AF822616

AFGEVOERD
LWBIB

Sem. Griekse Letterkunde
Blandijnberg 2
Gent

ALEXANDER OF APHRODISIAS
ON STOIC PHYSICS

PHILOSOPHIA ANTIQUA

A SERIES OF MONOGRAPHS ON ANCIENT PHILOSOPHY

EDITED BY

W. J. VERDENIUS AND J. H. WASZINK

VOLUME XXVIII

ROBERT B. TODD

ALEXANDER OF APHRODISIAS ON STOIC PHYSICS

LEIDEN
E. J. BRILL
1976

Sem. Griekse Letterkunde
Blandijnberg 2
Gent

ALE/5

ALEXANDER OF APHRODISIAS ON STOIC PHYSICS

A STUDY OF THE DE MIXTIONE
WITH PRELIMINARY ESSAYS,
TEXT, TRANSLATION AND COMMENTARY

BY

ROBERT B. TODD

AFGEVOERD
LWBIB

LEIDEN
E. J. BRILL
1976

03 JANUARI 1977
4.256

ISBN 90 04 04402 7

Copyright 1976 by E. J. Brill, Leiden, Netherlands

All rights reserved. No part of this book may be reproduced or translated in any form, by print, photoprint, microfilm, microfiche or any other means without written permission from the publisher

PRINTED IN THE NETHERLANDS

To My Parents

CONTENTS

PART THREE

THE *DE MIXTIONE*

ACKNOWLEDGEMENTS

I would like to thank Professor D. J. Furley for his advice and criticism at various stages in the preparation of this work, and Professor A. A. Long for commenting on an earlier version of Part II. Needless to say, the final product is my responsibility alone.

I wish to thank the Committee on Research at the University of British Columbia for supporting my research with several summer grants.

I record with gratitude that this book has been published with the help of a generous grant from the Humanities Research Council of Canada, using funds provided by the Canada Council.

And finally I thank my wife Barbara who typed, edited, and did so much else to make it all possible.

Crouch End,
February 1975.

PREFACE

The importance of Alexander of Aphrodisias in the Aristotelian tradition in Western philosophy is well established. This reputation however rests almost exclusively on his very influential interpretation of Aristotle's doctrine of the active intellect. The subject of the present study, the *de mixtione*, is a treatise in which he deals with the philosophically less important topic of the mixture of physical bodies. My aim is to show that both as an exposition of Aristotelian thought and as an extended discussion of Stoic physics it offers an excellent opportunity to observe the development of Peripatetic scholasticism in the face of ideas developed in post-Aristotelian philosophy. In this way I shall try to establish the largely unacknowledged importance of Alexander's contribution to the Greek philosophical tradition.

Alexander is still unfortunately a relatively obscure author and so I have devoted Part One of this study to a basic description of his works and a preliminary attempt to place him in his intellectual milieu. His philosophical creativity, as this essay will show, has greater rein in his short treatises than in his monumental commentaries, and it is from these works that his relation to other philosophical schools can best be gauged. Like his *de fato* the *de mixtione* is basically an attack on the Stoics, but it also contains a great deal of important source material and some constructive criticisms of Stoic physics. Much of this I shall evaluate in a commentary in Part Three, but these aspects of the work must also be seen in the light of similar contributions by our other sources for Stoic physics as well as Alexander's own overall relation to Stoicism. For this reason in Part Two I survey the latter before undertaking an extended examination of Alexander's exposition and critique of the Stoic theory of total blending (κρᾶσις δι' ὅλων), the main subject of the *de mixtione*.

Part Two as a whole will, I hope, contribute something to the almost untouched work of evaluating in detail our sources for Stoic physical theory. Source material for much of Stoic philosophy represents the residue of a complex and frequently polemical tradition, and while the character of this tradition can never be established in full some pattern can be discerned in it by the close analysis of specific doctrines or related groups of doctrines. I shall argue that through applying this method to the small but complex body of evidence for the Stoic theory of total blending the accepted interpretation of this doctrine and its place in the Stoic system must be radically revised.

My whole discussion will naturally be limited because of the close coherence of Stoic philosophy in which no single doctrine can be satisfactorily interpreted in isolation. By the terms of the present work I cannot apply, either in Part Two or in the Commentary, the same method of detailed analysis to the evidence for all the physical theories related to total blending as to the evidence for that doctrine itself. Within these limitations, however, I shall aim to provide a complete enough interpretation to set Alexander's account of total blending in the *de mixtione* in clear perspective (in Part Two, sect. ii), and also to show our author's place in the tradition of criticism and polemic that surrounded this doctrine in antiquity (in Part Two, sect. iii). As such this discussion can stand as a preliminary essay for a more systematic study of the sources for Stoic physical theory.

In Part Three I deal exclusively with the *de mixtione*. The translation is based on the text of Bruns' critical edition of 1892. The Greek text that I have included at the suggestion of the editors of "Philosophia Antiqua" is designed solely as a complement to this translation and the subsequent commentary and is no substitute for the new critical edition that Elio Montanari's recent study of the manuscripts (*Accademia Toscana* XXXVI [1971]) has shown to be necessary. The commentary is designed to complement the two preliminary essays by discussing in detail Alexander's methods as a Peripatetic scholastic both in his treatment of Stoicism and in his exegesis of Aristotle in the later chapters of the work. It will probably be more convenient to read the translation, or at least the outline, of the *de mixtione* before reading Part Two as this itself serves as a commentary in several places.

This work is designed to be as accessible as possible to the Greekless reader though of course much of the related material remains untranslated. Apart from textual matters in Part Three such a reader should have difficulty only with some detailed points of language dealt with in footnotes and in a few places in the commentary.

The manuscript was essentially completed by the autumn of

1973. I was unable to take account of Lutz Bloos' study *Probleme der Stoischen Physik*, Hamburg, 1973, nor was I aware of the 1971 New York dissertation by Bethia S. Currie, *God and Matter in Early Stoic Physics* (*Dissertation Abstracts* 32 [1972]). Also bearing on several of the aspects of Stoic physics that I discuss is Michael Lapidge's article, "ἀρχαί and στοιχεῖα: A Problem in Stoic Cosmology," *Phronesis* 18 (1973) 240-278, that also appeared after my work was complete. There are general surveys of Stoic physics in two recent introductory works: A. A. Long, *Hellenistic Philosophy*, London, 1974, and F. H. Sandbach, *The Stoics*, London, 1975. Finally, although not related to any of the main themes of this work, P. L. Donini's *Tre Studi sull' Aristotelismo nel II secolo d.C.*, Turin, 1974, contains important new contributions to the study of Alexander of Aphrodisias, and the first volume of Professor Moraux's *Aristotelismus bei den Griechen* (Berlin, 1974) has begun the authoritative version of the history that I have sketched briefly in Part One.

NOTE ON REFERENCES: ABBREVIATIONS

I have generally referred to books and articles in an abbreviated form; with the exception of encyclopaedia articles, standard works, and some items irrelevant to the main subjects of this study, this will be elucidated by a full citation in the bibliography. Among the references to ancient works note especially: *Mant.* = Alexander *de anima libri mantissa*, *Quaest.* = Alexander *Quaestiones*, both works in Vol. II of *Supplementum Aristotelicum*. The Aristotelian commentators are referred to by the title of the relevant Aristotelian work and the page and line number of the Berlin Academy edition. Other abbreviations:

AGPh	*Archiv für Geschichte der Philosophie*
AJP	*American Journal of Philology*
BICS	*Bulletin of the Institute of Classical Studies*
CAG	*Commentaria in Aristotelem Graeca*
CP	*Classical Philology*
CQ	*Classical Quarterly*
DK	H. Diels-W. Kranz, *Die Fragmente der Vorsokratiker*
GGA	*Göttingen Gelehrte Anzeiger*
JHP	*Journal of the History of Philosophy*
JHS	*Journal of Hellenic Studies*
JP	*Journal of Philology*
JTS	*Journal of Theological Studies*
LSJ	Liddel-Scott-Jones, *A Greek-English Lexicon*
MH	*Museum Helveticum*
PG	*Patrologia Graeca*
PIR²	*Prosopographia Imperii Romani* (2nd ed. where available)
RE	Pauly-Wissowa, *Real-Encyclopädie der Classischen Altertumswissenschaft*
REG	*Revue des Études Grecques*
REL	*Revue des Études Latines*
RM	*Rheinisches Museum*
SA	*Supplementum Aristotelicum*
SVF	*Stoicorum Veterum Fragmenta*
TAPA	*Transactions and Proceedings of the American Philological Association*
ZNTW	*Zeitschrift für Neutestamentliche Wissenschaft*

PART ONE

ALEXANDER OF APHRODISIAS—AN INTRODUCTION

A study of a work by Alexander of Aphrodisias must be prefaced by some general introduction to the author, for he has not been comprehensively studied but only surveyed in several encyclopaedia articles.[1] This chapter will attempt to enlarge on such basic introductions by presenting Alexander in the context of the Peripatetic school of his age. He was the holder of a chair in Peripatetic philosophy at Athens [2] sometime between 198 and 209 A.D.,[3] and although this is our only known biographical fact,[4] we can still probe his relation to other Peripatetics of the preceding and succeeding generations. Though this task has not been previously attempted, and even now the necessary scholarly aids have not been established,[5] I shall make a start on setting Alexander in

---

[1] See Bibliography II-i for these, and other general presentations.

[2] That he taught at Athens is implied (1) by the public chair he held, almost certainly one of those established by Marcus Aurelius in 176 (p. 6 and n. 29 below), and (2) by the reference to Aristotle's statue at Athens at *Met.* 415. 29-31. Though this is admittedly not a hard fact, I cannot follow Lynch (*Aristotle's School*, p. 214) who seems fairly certain that Alexander was appointed elsewhere, and is also certain (p. 193 n. 34) that he at least worked in Rome since Athens probably lacked "the kind of library facilities" a commentator would need. This is highly conjectural.

[3] The tradition that sets the termini for Alexander's tenure as 198 and 211 on the basis of the dedication of the *de fato* to Septimius and Antoninus (*de fato* 164. 3ff.) would seem to be in error. The earliest source to which I have traced this is J. A. Fabricius, *Bibliotheca Graeca* (Hamburg, 1716), II, 274 (also IV, 63), followed by J. Brucker (*Historia Critica Philosophiae* [Leipzig, 1742], II, 481), C. G. Zumpt (*Abh. d. Akad. d. Wiss.*, Phil.-Hist. Klasse, Berlin, 1843, p. 98), Zeller (p. 778), Ueberweg-Praechter (p. 564), and Gercke (*RE* I-i, col. 1453). Under the influence of these handbooks it has become dogma. Yet it must be false since Geta was created Augustus in 209 (see S. N. Miller, *Cambridge Ancient History*, XII, 41) and we know of an inscription indicating that the news had reached Athens by December 209 or January 210 (A. Birley, *Septimius Severus*, p. 264). Is there any reason to assume that Alexander would have omitted Geta thereafter?

[4] We cannot even be certain that his Aphrodisias is the major Carian city which has been extensively excavated in recent years, though in view of its well-developed cultural life (see L. Robert, *Antiquité Classique* 35 [1966] 397-432) it probably was.

[5] The main requirements are (1) a collection of the fragments of the pre-Alexandrian Peripatetics, of whom only those of Aristocles (H. Heiland,

Sem. Griekse Letterkunde
Blandijnberg 2
Gent

some historical perspective by offering a brief history of Peripatetic philosophy from roughly 160 to 230 A.D.[6]

I. SOME PRE-ALEXANDRIAN PERIPATETICS

The later years of the Roman Republic, as is well known, mark a watershed in the history of Aristotle's influence in antiquity. His school-texts—the so-called esoteric works—were at that time, if not actually recovered after two centuries of burial, at least made more widely available for study.[7] In this period they were made the object of a scholarly treatment and critical discussion with which we associate the names of Andronicus of Rhodes, Boethus of Sidon, and later, in the early decades of the Roman Empire, Nicolaus of Damascus and Xenarchus of Seleucia.[8] It is solely in virtue of this work that these figures are termed Peripatetics. We know little about their wider philosophical beliefs, if any, and nothing about the relation that they might have borne to the Peripatetic school at Athens.[9] It is not until the early part of the second century that further evidence appears of the significant continuation of this scholarly work. To this period belong Adrastus

---

Aristoclis Messenii Reliquiae) and Herminus (H. Schmidt, *De Hermino Peripatetico*) are available in the unsatisfactory form of dissertations, and (2) a collection of the fragments of Alexander's lost works, principally from the quotations by later commentators. Only a sample of the latter exists (P. Moraux, *Alexandre d'Aphrodise*, Appendix III). Moraux's promised history of Aristotelianism from Andronicus to Alexander will help to remedy these deficiencies; for brief indications of its range see Moraux, "Einige Aspekte des Aristotelismus," and *D'Aristote à Bessarion*, pp. 13-40.

[6] The period from Adrastus and Aspasius (c. 100 A.D.) to Alexander is covered in Zeller, pp. 776-788, and Ueberweg-Praechter, pp. 556-564, and surveyed briefly by Usener, *Kleine Schriften*, III, 206-209.

[7] On the problem of the fortune of these texts in the period between Theophrastus and Andronicus see Düring, *Aristotle in the Ancient Biographical Tradition*, ch. 18 with literature.

[8] On this period see Zeller, pp. 619-631, Ueberweg-Praechter, pp. 556-565, and Brink, "Peripatos," *RE* Suppl. Vol. VII, cols. 938-347. More recent literature to be noted: the studies of M. Plezia on Andronicus (*De Andronici Rhodii studiis aristotelicis*); Moraux, "Xenarchos (5)," *RE* Suppl. Vol. IXA; and H. Drossaart-Lulofs' edition of Nicolaus' Περὶ φιλοσοφίας. Lynch (*Aristotle's School*, ch. VI) has now argued that the Peripatetics of this period in no sense continued the work of the original Lyceum, which he believes ended when Sulla sacked Athens in 86 B.C.

[9] The evidence on Andronicus is unclear, and he may have been διάδοχος at Athens (K. Brink, *RE* Suppl. Vol. VII, cols. 938-940); but strongly against this is Lynch, *Aristotle's School*, pp. 203-204.

of Aphrodisias and Aspasius, the latter the only Aristotelian commentator before Alexander of whom there are any complete works extant. Both wrote commentaries on various logical, ethical, and physical treatises, and Adrastus, like Andronicus, studied the problem of the order of the Aristotelian writings.[10] Again we do not know of any connection between them and the school at Athens.[11]

In attempting to trace further the continuity of the second-century Peripatetic school we must turn to the polymathic figure of Galen, for he provides evidence both of these scholars and of a wide range of less significant figures. In his student days at Pergamum he appears to have heard the lectures of a student of Aspasius,[12] who we know from Simplicius was the teacher of Herminus,[13] who in turn was the teacher of Alexander of Aphrodisias.[14] From an Arabic source it seems possible that Galen wrote a treatise on the Aristotelian doctrine of a first mover directed particularly against this Herminus.[15] This all suggests that a widely-travelled savant such as Galen would naturally encounter either personally or intellectually these Peripatetic scholars, a good indication of their prominence. Also it was to be expected that with his interest in logic Galen would naturally gravitate to philosophers who intensively studied the Aristotelian logical works, and in general placed logic at the forefront of their system of education.[16] Since he died in 199 Galen's life overlapped with that of Alexander of Aphrodisias, but in the latter's extant works there is no evidence of anything except a limited intellectual contact between them.[17]

---

[10] On their writings see below, pp. 12-13.

[11] Ueberweg-Praechter (pp. 664-665) lists Xenarchus, Aspasius, and Herminus tentatively as "successors" (διάδοχοι). But this is pure guesswork, since whether or not Lynch (cf. n. 8) is right, we do not know whether the Peripatetic school had such heads in this period.

[12] V. 42 K, an event to be dated to 145-147 A.D. For his study of the commentaries of Aspasius and Adrastus on the *Categories*, cf. XIX. 42 K (*de libr. propr.*, p. 119 Mueller).

[13] *de caelo* 430. 32-33, and 431. 10-11 imply this.

[14] Simplicius *de caelo* 430. 32-33 (Ἑρμίνου . . . ἤκουσα).

[15] The evidence is complicated (see S. Pines, *Isis* 52 [1961] 23).

[16] Both Galen XIX. 42 K(= *de libr. propr.*, p. 118 Mueller) and Alexander (*Top.* 69. 15, *Pr. An. I* 16. 16, and *Met.* 85. 10) knew Eudemus' logical treatise Περὶ Λέξεως (for fragments see F. Wehrli, *Schule des Aristoteles*, VII, frs. 25-29).

[17] *Top.* 549. 23-24 is the only reference to Galen where he is mentioned as an example of an intellectual authority. There are many examples of

Sem. Griekse Letterkunde
Blandijnberg 2
Gent

It appears, however, that Alexander criticised Galen's treatise on the first mover in an Arabic text attributed to him.[18]

Minor Peripatetics: Alexander of Damascus and Aristocles of Messenia

The Peripatetic school of the second century did not consist solely of scholarly commentators. In the pages of Galen we encounter some less influential figures. There is the Eudemus whom we know of only through Galen as the friend of his father, his mentor, and later, on his first visit to Rome in 161, the agency of his entry into aristocratic circles.[19] It is in such circles that we encounter the only Peripatetic before Alexander whom we know to have held a chair of philosophy at Athens. This is Alexander of Damascus who was present at anatomical demonstrations that Galen gave at the invitation of the consul Flavius Boethus in 163.[20] At this time Alexander was Flavius' teacher, and is described as knowledgeable about Platonism, but particularly inclined to Aristotelianism.[21] His behaviour at this event was eccentric and yet it may be instructive to examine it in detail. Galen planned to demonstrate the nature of breathing and speech by conducting anatomical experiments on goats and pigs. Beforehand he tried to establish a method for determining the results of these experiments. He himself would point to the evidence of the dissection (τὰ ἐκ

Alexander dealing with the same topics as Galen (e.g. the locus of the ruling-principle [ἡγεμονικόν] of the soul, *de an.* 94. 7-100. 17; or the relation of the soul to the physical constituents of the body, *de an.* 2. 10-11. 13; see Donini, *Atti d. Accad. di Torino* 105 [1971] 61-107 at 98-107) but it is as difficult to determine Alexander's precise knowledge of Galen as it is to establish the nature of his Stoic sources (see Pt. II-i *passim*).

[18] S. Pines, *Isis* 52 (1961) 21-54 *passim*, and N. Rescher and M. Marmura, *The Refutation by Alexander of Aphrodisias of Galen's Treatise on the Theory of Motion*. Also cf. R. Walzer, *Galen on Jews and Christians* at pp. 48 n. 3 and 74 n. 2 on evidence of responses to Galen by Alexander.

[19] References are most conveniently collected at *PIR*² E109. Eudemus is perhaps to be identified with the Peripatetic Euthydemus at Lucian *Hermotimus* 11 (V. Nutton, *Proc. Camb. Phil. Soc.* n.s. 18 [1972] at p. 58). On Galen's social progress at Rome note especially XIV. 612ff. K. For Galen's biography, cf. the convenient outline in Bowersock, *Greek Sophists*, ch. 5.

[20] XIV. 626-629 K. On the date cf. Ilberg, *Neue Jahrbücher* 15 (1905) 289, who provides further evidence on the political figures present, as does Bowersock, *Greek Sophists*, pp. 62-63 and 82-83.

[21] Galen XIV. 627 K.: . . . Ἀλεξάνδρῳ τῷ Δαμασκηνῷ, γιγνώσκοντι μὲν τὰ τοῦ Πλάτωνος, ἀλλὰ τοῖς Ἀριστοτέλους προσκειμένῳ μᾶλλον.

τῆς ἀνατομῆς φαινόμενα) while he hoped that Alexander, whom he refers to as the teacher of all of them would formulate the results (συλλογίσασθαι τὰ περαινόμενα).[22] The position that Alexander held in the company [23] is indicated by the agreement with which Galen's request was met, though also, as Galen points out, it was hoped that in this way the disputatiousness (φιλονεικία) to which the Peripatetic was prone would be avoided. This proved vain, since before Galen could even begin Alexander asked whether the doctor took it for granted "that we must rely on what appears through the senses" (τοῖς διὰ τῶν αἰσθήσεων φαινομένοις πιστεύειν ἡμᾶς δεῖν).[24] At this Galen withdrew saying only that he had not expected to meet such "crude sceptics" (ἀγροικοπυρρώνειοι).[25] Recalling the incident in more general terms in a treatise written some years afterwards Galen refers to Alexander as at that time dignified with the appointment of public teacher of Peripatetic philosophy at Athens.[26]

It is easier to comment on the historical than the philosophical import of this evidence. First, we see a Peripatetic philosopher

---

[22] συλλογίζεσθαι ("to formulate") does not seem to be used here in the technical Aristotelian sense of "to syllogise," so that Galen would not be hinting at Alexander's philosophy. It simply means to relate phenomena to a non-phenomenal event. As the eventual experiment showed (XIV. 629-630 K) this involved establishing a causal relation between the phenomena of the dilation and contraction of the thorax and the presence and absence of sound. Cf. a similar usage at XV. 694 K.

[23] The description of him as "teacher of all" (ἁπάντων διδάσκαλος) at XIV. 628 K could be a courtesy title. It need not imply that he was Galen's teacher as Bowersock, *Greek Sophists*, p. 60 n. 50 (following *PIR*² G24) takes it. It is certain only that he was the teacher of Flavius Boethus, and possibly Cn. Claudius Severus (v. n. 37 below).

[24] XIV. 628 K.

[25] Cf. VIII. 711 K, where it is used similarly as a term of abuse. Galen refers to certain sceptics and ἀπορητικοί who "deny that they have firm knowledge of their own experiences and are rightly called crude sceptics (ἀγροικοπυρρώνειοι)." These extreme sceptics are probably not to be associated with any of the established schools (n. 34 below). Gercke (*RE* I-i, col. 1453) rightly points out that this incident seems to have become enmeshed in legend among the Arabs. The *Kitāb Al-Fihrist* (10th cent. A.D.) reports that Alexander of Aphrodisias called *Galen* "Mauleselkopf" (as translated by A. Müller, *Die Griech. Philos. in der Arab. Überlief.*, p. 23), but as Müller himself noted (p. 57 n. 45), there is no evidence in Greek sources of relations between the two (cf. also nn. 17 and 18 above). On the identification of the Aphrodisian and Damascan Alexanders in Arabic sources see also G. Strohmaier, *Encyclopaedia of Islam*, IV, fasc. 63-64 (Leiden, 1973), 129-130.

[26] *De Anatom. Administrat.* I-i (II. 218 K): . . . τοῦ νῦν 'Αθήνησιν ἀξιουμένου λόγους διδάσκειν δημοσίᾳ.

moving in those circles where the intellectual currents of the age—philosophy, rhetoric, sophistry and medicine—blended.[27] Like Hadrian of Tyre, who was also present at the gathering, Alexander moved on, presumably assisted by his influential friends, to a teaching appointment at Athens.[28] It is fairly certain that he was the first holder of the relevant philosophical chair of the four established by Marcus Aurelius on his visit to Athens in 176.[29]

---

[27] This blend is well described in Bowersock, *Greek Sophists*, especially ch. 5, "The Prestige of Galen."

[28] On Hadrian see *PIR*² H4. Philostratus (*V.S.* 588) shows that Hadrian was appointed to a municipal chair of rhetoric, for at the time of Marcus Aurelius' visit to Athens in 176 which established imperial chairs (cf. next note "he *already* held (ἔκρατε μὲν ἤδη) the chair of the sophists."

[29] The evidence for the general endowment, which included a chair of rhetoric, is at Dio Cassius LXXII. 31. On the philosophical chairs see Philostratus *V.S.* 566 and Lucian *Eun.* 3. 8, and cf. Tertullian *Apolog.* 46. I agree with H. I. Marrou (*History of Education*, p. 568 n. 10) that Lucian's report does not prove that there were two chairs for each sect, nor indeed does Porphyry *Vit. Plot.* 20. 39-40. Alexander of Damascus is variously said to have held the chair in Peripatetic philosophy "around 270" (Zumpt, *Abh. d. Akad. d. Wiss.*, Phil.-Hist. Klasse, Berlin, 1843, p. 97; followed by Ueberweg-Praechter, p. 665), or to have been the first holder of the imperial chair (Christ-Schmidt-Stählin, *Gesch. der Griech. Lit.* II. 2, p. 833 n. 8; Gercke, *RE* I-i, col. 1452, who adds "vielleicht"). The matter can only be discussed on the basis of the dating of Galen's works. The first reference to Alexander being present at the anatomical demonstrations is in the "On Prognosis" (Περὶ τοῦ προγιγνώσκειν, Galen XIV K), a work written in 178 (I. Ilberg, *Neue Jahrbücher* 15 [1905] 288 n. 2), or late 177 (V. Nutton, *Proc. Camb. Phil. Soc.* n.s. 18 [1972] at p. 32 n. 1). The reference to him as professor at Athens is in the "Inquiries into Anatomy" (Ἀνατομικαὶ Ἐπιχειρήσεις, Galen II K), a work of uncertain date. Ilberg (*RM* 44 [1889] 209) is no more precise than 169-180. Though not impossible it would seem to be unlikely that in the former work of late 177-178 Galen would refer to Alexander without adding that he was also professor at Athens; thus the reference to him in this capacity was probably made after this date. διδάσκειν δημοσίᾳ ("to teach in public," Galen II 218 K) would then have to refer to his appointment to the imperial chair probably between 178-180, and it might still have been influenced by Herodes Atticus who died in 178 (cf. nn. 30, 31). How these imperial appointments affected the earlier tradition of a succession of headships is not known. The term "successor" (διάδοχος) is still used after 176 (Porphyry *Vit. Plot.* 15. 19), and Galen XIX. 50 K seems to imply that these were paid appointments such as Marcus endowed. Presumably the imperial chairs supplanted the old headships where they still existed (J. Walden, *Universities of Ancient Greece*, p. 102 n. 1), though recently Lynch (*Aristotle's School*, pp. 169-174) has made out a good case for there being no relation whatsoever between the two. Alexander of Aphrodisias certainly does refer to himself as a teacher (διδάσκαλος, *de fato* 164. 15), and only such teachers are mentioned by Dio Cassius (LXXII. 31). A similar emphasis on pedagogy can be found in Philostratus' description (*V.S.* 560) of Theo-

Since Herodes Atticus perhaps had some influence over these appointments [30] we have a further indication that Alexander's social connections were important.[31] The philosophical difficulties posed by this appointment are somewhat greater. What could the Peripatetic doctrines that Alexander taught at Athens have been if he could espouse the scepticism evinced at Galen's demonstration? The problem might be soluble in the light of historical considerations. Galen vividly describes the vicious backbiting that marked intellectual life at Rome at the time of his first visit.[32] As a newcomer he had entered aristocratic circles very rapidly after a sensational cure of his teacher Eudemus.[33] Conceivably the more established Alexander resented his intrusion, and there would be no

---

dotus, the first holder of the imperial chair in rhetoric (προὔστη . . . τῆς 'Αθηναίων νεότητος), while the pedagogical role of the professors forms the whole theme of Lucian's *Eunuch*; cf. also Alexander *de fato* 189. 4-5 on the Stoics as teachers. Finally in dedicating Book VIII of his *Onomasticon* to Commodus, Pollux complains of the pressures of teaching (ἡ συνουσία ἡ πρὸς τοὺς νέους). Whatever the relation the new chairs bore to the traditional positions of the successors, they probably introduced a new emphasis on public teaching, derived perhaps from the Stoic emperor's own sensitivity to pedagogy; cf. *Meditations* Bk. I *passim* for dedications to various teachers.

30 Philostratus *V.S.* 566; although, as Bowersock says (*Greek Sophists*, p. p. 49), this shows only that "Marcus took the advice of Herodes." Since the latter died in 178 the system of appointment referred to by Lucian (*Eun.* 3. 8) whereby a candidate was "judged by the vote of the best men" (δοκιμασθέντα ψήφῳ τῶν ἀρίστων) could have been in force soon after, and was possibly the system under which Alexander of Aphrodisias was appointed. Who these "best men" were is not known. Lesky's description (*History of Greek Literature*, English trans., p. 841) of the *Eunuch* as a depiction of "the struggle over the chair of philosophy at Athens (176) in all its deplorable details" must certainly be corrected.

31 This connection might have been the following: Favorinus of Arelate was the teacher of Herodes (Philostratus *V.S.* 490), and present at Galen's anatomical demonstration was Demetrius of Alexandria, who at that time was giving rhetorical demonstrations in the style of Favorinus (κατὰ τὴν ἰδέαν τῆς Φαβωρίνου λέξεως, Galen XIV. 627 K). Probably Favorinus was dead by 163 (C. P. Jones, *CQ* 17 [1967] 312 n. 5), but presumably Demetrius knew him in his lifetime. If he and Alexander were friends our philosopher might thereby have known the great sophist. A firmer speculation on the reasons for Alexander's advancement might be grounded in the fact that the enthusiastic Peripatetic Cn. Claudius Severus (Galen XIV. 613 K) knew Alexander (XIV. 629 K), and was possibly also present with Marcus Aurelius at Athens in 176 when the philosophical chairs were established (Philostratus *V.S.* 588).

32 XIV. 620-622 K where Galen complains of this, and a prevailing charlatanry, and cf. esp. XIV. 625 K.

33 See XIV. 612-613 K for his encounter with Flavius Boethus and Sergius Paulus (*consul ordinarius* for the second time in 168; see *PIR*[5] S377) at Eudemus' bedside; and XIV. 625 K on the development of his reputation.

better way to scotch an anatomical experiment than at least to suggest a sceptical pose; for among contemporary medical schools both the Empiricists and Methodists were opposed to dissection on precisely such sceptical principles.[34] Alexander's question is so general that it cannot be linked with the refined forms of scepticism that these schools adopted, but in the circumstances it would have been quite sufficient to constitute a root and branch attack on anatomical experimentation. This would leave us free to assume that in his less malicious moments Alexander was a more orthodox Peripatetic.

This is perhaps all that should be made of the incident. Yet it is worth spelling out the considerations that might have led Alexander to adopt scepticism as a philosophical position and not just for *ad hominem* purposes. There is first the general point that a philosopher's knowledge of Aristotle and ability to teach his works might not have entailed a philosophical agreement with all his doctrines. We have seen that in the first and second centuries Peripatetics were essentially philologists and scholars.[35] Alexander could presumably have followed in their train while being quite free to adopt an alien philosophical stance. This suggestion of a dichotomy between philosophy and scholarship might be more plausible if we had some actual evidence of Alexander having written commentaries or monographs, but apart from the citations in Galen his name is never mentioned by later commentators or

---

[34] On the Methodists' rejection of anatomy see L. Edelstein, "Methodiker," *RE* Suppl. VI (= *Ancient Medicine*, p. 180), and on that of the Empiricists cf. Edelstein, *Ancient Medicine*, pp. 269-270. The best presentation of their differing forms of scepticism can be found at Sextus Empiricus *P.H.* I. 236ff.

[35] In general the relation between an official philosophical affiliation and a doctrinal stance needs to be carefully measured at this period. Public inscriptions, for example, will refer only to a single affiliation (e.g. for Πλατωνικός see Tod, *JHS* 77 [1957] 134; or διάδοχος Στωϊκός at *I.G.* III. 661 and 1441), while philosophical writers, on the other hand, will note any eclecticism (e.g. Porphyry's description of one Trypho as "both a Stoic and a Platonist" [*Vit. Plot.* 17]). This gap may have been particularly wide among Peripatetics since there was a well-developed tradition of philosophical scholarship that carried no clear indication of broader philosophical commitment. So, for example, the Severus whom Marcus (I. 14) respected for his liberal and essentially Stoic political views could be the Peripatetic Cn. Claudius Severus (*Hist. Aug.* IV. 3. 3 and *PIR*[1] C808) father of Severus who was also a student of Peripatetic philosophy (Galen XIV. 613 K and cf. *PIR*[1] C811 and n. 31 above). There is no reason why it should be "remarkable" (*pace* Farquharson on *Med.* I. 14) for the same man to range so widely in the prevailing philosophical climate.

philosophers. Still, whether or not we accept this particular suggestion we are still forced, if we take Galen's anecdote seriously, to give him the epithet of "eclectic" and to determine the sense in which it applies to him.

There are three philosophical components to Alexander of Damascus: knowledge of the works of Plato, and of Aristotle, and possibly some general form of scepticism. The only part of his philosophical knowledge compatible with this scepticism would be his Platonism. In the second century the sceptical Academy still thrived, and Galen was quite capable of mentioning Academics and Pyrrhonists together.[36] Indeed the famous sophist Favorinus of Arelate seems to have combined the doctrines of both schools while still favouring the Peripatetics.[37] Perhaps Alexander was capable of a similar eclecticism, and there is the possibility that he knew Favorinus personally.[38] If he did resemble the sophist philosophically we should have to assume that his Aristotelianism was indeed a scholarly knowledge suitable for pedagogical purposes. Certainly if teaching of Aristotle was tending to the mechanical form of which we have some hints in Alexander,[39] and which we know it acquired in later centuries,[40] this would be an even more plausible suggestion.

---

[36] E.g. V. 60 K (= I. 47 Marquardt). Earlier we know that it was the sceptical Academy that Epictetus attacked (Arrian *Diss.* I-v and II-xx).

[37]. See W. Schmid, *RE* VI-ii, col. 2081, and Galen *de opt. doctr.* I (I. 41 K = pp. 82-83 Marquardt) on Favorinus' Academic scepticism. He counted himself an Academic (Aulus Gellius XX. 1. 9 and XXI. 21, with Galen *de opt. doctr.* I. 40 K = p. 82 Marquardt), though Plutarch (*Quaest. Conviv.* VIII. 734F) says Favorinus was an enthusiastic Peripatetic; cf. the description of his logical quibbling at Aulus Gellius V. 11. 8-14. Plutarch's description of Favorinus' Peripatetic tendencies should be compared with Galen's description of Alexander as "particularly tending towards Aristotle's doctrines" (cf. above, n. 21). The parallels between the two grow if we can assume that Alexander would have had to exhibit a similar enthusiasm in his capacity as the teacher of Flavius Boethus who was, according to Galen, a zealous Aristotelian (XIV. 612 K), like Cn. Claudius Severus (XIV 613 K), possibly another student of his (cf. XIV. 629 K, where he is one of those who reproach Alexander for his sceptical outburst).

[38] Note 31 above.

[39] *Pr. An. I* 8. 3f., 9. 5ff., and *Top.* 1. 3, and 125. 4-8. On the relation between the exposition of a work's goal (σκοπός) and value (χρήσιμον) and oral teaching in general see K. Praechter, *Byz. Zeit.* 18 (1901) 525-531. There is a hint that even earlier Alexander of Aegae (first century A.D. Peripatetic) was interested in the question of a work's goal; cf. Simplicius *Cat.* 10. 19ff. and Ueberweg-Praechter, p. 561.

[40] This is true of both the Athenian and the Alexandrian commentators

On the assumption then, that Alexander of Damascus' scepticism can be taken seriously, we can regard him as a philosopher sufficiently knowledgeable about Aristotle to acquire a teaching chair in his philosophy while probably espousing a sceptical Platonism. One reason for holding to this picture is the interesting contrast it provides with the better known character of another second century Peripatetic, Aristocles of Messenia.[41] Here we have a philosopher who is called a Peripatetic [42] yet has a knowledge of a more orthodox and in particular an antisceptical Platonism that probably has its roots in the earlier synthesis of Antiochus of Ascalon.[43] Traditionally Aristocles has been regarded as one of the teachers of Alexander of Aphrodisias which would conveniently date him to the period 170-200 A.D. during which Alexander of Damascus flourished. Recently however, Paul Moraux has thrown doubt on the tradition of textual emendation that rendered Aristocles the teacher of the great commentator.[44] It nonetheless remains certain that Aristocles belongs to the second century A.D. The list of his works shows that blend of philosophy and rhetoric so typical of the era;[45] and we find him quoted extensively by Eusebius who drew heavily on philosophers of this period.[46] It is therefore plausible to contrast him with our reconstruction of the philosophy of Alexander of Damascus. In each case we find a second century Peripatetic turning to Plato and relating him to scepticism. Aristo-

---

of the fifth and sixth centuries; see Praechter, *Byz. Zeit.* 18 (1901) 529-533, and on Proclus, A. J. Festugière, *M.H.* 20 (1963) 77-100.

[41] His fragments are collected by Heiland, diss. Giessen, 1925.

[42] The Suidas s.v., and frequently by Eusebius who quotes him most extensively; see Heiland, pp. 1-3.

[43] Heiland, pp. 102-104.

[44] *AGPh* 49 (1961) 169-182 where that tradition is copiously documented. Aristocles has only been made Alexander's teacher by the emendation of the Ἀριστοτέλης of various texts, notably *Mant.* 110. 4, to his name on the ground that as a reference to the Stagirite this was unintelligible. Cf. n. 53.

[45] The Suidas (s.v.) records an "Arts of Rhetoric" (τέχναι ῥητορικαί) and an "Is Homer superior to Plato?" (Πότερον σπουδαιότερος Ὅμηρος ἢ Πλάτων). Both titles suggest that, rather like Favorinus, Aristocles was a rhetor and a sophist; the latter seems to be a demonstration of the antiquarian kind that was so typical in the Second Sophistic. See Bowie, *Past and Present* 46 (1970) 4-10 on the use of such themes.

[46] He selected, in particular, philosophical texts of a rhetorical and polemical character, like those from Aristocles' *On Philosophy* (Περὶ φιλοσοφίας), or Atticus' attack on the Peripatetics, both of which present opponents in a one-dimensional aspect; see Mras, *Anz. d. Oest. Akad. d. Wiss. in Wien*, Phil.-Hist. Klasse 93 (1956) 209-217.

cles attacks, largely on the basis of arguments from the *Theaetetus*,[47] the scepticism of Protagoras as well as Pyrrhonism.[48] Whether any of this could have been provoked by Alexander of Damascus' possibly sceptical Platonism is a matter of conjecture, since we cannot be certain of Aristocles' precise date. It is however clear that his eclecticism is systematic, which is grounds enough for our suggestion that the same might be true of his more obscure contemporary. Like Alexander of Damascus Aristocles is not known to have written any commentaries or scholarly works on Aristotle. This is less surprising in his case, since if he was indeed influenced by Antiochus he would not have approached the Aristotelian texts with scholastic rigour. With Alexander we must assume, unless we are to allow for an eccentric form of pedagogy, something of a scholastic approach; he might simply have been versed in the Aristotelian works and have relied on the commentaries of his more productive scholarly predecessors, such as Adrastus and Aspasius.

Alexander's Teachers

In addition to Herminus,[49] there are two other known teachers of Alexander of Aphrodisias who must have been active in the latter third of the second century, if the epithet "teacher" (διδάσκαλος) implies personal contact. Sosigines [50] was a wide-ranging scholar. As a typical Peripatetic he dealt with logic, also wrote a treatise on vision in at least eight books,[51] and was particularly interested in astronomical problems. The other teacher has been only recently identified. He is Aristotle of Mytilene, mentioned by Galen as one of the most distinguished contemporary Peripatetics in a treatise written in the 190's.[52] This, in Moraux's view,[53] makes

[47] Fr. 4 Heiland (= Eusebius *P.E.* XIV. 19. 766b ff.).
[48] Fr. 6 Heiland (= Eusebius *P.E.* XIV. 18. 758c ff.).
[49] Cf. p. 3, nn. 14, 15 above.
[50] See Rehm, *RE* Zw. Rh. III (1929) 1157-1159 for details, and on the distinction between him and Julius Caesar's advisor, see Th. Martin, *Annales de la Faculté des Lettres de Bordeaux* 1 (1879) 174-187. Alexander himself refers to him as his teacher (ὁ διδάσκαλος ἡμῶν, *Meteor.* 143. 13).
[51] Περὶ Ὄψεως, Alexander *Meteor.* 143. 14.
[52] Cf. Περὶ Ἐθῶν, Müller, *Script. Min.*, II, 11. 4-5, and see Moraux, *AGPh* 49 (1967) 177. He is termed "a leading figure in Peripatetic scholarship" (ἀνὴρ πρωτεύσας ἐν τῇ Περιπατητικῇ θεωρίᾳ).
[53] See *AGPh* 49 (1967) 176-182 for the full argument. This revised Moraux's earlier view (*Alexandre d'Aphrodise*, ch. 4) that the Aristotle referred to in various Alexandrian texts was the Stagirite, which had been criticised by F. Trabucco, *Acme* 11 (1958) 119-126.

him the Aristotle referred to as Alexander's teacher in several texts and renders the emendation of this name to Aristocles unnecessary. If this is correct his philosophical views are represented by a passage of the *de intellectu* dealing with the doctrine of "the external intellect" (νοῦς θυράθεν).[54]

II. THE ALEXANDRIAN CORPUS [55]

The purpose of this description of the Peripatetic school in the generation before Alexander has been to place his own contribution in perspective. It is clear now that a strong tradition of scholarly exegesis was established before him, and that he was linked to it through his teachers Herminus and Sosigines. There are also some indications of a less rigorous tradition represented by Alexander of Damascus and Aristocles of Messenia in which philosophical eclecticism was combined with an affiliation to the culture of the Second Sophistic. Alexander's contribution was to be entirely the continuation of the former tradition [56] and its elevation to new heights, primarily through the composition of a series of commentaries on Aristotelian works.

The Commentaries

The precise form of pre-Alexandrian philosophical commentaries is not known to us,[57] the only extant example being the somewhat

---

[54] *Mant.* 110. 4-112. 5 Bruns; cf. n. 44.

[55] See Pt. II of the bibliography for additional literature on the works and subjects discussed in this section.

[56] This is another reason, quite apart from those adduced by Moraux, why Alexander might not have been the pupil of Aristocles of Messenia. Nothing however is known of the contribution of Aristotle of Mitylene to the specifically scholarly tradition, and therefore his influence on Alexander as against that of the known scholars, Herminus and Sosigines, might have been slight.

[57] Pre-Alexandrian commentators are quoted rather sparingly, both by Alexander and later commentators. For example, Andronicus is only mentioned in Alexander's extant works at *Pr. An. I* 160. 32, and Düring (*Biographical Tradition*, p. 416) believes that he drew his knowledge about the arrangement of Aristotle's works from Adrastus, who wrote on this subject (Simplicius *Cat.* 16. 1-2). Boethus and Xenarchus are mentioned by name only at *Mant.* 151. 8 in a text of dubious authenticity. The very nature of textual commentary makes reference to earlier authorities otiose unless an important point arises, such as a textual reading; cf. the references to Aspasius at *de sensu* 10. 2 and *Met.* 59. 6 and 379. 3. For these reasons it is difficult to measure the originality of a commentator, and thus while Alexander's achievement

paraphrastic commentary on the *Nicomachean Ethics* by Aspasius.[58] We know that something like this technique was practiced in both the Platonic and Aristotelian schools as early as the first generation after their foundation,[59] but it was not developed to any extent until the first and second centuries A.D. To this period belong several commentaries on the *Timaeus* by Platonists,[60] as well as the works of Peripatetics mentioned above. We cannot be certain of their form. They might have been epitomes, or more elaborate paraphrases, or text by text commentaries like most of Alexander's works. We are rather better informed about the range of Peripatetic commentaries, and we find that an interest in the logical works predominated, epitomised by the celebrated description of Herminus in Lucian's *Demonax*.[61] There are references to commentaries on the *Physics* by Aspasius and Adrastus, and one on the *de caelo* by the former.[62] Both also wrote on the *Nicomachean Ethics*;[63] only Aspasius is known to have commented on the *Metaphysics* before Alexander.[64] Sosigines seems only to have written mono-

---

was no doubt great we cannot afford to gauge his debt to his predecessors solely from the *indices nominum*.

[58] *CAG*, XI-i. Only that on Books I-IV, part of that on Book VII, and that on Book VIII are extant.

[59] Crantor was the first of many to write a commentary on the *Timaeus* (Von Arnim, *RE* XI, col. 1586), and Eudemus seems in effect to have written a commentary on some of the logical works and the *Physics*; note the texts at Wehrli, *Schule des Aristoteles*, VII, frs. 9-24 and 31-123.

[60] These were composed by the more orthodox Platonists Taurus, Atticus, and Severus, and there is also a record of one by the Peripatetic Adrastus. Alexander is known to have criticised Taurus' commentary—see Praechter, *RE* VI-i, col. 68.

[61] *Demonax* 56, where he is said to deserve δέκα κατηγορίαι (ten categories = ten indictments). Simplicius' proem to his commentary on the *Categories* is a rich source of evidence on earlier studies of this work; cf. the *index nominum*. For our purposes we need note that there were commentaries by Aspasius, Adrastus, and Herminus. Alexander quotes from Herminus on some texts in the *Topics* (569. 3-5, 574. 26) and *Prior Analytics I* (72. 27, 89. 34ff., 91. 21ff.). Aspasius appears to have commented on the *de interpretatione*.

[62] Cf. Simplicius' commentaries to these works, *index nominum*.

[63] For Aspasius cf. n. 58; for Adrastus, Athenaeus XV. 673.

[64] He is quoted by Alexander in his commentary; cf. n. 57 above. This limited range of Aristotelian works on which commentaries were written illustrates Düring's point that it was "ein sehr zusammengeschrumpter Aristoteles, für den man sich interessiert" (*Antike und Abendland*, 4 [1954] 124) in the period from Andronicus to late antiquity. The indices show that Alexander had a knowledge of the biological and zoological works, as well as the *Ethics*. It has also been shown that he had a direct acquaintance

graphs on questions in logic, optics and astronomy, and perhaps thereby to have influenced Alexander's contributions in this genre.

We know that Alexander commented on all of Aristotle's logical treatises with the exception of the *Sophistici Elenchi*,[65] though only his works on Book I of the *Prior Analytics* and on the *Topics* are extant. Whether in this he surpassed in volume the work of his predecessors is difficult to gauge; for, as we have seen, the work of earlier commentators is often assimilated by their successors without attribution.[66] Certainly Alexander exhibited very clearly that concern for logic that was regarded as part of the vulgar image of a Peripatetic of his age.[67]

He also wrote commentaries on the four principal physical treatises of Aristotle, the *Physics, de caelo, de generatione et corruptione*,[68] and *Meteorologica*. Only his commentary on the latter, a rather paraphrastic work, is extant. Extensive remains of the first two works are preserved by Simplicius in his commentaries on them, while, as this study will demonstrate in part,[69] some of his commentary on the *de generatione et corruptione* can be reconstructed from Philoponus' commentary, for Alexander's *de mixtione* has close affinities with the Aristotelian treatise and can therefore be used as a guide to the Alexandrian content of Philoponus' work.

Alexander, as far as we know, is the first Peripatetic to have dealt extensively with the psychological treatises of Aristotle.

---

with the esoteric works—see P. Wilpert, *Hermes* 68 (1940) 369-396. There is some evidence that the range of commentaries was deliberately extended by Michael of Ephesus, the twelfth century Byzantine scholar, at the instigation of Anna Comnena; see R. Browning, *Proc. Camb. Phil. Soc.* n.s. 8 (1962) 6-8. For Michael's works see *CAG*, XXII.

[65] That on *Soph. El.* in *CAG*, II-iii is spurious—see Wallies, *Praefatio*. For evidence of these other commentaries see Zeller, p. 790 note. For literature, see Bibliography II-2.

[66] Cf. above n. 57.

[67] In addition to the description of Herminus in the *Demonax* (n. 61 above) see Tertullian *de praescriptione* VII. 6, Justin *Dialogue* II. 3, and Eusebius *Hist. Eccl.* V. 28. 13-14 (on which cf. Walzer, *Galen on Jews and Christians*, pp. 76-78). The latter describes the heresy that arose from applying Peripatetic logic to scripture. From an earlier period we may note the ridicule by one of the interlocutors at Plutarch *Quaest. Conviv.* IX. 743D directed against Menephylus the Peripatetic about the apparent use of the Muses in logical exercises. Cf. also Seneca *Ep.* 45. 6-10, Epictetus IV. 4. 14, and Marcus Aurelius *Med.* I. 17. 9 for hostility to Peripatetic logic-chopping from a non-scholastic Stoic viewpoint.

[68] For evidence see Zeller, p. 790.

[69] See Comm. on *de mixt.* 16 *passim*.

He wrote a major commentary on the *de anima* that is lost but quoted from principally by Philoponus,[70] and there is an extant work on the *de sensu*.[71] Conceivably he was led to this area by his teacher Sosigines who wrote an *On Vision*.[72]

The only other extant commentary is that on the *Metaphysics* of which only Books A-D are genuine.[73]

If we cannot be certain that Alexander's commentaries were more wide-ranging than those of any single predecessor, his extensive influence on many subsequent commentaries is a strong indication that he certainly surpassed them in quality. It is very difficult to judge the extent to which these works sprang from his pedagogical activity. The text by text discussions are much less formally constructed than some of the stylised products of the later Alexandrian school,[74] but it is still a fair presumption that these commentaries must in some way have complemented his teaching. We can distinguish in them roughly three elements. First, there is the elucidation of the text and a basic gloss on it; Alexander gives evidence of having consulted several manuscripts (ἀντίγραφοι) to establish a text,[75] as well as having utilised the works of his predecessors. Second, there is a more expansive paraphrase that is offered particularly where the meaning of the text is problematic. Thirdly, there is the introduction of parallels from other philosophical schools, either because there is a reference to them in the text, or because, as in the case of the Stoics, their doctrines have a relevance to the issue under consideration.[76]

---

[70] Evidence at Zeller, pp. 790-791.

[71] This is probably a later work as it is not referred to elsewhere; see Wendland, *CAG*, III-i, *Praefatio*, p. v. The reference to another discussion of memory and reminiscence at *de an.* 69. 20 may be to a commentary on Aristotle's *de memoria*—Alexander may therefore have commented on the whole of the *Parva Naturalia*.

[72] Above, n. 50.

[73] See Moraux, *Alexandre d'Aphrodise*, pp. 14-19 for a summary of theories about the authorship of the commentary on Bks. E-N.

[74] Cf. above, nn. 39 and 40.

[75] This is evident from *Met.* 75. 26, 104. 20-21, 145. 21, 356. 34; *Pr. An. I* 144. 5, 151. 15, 210. 30, 304. 14; *de sensu* 101. 4; and *Meteor.* 5. 29.

[76] These distinctions are not rigid, and are only loosely comparable to the *glossa*, *commentium* or *expositio*, and *paraphrasis* of mediaeval commentators. Sophonias, the fourtheenth-century Byzantine commentator, has some interesting comments on the superiority of paraphrases to running commentaries (*de an.*, pp. 1-2 at *CAG*, XXIII-i).

The Minor Treatises

Clearly the commentary (ὑπόμνημα) is a limiting scholarly form. In its concentration on specific texts it inhibits adequate discussion of problematic issues, or a broad enough treatment of opposed views. In particular it limits the exegesis of wide ranging questions where texts from a variety of treatises have to be synthesised. It seems that in order to meet these deficiencies Alexander composed several monographs (συγγράμματα).[77] The most important is his long essay, the *de anima*, an interpretative paraphrase of the Aristotelian work. Here Alexander presented controversial views on the formation of the soul and the nature of the intellect in a continuous fashion that seems designed for communication to a wider public than would have had access to the commentaries. Two other essays seem to have been a response to pressures from the contemporary philosophical culture, the *de fato* and the lost treatise, the *de providentia*, where he dealt with questions that are not directly treated by Aristotle and that therefore required the development of his doctrines in new directions. The *de fato* is a long critique of the Stoic doctrine of fate and the establishment of a parallel Peripatetic theory, while the work on providence, as far as we can determine from the fragments, attempted to integrate a theory of providence with Aristotelian cosmology. Both contributed to a long standing philosophical debate on these questions, while the latter could have been a response to one of the standard Christian criticisms of Peripatetic negativity about divine providence.[78]

---

[77] The complementary relation of treatises (συγγράμματα) and commentaries (ὑπομνήματα) seems to have developed among the Alexandrian commentators on Homer; see R. Pfeiffer, *History of Classical Scholarship*, p. 213, and cf. *Schol. in Il. Vet.* A, B. 111. The distinction is used by commentators who refer to Alexander's critique of the Stoic theory of mixture in both the σύγγραμμα (the *de mixtione*) and his commentaries; cf. Themistius *Phys.* 104. 18-22, and Simplicius *Phys.* 530. 14-16, and p. 91 below. It seems clear that the σύγγραμμα is a polished work (Alexander *Top.* 259. 5-6) bearing a title, while the ὑπομνήματα are without titles, and can often—it appears—be composed in haste. This was the case, for example, with Galen's dictation of the anatomical experiments interrupted by Alexander of Damascus (cf. above, p. 4-57 and Galen XIV. 630 K).

[78] On this whole topic see A.-J. Festugière, *L'idéal religieux*, Excursus C. Further on the *de providentia* see Commentary on *de mixt.* 10. 223. 6-14. Alexander followed the middle-Platonist and Christian attack on the Stoic doctrine of fate: cf. Clement *Strom.* I. 83. 5, Albinus *Didask.* 26, 179 Hermann, and Alexander *de fato* 26-30 *passim* but especially 196. 25-26, 197. 1-2

Since the *de mixtione* is the object of this study I shall not describe it at any length. It seems to be a complement to Alexander's other monographs in its critique of the physical theory that underlay the Stoic doctrines of God, fate, providence and the soul,[79] as well as affording the opportunity for a continuous critique of the theory of mixture along lines we know that Alexander followed in his commentaries on some of the physical treatises of Aristotle. It also justified a separate discussion of two problematic Aristotelian theories, those of mixture and organic growth.

It will be clear that there was a strong polemical element in these treatises (συγγράμματα).[80] Alexander's main opponents are the early Stoics, principally Chrysippus,[81] and here, as we shall see, he joins Platonists and Christians almost to the letter in some of his criticisms.[82] In this anti-Stoicism we can also observe his general affinities to the contemporary philosophical culture; for while he attacks the Stoics in the same manner as some of his contemporaries, he also defends the Peripatetics on the very grounds on which they were attacked by both pagan and Christian. This is true not only of his work on providence, but in his defence of the study of logic,[83] and possibly in his explications of Aristotelian ethics.[84]

---

206. 1, 206. 28-30 and 209. 20-210. 3 where similar wording is used, which suggest that he shared an anti-Stoic *fons communis* (see Lilla, *Clement of Alexandria*, pp. 50-51, who notes the first two passages but neglects Alexander). The proximity of this whole issue to contemporary philosophical culture is indicated by the fact that the *de fato* is the only work of Alexander's that Eusebius quotes (K. Mras, *Anz. d. Oest. Akad. d. Wiss. in Wien*, Phil. Hist. Klasse, 93 [1956] 215).

[79] *de mixt.* 227. 5-10 makes this clear.

[80] At *Phys.* 489. 20-22 Simplicius refers to Alexander's refutation of the view that infinite body can be in place "in the work written against the Epicurean Zenobius" (ἐν τοῖς πρὸς τὸν Ἐπικούρειον Ζηνόβιον ἀντιγεγραμμένοις). If this is in fact a treatise, it is another polemic to be added to his works, but the reference may be to a discussion in a commentary or a scholium. Alexander's general critique of Epicurean physics can be traced in *de mixt.* 2, *Quaestiones* III. 12, and Alexander ap. Philoponus *de gen. et corr.* 12. 6-25. As for other treatises, *Pr. An. I* 125. 30-31 seems to refer to an independent discussion of syllogisms, but the treatise "On Spirits" (Περὶ δαιμόνων) referred to by Michael of Ephesus *Parv. Nat.* 83. 27 and 84. 26 is almost certainly spurious; see Cranz, *Catalogus* II, pp. 420-422.

[81] See further Part II-i on this and the general topic of Alexander's anti-Stoicism.

[82] In addition to n. 84, cf. Commentary on *de mixt.* 11. 225. 5-18, 226. 24-30.

[83] *Pr. An. I* 4. 30ff. and 6. 8ff. are both in effect an apologia for the study of logic.

[84] Lilla, *Clement of Alexandria*, pp. 68-71 shows that the idea that "virtue

The Mantissa and Quaestiones

In addition to this balanced achievement of commentaries and monographs, there are several works of dubious authenticity attributed to Alexander. Most of these can be disregarded as spurious since they are totally unrelated to his main works.[85] There are however two collections that do bear more closely on the themes of the major treatises. The first of these is appended to the *de anima* as, in effect, a second book. It consists of a series of separate texts of which the opening five deal directly with psychology, presumably the justification for its presence there. The remaining twenty treat a variety of topics: the largest group deals with questions of vision and light, while others cover problems in physics and ethics, the final four being a contribution to the Peripatetic doctrine of fate and providence. The collection has been termed rather slightingly the *Mantissa* by its editor Bruns. Its most important text is the *de intellectu*, a restatement of Alexander's interpretation of the Aristotelian doctrine of intellect which had an independent fortune in the Arabic and mediaeval worlds. Until Moraux's study it was regarded as a complement to the *de anima*, but since he has pointed up significant differences between the two, its authenticity is at least questionable, though not all would follow him in declaring it spurious.[86] The remainder of the texts obviously form a heterogeneous collection. They are on the whole polemical and written in a very stylised fashion. The main opponents are the Stoics, but more than this there is a

---

is self-sufficient with regard to happiness" is a commonplace in Stoicism, middle-Platonism and Clement. The Peripatetic denial of this against the Stoics must also have been a commonplace; cf. Aulus Gellius XVIII-i, and Lucian *Eun.* 3. It is therefore interesting to note that *Mant.* 159. 15-168. 26 presents arguments for the Peripatetic position. Whether this work is Alexandrian is questionable (cf. below), but that it belongs in the context of second and third century philosophical culture seems undeniable.

[85] These are: *de febribus* (Περὶ Πυρετῶν), ed. J. Ideler, *Physici et Medici Graeci* (2 vols., Berlin, 1841-1842), I, 81-106; and the four books of *Problemata*, Bks. I and II at Ideler, *op. cit.*, I, 3-80, Bks III and IV, ed. U. C. Bussemaker, *Aristoteles, Opera Omnia*, IV (Paris, 1857), and ed. H. Usener (Berlin, 1899).

[86] See Moraux, *Alexandre d'Aphrodise*, especially ch. 3. F. Trabucco, *Acme* 11 (1958) 117-118, follows Bruns (*SA*, II-i, *Praef.*, p. xii) in regarding it as one of the authentic treatises of the *Mantissa*. His view is accepted by P. Merlan, *Monopsychism, Mysticism, Metaconsciousness*, p. 46 n. 3. Recently B. C. Bazán, *Rev. Phil. de Louvain* 71 (1973) 468-487, has made an impressive case for its authenticity.

consistent anti-materialism running through the majority of these texts that gives them an underlying unity. Since the form of this anti-materialism is basically that taken by Alexander's major critique of the Stoic theory of mixture, I shall try in my discussion of that critique in Part Two below [87] to throw some light on this part of the collection. This will not in any sense demonstrate authenticity, but will simply show that some of the texts bear a systematic relation to authentic works of Alexander.

The second questionable collection is a heterogeneous group of texts known as the *Quaestiones*,[88] arranged in four books. The fourth, the 'Ηθικὰ Προβλήματα ("Problems in Ethics") is the only one that accurately reflects its title. The other three, though termed "problems and solutions of physical questions," in fact contain some sixty-nine items not all of which are problems, while only some twenty-four are on physics, whereas seventeen deal with psychology, eleven with logic and metaphysics, and six with questions of fate and providence. Certain patterns can be discerned in this maze. The texts on psychology seem to be a series of notes related probably to Alexander's lost commentary on the *de anima*,[89] while a similar series deals with issues in the *de generatione et corruptione*.[90] Again some have a very simplistic character which suggests that they may be students' notes, or some form of pedagogic device, while others show a close connection with themes from authentic works of Alexander.[91] As with the texts of the *Mantissa* this is perhaps as far as one can go in dealing with the problem of their authenticity.

Alexander's Successors: His Influence

The attribution of these works to Alexander is obviously a compliment to his preeminence in the Peripatetic school. We do

---

[87] See especially p. 81 and n. 234.

[88] Quaestiones, of course, became one of the major genres of mediaeval Aristotelian scholarship, and perhaps earlier, of Arabic scholarship, since many extant Arabic works attributed to Alexander are in this form; see A. Dietrich, *Nachr. Gött. Akad.* 1964, pp. 93-100.

[89] Moraux (*Alexandre d'Aphrodise*, pp. 22-24) thinks that this collection may have been paralleled in the "notes on logic" (σχόλια λογικά) which Alexander refers to at *Pr. An. I* 250. 1-2, or the collection of explanations of readings (λέξεις) from the *de sensu* quoted by a scholiast on *Quaest.* I. 2 (see Bruns, *SA*, II-ii, p. ix).

[90] Manifestly II. 22, and probably I. 5, I. 15, I. 19, and III. 5.

[91] This is the case with *Quaestiones* II. 12 which I have discussed at *Philologus* 116 (1972) 293-305.

not know whether they were written directly under his influence or in the Peripatetic school of a later generation. All that is probable is that they did emanate from the school, as organised in Athens or elsewhere, since they are dominated by a polemical or apologetical tone. Alexander's claim to them might be an indication of the extent to which he dominated his successors. Of these very little is known, and we must be guided mainly by guess-work. An Ammonius is referred to by Philostratus [92] as the "best-read" (πολυγράμματος) man he knew, and he and his contemporary Ptolemaeus are described by Longinus in Porphyry's life of Plotinus as "very learned" (φιλολογώτατοι) with Ammonius being the superior of the two.[93] These qualities might imply that they carried on into the 220's and 230's the scholarly work that Alexander of Aphrodisias had so strikingly advanced. Longinus also reports [94] that neither wrote any technical works but only "poems and show-speeches" (λόγοι ἐπιδεικτικοί), which suggests a return to that coalescence of philosophy and rhetoric of which we saw indications in Aristocles' works, and that at least seemed implied by the social circles in which Alexander of Damascus moved.[95]

Alexander of Aphrodisias' major influence on later Greek philosophy was to be reserved not for the Peripatetic school, of which he is the last significant figure, but for the long tradition of Platonist commentators on Aristotle. He was read along with Aspasius and Adrastus by Plotinus,[96] whom he to some extent influenced. The new tradition was inaugurated with Porphyry's introduction and commentary on the *Categories*,[97] and was continued in the Platonist schools of Athens and Alexandria in the fifth and sixth centuries. It is from the works of these commentators that so much of Alexander's lost commentaries remains to be retrieved.

---

[92] *V.S.* 618. On him see *RE*, I-i, cols. 1862-1863 where the guess is made that both were disciples of Alexander.

[93] *Vit. Plot.* 20. Ptolemaeus cannot be identified (*pace* A. Dihle, *RE*, XXIII-ii, col. 1860) with the critic of Dionysius Thrax described by Sextus Empiricus *A.M.* I. 60-61, if Sextus is to be dated ca. 100 A.D. (see F. Kudlien, *RM* 106 [1963] 251-254).

[94] Ap. Porphyry *Vit. Plot.* 20.

[95] See above, pp. 6-8. Cf. also *de fato* 165. 5-7 where Alexander assures the emperors that his philosophical treatise is not designed as a demonstration (ἐπίδειξις) but as an instructive inquiry (ἐξέτασις, διδασκαλία) of the subject.

[96] Porphyry *Vit. Plot.* 14. On Alexander and Plotinus see the literature at Bibliography II-4.

[97] At *CAG*, IV-i.

PART TWO

ALEXANDER OF APHRODISIAS AND THE STOIC THEORY OF TOTAL BLENDING

I. ALEXANDER AND STOICISM: A PRELIMINARY SURVEY

Stoics and Peripatetics had close affinities, especially in ethical theory, in the period of philosophical eclecticism represented by Antiochus of Ascalon and reflected in Cicero's philosophical works,[1] or in Arius Didymus' ethical surveys. It is difficult to judge whether such syncretism continued to be evident in the Peripatetic scholasticism of the first two centuries of the Roman Empire described in the preceding section. [2] In certain respects the two schools were polarised in the standing debate on the relative merits of their different doctrines of categories,[3] and they seem also to have clashed over the different argument-forms recognised by their logical systems.[4] Apart from this, and some standard debates on ethics,[5] there is little evidence of Peripatetic attempts to criticise or philosophically interact with Stoicism. It might then have been unusual for Alexander of Aphrodisias to pay greater attention to this school than any other; perhaps just as he extended Peripatetic scholasticism beyond any of its previous achievements he may also have been the first to take it into this particular area. But with so little evidence of his predecessors we cannot be certain. What

---

[1] Cf. *De Fin.* iv. 26. 72 and Reid's note on *Acad. Post.* I. iv. 17. 1. Some recent scholarship has reemphasised these on general philosophical grounds: see A. A. Long, *BICS* 15 (1968) 72-75, J. M. Rist, *Stoic Philosophy*, chs. 1-3.

[2] See Pohlenz, *Die Stoa*, I, 355-357 on this and the whole topic of the Peripatetic treatment of Stoicism.

[3] L. A. Cornutus, the younger Seneca's protégé, wrote a commentary on Aristotle's *Categories* which was known to Simplicius (*Cat.* 64. 24 etc.) and therefore may well have been known by Alexander.

[4] See I. Mueller, *AGPh* 51 (1969) 173-185, an article based on material in Alexander's commentaries on *Pr. An. I* and the *Topics*.

[5] E.g. Plutarch's *de virtute morali* is an essentially Peripatetic attack on Stoicism, possibly inspired by Andronicus. Cf. also the discussion of emotions at Aspasius *N.E.* 44. 10-45. 22 where the views of the Stoics and Andronicus are contrasted. At *REL* 45 (1967) 396-419 P. Grimal has discussed Seneca's critique of Aristotle in the *de vita beata*.

is obvious is that particularly in his treatises Alexander chose both to criticise and assimilate Stoic doctrines as he sought to establish interpretations of Aristotle's thought. This suggests at least a highly individual response to Stoicism among members of his school, and one which must be taken account of before analysing texts in which he deals with a specific Stoic doctrine.

First we must consider the general form of Alexander's references to the Stoics. In Von Arnim's collection he is cited some one hundred and fifty seven times in the two volumes of Chrysippean texts (*SVF* II and III) [6] although he mentions this particular Stoic by name in only two contexts.[7] This was justified by the editor on the grounds that Alexander encountered that form of Chrysippean Stoicism upheld by the scholastic Stoics in the first two centuries of the Roman Empire.[8] Yet such an unqualified claim about a putative Stoic source is unsatisfactory when it is not derived from the internal perspective of the author himself. Von Arnim's brief preliminary discussion of Alexander did nothing to refine it, nor can his device of typographical variations be said to satisfactorily indicate the differing value of the material collected on Chrysippus from Alexander and other sources.[9] Clearly an author's approach

---

[6] By Von Arnim's numbered citations there are 48 from the commentaries and 109 from all other works. This is only a rough guide since often more than one passage is quoted in a single citation.

[7] *de mixt.* 213. 7; 216. 8, 9, 14; 218. 8, 9; and *Pr. An. I.* 177. 25 and 180. 36 (cf. note 12 below). Other Stoics referred to by Alexander are Antipater of Tarsus with regard to some of his logical theories (*SVF* III 24, 26, 27) and, very briefly, Sosigenes, a pupil of Antipater, à propos his theory of mixture at *de mixt.* 216. 12.

[8] *SVF I, Praef.*, p. xvii: "Nam tres illi auctores [Alexander, Plutarch, Galen] . . . scholasticos sui temporis Stoicis impugnant."

[9] *SVF* I, p. v. Three categories are established here: (1) fragments from Chrysippean works; (2) placita, either bearing his name or closely related to him; (3) a very wide class of admittedly varying value termed "quae aliquo modo ad cognoscendam Chrysippi doctrinam utilia viderentur." The latter assumes a Chrysippean orthodoxy which is not always recognised; see next note and cf. Galen *De Plac. Hipp. et Plat.* III. 4, IV. 5, pp. 279-280, 362-363, 376 Müller, where the ignorance of Chrysippus' works by contemporary Stoics is noted. On the other hand this evidence has to be employed where its congruence with Chrysippean doctrine can be recognised, and Gould (*Philosophy of Chrysippus*, p. 1) is too drastic in excluding such passages since they only "[purport] to convey general Stoic doctrine": a general *reference* is not the same as "general doctrine"—cf. A. C. Pearson, *JP* 30 (1907) 219 on this. Again it is not sufficient to assume that Chrysippus' name in a text makes a doctrine *ipso facto* Chrysippean; thus von Arnim's class 2 needs careful scrutiny. Gould (*loc. cit.*) overlooks this in terming all

to Stoicism will be conditioned by the sources available to him, but when, as is the case here, these are not extant, we need to analyse carefully that author's own treatment of this philosophy in the different contexts of his works before speculating on his possible sources. Again, even though Von Arnim's claim is usually justified that Chrysippean doctrines are most in evidence in Alexander's works,[10] we still need again to evaluate carefully his reports and not accept their validity simply because of this very general truth. The following is only a sketch of the prolegomenon necessary to a selection of evidence on Stoic philosophy from the Alexandrian corpus. It will try to identify the main categories that such a discussion should follow and describe some of the central problems; only in the second and third parts of this essay, and later in the commentary on the *de mixtione,* do I undertake the detailed criticism that a comprehensive inquiry should also contain.

Of fundamental importance to any account of Alexander's relation to Stoicism is the manner in which he introduces his references to this school. Where he refers to earlier Peripatetics his regular practice is to add a further reference to a specific work, presumably because he possessed a copy or had reliable knowledge of its contents.[11] However in only one of the two contexts in which he refers to Chrysippus by name does he follow this practice, mentioning but not quoting from a treatise "On the Universe" (Περὶ κόσμου).[12] Otherwise, in all his discussions of the Stoics the

such texts "fragments"—e.g. Diog. Laert. VII. 54 (= *SVF* II 105), category 2 for Von Arnim, is a fragment for Gould (p. 63).

[10] Cf. however *de an.* 68. 15-16 where Alexander argues in general terms that a presentation (φαντασία) cannot be an imprint in the soul because a large number of simultaneous imprints will be indistinct. Sextus (*A.M.* VII. 228 = *SVF* II 56) reports that this was Chrysippus' criticism of Cleanthes! If Alexander is criticising the Stoics at this point he has taken a non-Chrysippean view as orthodoxy.

[11] The indices to Alexander's works under Eudemus and Theophrastus demonstrate this amply enough. Clearly references to earlier commentators (e.g. to Herminus in the commentary on the *Topics*) are not attributed to a given work since it was simply a similar commentary to that being written.

[12] *Pr. An. I* 180. 36. The subject is the preservation of numerical identity in a recreated cosmos after the conflagration; cf. the texts from the whole context at *SVF* II 624. It is questionable whether Alexander actually possessed this book, since he never quotes from it again, although it deals with the topic of fate (see the further quotation by Stobaeus at *SVF* II 913) which Alexander dealt with in the *de fato.* Also in his earlier reference to Chrysippus at *Pr. An. I* 177. 25 (not in *SVF*) on the same general topic Alexander does not include a reference to this book title.

school is referred to collectively, in the manner of most modern scholars, or more often still without any form of identification. By this token it might be reasoned that Alexander did not possess any treatises by early Stoics and was totally reliant on some form of secondary material. But this would not be decisive, since Alexander may have more freely paraphrased and abstracted from Stoic treatises than from those of his own school. However it is clear that he had access to certain doxographical reports of Stoicism,[13] and in many of his formulaic reports of Stoic ethics could well be drawing on the kind of Stoic handbook of definitions that we know his contemporary Origen employed.[14] Again he may have known the works of contemporary Stoic scholasticism best represented for us by Hierocles' summary of ethical doctrines preserved on papyrus.[15] It seems in fact quite probable that Alexander had a variety of sources at his disposal that we cannot, like Von Arnim,[16] collectively characterise with any confidence: this uncertainty not only arises from the lack of independent source material but from the very framework in which Alexander couches his discussions of Stoicism.

The mode of reference that we have just reviewed is one indication of this. The Stoics are frequently described collectively or tacitly simply because they are being identified within the broader context of a discussion of an Aristotelian subject. So, for example, in Alexander's *de anima* "the Stoics" are said to believe that the soul is a harmony since it is a pneuma composed of fire and air.[17] Now the identification of soul as pneuma is orthodox Stoicism,

---

[13] See below, pp. 55-65.

[14] R. Cadiou, *REG* 45 (1932) 271-285, demonstrates Origen's use of a Stoic book of definitions by one Hierophilus, and E. Klostermann, *ZNTW* 37 (1938) 54-61 lists standard definitions in Origen and Clement. Alexander's definition of ἡδονή at *Top.* 181. 4 (= *SVF* III 434) corresponds to those listed by Klostermann. Cf. also the definition of ἔρως at *Top.* 139. 24-25 (= *SVF* III 722) with those of Diogenes Laertius (*SVF* III 396, 716) or Stobaeus (*SVF* III 395, 717), or that of ἀρετή at *de fato* 211. 17 (= *SVF* III 283) with Stobaeus (at *SVF* III 262).

[15] *Hierokles: Ethische Elementarlehre*, ed. H. Von Arnim; pp. 11ff. of the introduction to this work outline its place in this scholastic movement. See Comm. on *de mixt.* 1. 213. 2-13 and 4. 217. 36 on its relevance to our study.

[16] Also cf. Bruns, *S.A.*, II-i, *Praef.* p. xvi, who dogmatically claimed that Alexander's many tacit references to the Stoics indicated that he possessed no original works by early Stoics at all.

[17] 26. 15-17. Here they are grouped with the Epicureans on the grounds that both held that the soul was composite.

but its characterisation as a harmony is a distortion derived from Aristotle's discussion of this theory of the soul in his *de anima* A4 which Alexander is drawing on here, and is designed to facilitate criticism from an Aristotelian standpoint.[18] This random example is very typical and could be multiplied particularly from Alexander's commentaries where the Stoics are referred to only *en passant* and entirely within the context of a discussion of a specific Aristotelian text. This is notably the case with the numerous formulaic references to Stoic ethical doctrines in the commentary on the *Topics*.[19] In principle it is difficult to decide whether such references are the result of Alexander's own abstraction of Stoic ideas from the works of early or contemporary Stoics,[20] or whether they are taken from some independently developed handbook in which they would already be presented in a formulaic fashion, particularly when early Stoics themselves seem to have written such works.[21] Rather than continue these speculations we are better employed for the moment in trying to characterise accurately the complexity of Alexander's relation to Stoic philosophy as we find it in his works.

Our best evidence here is his treatment of the Stoics in his essays the *de anima, de fato,* and *de mixtione.* In these monographs his Aristotelian exegesis is untrammelled by an accompanying text, while he can deal with his major philosophical opponents in much broader terms than in the commentaries. For example, one major approach that he adopts is to identify the general principle of a particular Stoic doctrine and criticise this rather than the doctrine itself and its ramifications. An extreme case of this occurs in the

---

[18] See e.g. *SVF* II 897; Alexander is also concerned to distinguish this theory from this own theory of the soul as a form supervening on the mixture of the elements; on this see Donini, *Att. d. Accad. di Torino* 105 (1971) 61-107 at 82-92.

[19] *SVF* III 62, 67, 147, 152, 434, 594, 711, 722. The Epicureans are treated similarly, e.g. *Top.* 14. 7, 19. 13 and 139. 23. Cf. also *SVF* II 329 where Alexander's critique of the Stoic category of τὸ τί arises from the immediate context in the *Topics*.

[20] As far as I know only at *de fato* 18 does Alexander clearly refer to contemporary Stoic works, but this is in such general and discursive terms that nothing of his general practice can be gleaned.

[21] E.g. Diog. Laert. VII. 199 refers to Chrysippean works of ethical definitions (ὅροι). Also cf. *SVF* I, p. xli on the influence of a Chrysippean summary of ethics on Diogenes Laertius. On Chrysippus' role in the tradition of *florilegia* see K. Horva, *RE* Supplbd. 6 (1935) 80-81.

de fato where the bulk of the treatise is a polemic against "those who say that everything occurs according to fate," a group that is not even once specifically identified as Stoic.[22] Again, as we shall see, although the *de mixtione* contains an extensive exposition of the Stoic theory of blending, it is at its core an attack on "those who say that body goes through body," a group to which the Stoics are as it were affiliated in virtue of their theory of blending.[23] The *de anima* contains a briefer attack on the general thesis that soul is body,[24] while the fragmentary *de providentia* probably contained a polemic against the general notion of an immanent deity.[25]

Now this technique of isolating a general thesis as the kernel of an opponent's thought is a common enough practice in ancient as well as modern philosophy. Alexander may have been particularly influenced by Aristotelian historical method which rested on such general identifications of the views of earlier thinkers largely in terms derived from his own system.[26] But in the commentator's case these generalisations serve only to provide a basis on which his opponents views may be criticised: they polarise his opposition and allow him to introduce criticisms which can equally employ Aristotelian and more general ideas.[27] This is particularly the case with the series of texts in the *Mantissa* which attack a range of general theses on psychology, the theory of vision, and ethics.[28] Now although this method may be typical, it poses considerable problems when we try to establish firm evidence for Stoic doctrines, for generalisation is itself a distortion, and it is difficult to learn much from ensuing criticisms.[29] Again this method isolates Stoic

---

[22] A. A. Long, *AGPh* 52 (1970) 247-268 deals at length with this feature of the work in its early chapters.

[23] In sect. iii of this essay I try to demonstrate that this is a classification independently developed in the Peripatetic tradition.

[24] 17. 15-18. 10. "The Stoics" holding this thesis are however identified in explicitly Aristotelian terms as [οἱ λέγοντες] πᾶν σῶμα, ἢ ὕλην ἢ ἐξ ὕλης, εἶναι (17. 16).

[25] On this see Comm. on *de mixt.* 11. 226. 24-30.

[26] Without offering a general verdict on its merits, this is, I hope, an indisputable result of the researches of H. Cherniss, *Aristotle's Criticism of Presocratic Philosophy*, and J. MacDiarmid, *HSCP* 61 (1953) 85-156.

[27] In the polemic against corporeal qualities at *Mant.* 122. 16-125. 4, for example, 123. 36-124. 9 uses the concepts of matter and quality while 124. 9-20 presents a set of general anti-materialist arguments.

[28] Some of these are discussed at nn. 234, 243 below.

[29] Alexander *Quaest.* III. 12 which sustains the general thesis "that being

doctrines from the system to which they belong and makes the task of reconstructing their interrelation difficult and often highly speculative.[30] Finally, there is the encouragement that this approach may give to the assimilation of an opponent's terminology and ideas; when a philosopher considers only the general character of an opponent's thought broad similarities with his own can be more easily recognised.

This latter aspect, whatever its cause, needs special comment. Like many of his contemporaries Alexander freely borrowed the philosophical vocabulary of the Stoics and others to describe concepts peculiar to his own system. In many cases this can be quite innocuous where the alternative term simply functions as a description without any philosophical significance.[31] However there are cases where such borrowing involves a significant modification in Alexander's presentation of Aristotelian thought, and consequently affects his view of related Stoic doctrines. A good example of this is his ready assimilation of the language of the Stoic doctrine of common notions which, though in the commentaries serving merely as a terminological variant for (contradictorily) both "general assumptions" (ὑπολήψεις, ἔνδοξα) and axioms, is used in the *de mixtione* and *de fato* against the Stoics. Their doctrines of fate and mixture are then said to be against these common notions, while the corresponding Aristotelian doctrines are said to obey them.[32] Again the very definition of fate in "Aristotelian" terms in the *de fato* 6 as nature *qua* efficient cause is the result of an assimilation of the broad Stoic framework [33] for dealing with this question, as were very probably some aspects

---

is not infinite" (ὅτι μὴ ἄπειρον τὸ ὄν) has a long attack, in the manner of the texts in the *Mantissa*, against various theories of the infinite. At *SVF* II 536 Von Arnim, rather misleadingly I think, cites 101. 10-15 and 106. 10-13 from this text when it is clear that, despite the generality of the arguments, Alexander's opponents are Epicureans (cf. 104. 11-13 on infinite worlds, and cf. 101. 10-16 with Epicurus *Ep. ad Hdt.* 41). His error is indicative of the dangers of using these highly general polemics as evidence.

[30] On this cf. pp. 72-73 below.

[31] For a discussion of one such case see my article on ἐπιτηδειότης at *Acta Classica* 15 (1972) 25-35.

[32] Cf. *de mixt.* 232. 34-233. 1 and *de fato* 182. 20-22. On the whole topic of Alexander's assimilation of these terms see my discussion at *Symb. Os.* 48 (1973) 62 nn. 80-85.

[33] On this see G. Verbeke, *AGPh* 50 (1968) 81. Cf. Comm. on *de mixt.* 11. 225. 20-27.

of Alexander's complementary definition of providence.[34] These two concepts were developed by the Stoics: the commentator then offers a Peripatetic response.

There are other less obvious examples. In the *de anima* a genetic account of the formation of universals is modelled after a Stoic account of concept formation.[35] Again in the same treatise we find an account of the relation between imagination (φαντασία) and action systematised beyond anything in Aristotle.[36] Briefly the scheme is that assent (συγκατάθεσις) to imagination is "in our power" (ἐφ' ἡμῖν) because assent is preceded by deliberation and is therefore rational. Thus we follow the impulse to action (πρᾶξις) freely because we may withdraw our assent if we wish. As it stands this is a Stoicised version of Aristotle's view that choice (προαίρεσις) is a "deliberative impulse for the things in our power" (βουλευτικὴ ὄρεξις τῶν ἐφ' ἡμῖν).[37] However in the *de fato* this becomes the basis on which the Stoics are criticised, when Alexander claims that for them all human action is determined in the sense that assent to action and impulse are identical with "what is in our power" (τὸ ἐφ' ἡμῖν) with no deliberative process to ensure freedom by allowing for the rejection of an action.[38] That is to say, the Stoic theory of fate cannot preserve freedom as conceived of within a Peripatetic schematisation itself based on Stoic concepts. The pertinence of this criticism aside [39] it is in effect claiming that a Stoic doctrine is deficient because it does not deploy its own concepts in the same way as an assimilator who has isolated them from their own system. In structure it is exactly the same kind of critique as that based on the assimilation of the Stoic common notions.

This very brief sketch can do no more than suggest the complexity

---

[34] See Comm. on *de mixt.* 225. 18-226. 10. Further on the *de providentia* see Comm. on *de mixt.* 10. 223. 6-14.

[35] Cf. 83. 2-8 with Aetius *Plac.* IV. 11. (= *SVF* II, p. 28. 12-19). This was noted by Moraux (*Listes Anciennes,* p. 5 n. 16).

[36] 72. 13-73. 13. This passage deserves more detailed discussion and I hope to return to it.

[37] *N.E.* 3. 1113a 11-13. Cf. Long, *BICS* 15 (1968) 81.

[38] This is the burden of *de fato* 11-14. Long has discussed it in detail (*AGPh* 51 [1970] 258-265), but did not note how it is in large measure either conditioned by or reflected in the assimilation undertaken at *de anima* 72-73. Cf. also *Quaest.* III. 13, a more scholastic version of the latter. On a broadly similar critique of the Stoics in Plotinus see Graeser, *Plotinus and the Stoics,* pp. 48-49.

[39] On this see Long, *AGPh* 51 (1970) at pp. 258-262.

of Alexander's relation to Stoicism. He is clearly an author who has set a considerable distance between himself and his sources, whatever they were; for he has integrated his knowledge of Stoicism, as we would expect, into the framework of his exegesis of Aristotle, and often to the point where it is no longer doctrine to be rejected but an essential part of his own philosophical contribution. He is perhaps unique in the extent to which he has thoroughly assimilated Stoic thought, though many of his individual criticisms and patterns of assimilation can be paralleled from other authors.[40] These parallels must also be investigated if we are to establish our source's place in the general tradition of evidence on particular Stoic doctrines. The next section of this essay will therefore attempt to establish Alexander's significance as both source and critic in the context of the variegated evidence for the Stoic theory of total blending. Similar research could be conducted into many of our major sources for Stoic philosophy [41] and I would hope at least to establish the value of the method employed, if not of all the conclusions reached, in the following discussion.

II. THE STOIC THEORY OF TOTAL BLENDING

A Reconstruction from the Sources (*apropos de mixtione 3 and 4*)

The evidence for the fundamental doctrines of Stoic physics is almost entirely provided by authors of the first and second centuries A.D. The doxographical tradition, Philo, Plutarch, Galen, Sextus Empiricus, Diogenes Laertius, and Alexander of Aphrodisias are our major sources for such doctrines as body, place, time, infinite division, the elements and principles, and those of pneuma and total blending (κρᾶσις δι' ὅλων). The attempt to reconstruct any of these doctrines on the basis of the sources imposes two main

[40] His use of the common notions is comparable to that of Plutarch in the *de communibus notitiis*, though cf. my comments at *Symb. Os.* 48 (1973) 63 n. 88. Many other examples are collected in the Commentary; see especially on 6. 219. 9-18 and 11. 225. 5-18, 226. 24-30.

[41] Recent studies such as M. Spanneut's *Le Stoïcisme des pères de l'Église* and D. Babut's *Plutarque et le Stoïcisme* deal primarily with the influence of Stoicism on positive philosophical activity and not the effect that such assimilation has on an author's reports and criticisms of Stoic doctrines. A recent article by L. Roberts, "Origen and Stoic Logic," *TAPA* 101 (1972) 433-444 is however a good example of how an appreciation of a non-Stoic author's use of Stoic philosophy can materially assist the interpretation of specific doctrines.

requirements: to determine, as far as possible, the tradition by which our evidence is formed;[42] and, since most of these sources are anti-Stoic, to take careful account of their biases and to analyse their criticisms and polemics. Since the quality and quantity of evidence varies considerably from one doctrine to another there cannot be an invariable method for carrying out this programme. In the present case I have chosen first to formulate an interpretation of the theory of total blending that originates in those texts which come closest to being fragments, and then to test it in the context of a varied group of secondary reports.[43] Finally in section (iii) I shall discuss the principal ancient critique of the theory of total blending in the light of what should by then be a well grounded interpretation.

The Primary Sources—A Preliminary Interpretation

There are two texts which though not indubitably fragments at least provide evidence that Chrysippus dealt with the theory of mixture in specific works.[44] Since they are crucial to our interpretation they need to be quoted in full.

---

[42] Or, to use K.-H. Abel's expression (*Gnomon* 44 [1972] 645-651 at p. 646), we must deal with "der überlieferungsgeschichtlicher Aspekt." Abel's whole review touches on many of the general problems with source material raised in the course of this essay. See also H. J. Mette's review of Pohlenz's *Die Stoa*, *Gnomon* 23 (1951) 30-39 where two examples are given of inquiries based on a detailed analysis of the sources.

[43] I am excluding any attempt to establish an independent account of this theory's historical background, although comparisons, particularly with the Aristotelian theory of mixture, will be unavoidable. Such accounts, so prevalent in the literature on Stoic physics, must be derived from the reconstruction of doctrines from the sources. I am also excluding the use of texts which show the influence of the theory of blending. Such material can often assist interpretation (cf. my discussion at *Symb. Os.* 48 [1973] 60-63) but in the present case it is so parasitic on extant sources that we can neglect it. It is largely to found among the Christian Fathers (see H. Wolfson, *Philosophy of the Church Fathers*, I, ch. 16, pt. 3, and M. Spanneut, *Le Stoïcisme des Pères de l'Église*, index under κρᾶσις δι' ὅλων and μῖξις) and was for the most part derived from the standard tripartite classification of mixtures discussed at pp. 49-65 below.

[44] Reports that Zeno held that substances and qualities are totally blended (Galen at *SVF* I 92 [cf. p. 59 below]) or that body went through body (Hippolytus at *SVF* II 496, Themistius at II 468) are in very general terms and not substantiated elsewhere. Certainly the doctrine of a cosmic pneuma was Chrysippean (cf. n. 61 below) and, as we shall show, there is no reason to believe that total blending had any meaning except in relation to it.

Diogenes Laertius VII. 151 (= *SVF* II 479).
καὶ τὰς κράσεις δὲ διόλου γίνεσθαι, καθά φησιν ὁ Χρύσιππος ἐν τῇ τρίτῃ τῶν Φυσικῶν, καὶ μὴ κατὰ περιγραφὴν καὶ παράθεσιν· καὶ γὰρ εἰς πέλαγος ὀλίγος οἶνος βληθεὶς ἐπὶ ποσὸν ἀντιπαρεκταθήσεται, εἶτα συμφθαρήσεται.

And blends occur through and through [totally] [45] as Chrysippus says in the third book of his *Physics* [46] and not by surface contact and juxtaposition; for a small amount of wine thrown into the sea will be extended [through it] to a certain degree [47] and then blended with it.[48]

Plutarch *de communibus notitiis* 37. 1078E (= *SVF* II 480).
καὶ ταῦτα προσδέχεται Χρύσιππος εὐθὺς ἐν τῷ πρώτῳ τῶν Φυσικῶν Ζητημάτων, "οὐδὲν ἀπέχειν φάμενος, οἴνου σταλαγμὸν ἕνα κεράσαι τὴν θάλατταν," καὶ ἵνα δὴ μὴ τοῦτο θαυμάζωμεν, "εἰς ὅλον, φησί, τὸν κόσμον διατενεῖν τῇ κράσει τὸν σταλαγμόν."

And Chrysippus made the following claim right at the beginning of the first book of his *Physical Inquiries*,[49] saying

[45] διόλου here is unique in a physical sense. LSJ cite it in the sense of "altogether" but at Aristotle *Poet.* 1460a 8 (the only passage cited) in a temporal sense. The commoner δι' ὅλων might well be read here; cf. note on the text of *de mixt.* 221. 33.

[46] This book-title is mentioned only by Diogenes, and this is the only reference to Book III. Book I is cited as a source for central physical doctrines (the principles [VII. 139], the elements in the cosmogony [VII. 135, 142] and the doctrine of matter [VII. 150]) while Book II dealt with epistemology and phychology (cf. VII. 54, 55, 57 and 157). We shall show that total blending has at least some relation to the former set of doctrines.

[47] Translators take ἐπὶ ποσόν in a temporal sense: "for a while" (Hicks, Loeb ed.), "aliquamdiu" (Cobet, Paris [1878]). However if we take the phrase as describing the limit to which the wine is extended, the sense in which extension is a physical precondition of blending is much clearer. There are several closely related expressions in the descriptions of extension in *de mixtione* 4 that must clearly be taken in a physical sense: ἐπί τι ἐλθεῖν μέγεθος (217. 18), ἐπὶ τοσοῦτον χεῖσθαι (217. 30), εἰς τὴν ἐπὶ τοσοῦτον ἔκτασιν (217. 32). Cf. also *de mixt.* 6. 220. 18 (ἐπὶ πλεῖον . . . χεόμενα) and 220. 20 (τὴν ἐπὶ πλέον . . . χύσιν).

[48] Literally συμφθείρεσθαι means "to be destroyed along with" and elsewhere "joint-destruction" is a synonym for fusion (σύγχυσις), a blend in which constituents form a new compound from which they are irretrievable (see pp. 50-51 below). Either this is an error, or as I suspect (with Hicks, Loeb ed.) the term in fact means a blend in which constituents do not lose their identity; cf. p. 69 n. 188 on Plutarch's use of συγχεῖν to mean "blend."

[49] Plutarch (*Sto. Rep.* 1053F, *de comm. not.* 1084C) quotes the same book again. The former passage deals with the doctrine of ἕξις which is, as we shall see, relevant to our interpretation of total blending (see pp. 36ff. below).

> that there is nothing to prevent a single drop of wine blending with [tempering] [50] the sea, and so that we may not be shocked by this, says that in the blending the drop will extend through the whole universe.

These are the only places in which the theory of blending is reported in association with Chrysippus' name and a book title; thus although Diogenes often epitomises Stoic doctrines, and Plutarch paraphrases very loosely, both texts justify close independent scrutiny.

Diogenes first describes blending as occurring totally (δι' ὅλων) rather than by juxtaposition (or, as it is often termed, chemical combination). Now in itself this does not explain the nature of this totality. Aristotle also distinguished blending from the juxtaposition of corpuscles,[51] but provided an account of blending as the unification of substances into a compound in which they were present in potentiality.[52] The only indication of a comparable Stoic account is given by the second clause in Diogenes' description which explains that before a wine drop is blended with the sea it is extended through it. The particle introducing this clause (γάρ) is explanatory and therefore we must conclude that a blend is total only insofar as a small body pervades a larger one; that is, totality is reached when the former attains the total volume of the latter. Now such a conception is certainly compatible with other known uses of ἀντιπαρεκτείνεσθαι, the verb used to describe this process of extension. For example it is applied to cavalry being extended in line with a wall, or more conceptually of a point being extended to equal a line:[53] in each case the smaller entity undergoes the process of extension so as to become equal to the larger. In the

[50] "κεράσαι singulariter dictum sed vix sollicitandum" (Pohlenz, *ad loc.* in the Teubner ed.). "Tempering" preserves grammar but means nothing more than "blend with."

[51] *de gen. et corr.* A10. 328a 5-10.

[52] *de gen. et corr.* A10. 327b 22-31; see Comm. on *de mixt.* 15 *passim* for a discussion of this theory.

[53] Respectively, Josephus *B.J.* 3. 7, and Sext. Emp. *A.M.* III. 32-33 and IX. 384-385. Cf. also *A.M.* IX. 260 where a part is said to extend to equal a whole; Iambl. *In Nic. Arithm. Introd.* p. 13. 17 Pistelli a propos the division of a monad which would be extended (ἀντιπαρεκτείνεσθαι) to infinity; infinite division of a monad where ἡ ἐπ' ἄπειρον αὔξησις ἀντιπαρεκτείνηται; and Greg. Naz. *Or.* 43, p. 852e which refers to a human ἀντιπαρέκτασις towards God. In all these cases the emphasis is on a single entity's extension in relation to a larger one.

description of total blending this usage is enlarged [54] to include the case of a small volume of liquid extending *through* (διόλου), rather than alongside, a larger volume, and similarly becoming equal to it.

Now just on the basis of this text, and the complementary description in Plutarch,[55] we can uncover some of the logic behind this theory. Blending occurs only if there is an equality in volume between the two constituents, but such equality is the result of a process—ἀντιπαρέκτασις—undergone by one of the constituents and not an inherent property of both prior to their blending. In this way the theory maintains the traditional principle taken over by Aristotle that blending can only occur between equal volumes,[56] while denying Aristotle's claim that if there is initially a significant inequality the larger will supervene on the smaller. Chrysippus in fact seems to have deliberately reversed the example by which this claim was expressed, that "a drop of water will not mix with innumerable measures of water, for its form is dispersed and changes into the mass of water." [57]

The compound finally formed by a total blend is clearly one in which the constituents will be actually present together. No allowance is made, as in Aristotle's theory, for the emergence of a separate compound over and above the constituents and in which they will be present in potentiality.[58] Instead equal volumes must retain their identity while occupying one another's volume; or, put simply, two bodies must be in the same place.

So we begin to arrive at the paradoxicality of this theory which our ancient sources ridicule at length.[59] Certainly they were justified

---

[54] This is the only case in which it describes an interrelation between physical bodies, a good indication that it is a Stoic technical term.

[55] The quotation comes at the end of a polemic against total blending which deals almost entirely with the blend of unequals; further on its whole context see pp. 66-67 below.

[56] See *de gen. et corr.* A10. 328a 29-33, and cf. Comm. on *de mixt.* 14. 231. 3. On the Presocratic tradition of "equalitarian physics" followed by Aristotle see G. Vlastos, *CP* 62 (1947) 175. As far as the term κρᾶσις goes this is a natural requirement, for it means the tempering or evening-out of one thing by another; cf. LSJ, κεράννυμι I. 3, II, and III.

[57] *de gen. et corr.* A10. 328a 27-28.

[58] See *de gen. et corr.* A10. 327b 22-27, and 328a 10-11. On the uniformity (ὁμοιομερές) of a blend see Comm. on *de mixt.* 7. 221. 9-10, and on the constituents being in potentiality cf. on *de mixt.* 15. 231. 16-19. In this account constituents do not, of course, occupy one another's place.

[59] See sect. III *passim* for a detailed discussion of this critique.

if it is correct to take this theory in isolation from the Stoic physical system. But as with so many Stoic doctrines we can advance our understanding by attempting to establish the relation of this theory to other theories in the same area. Scholars have recognised this, and the most widely held view is that total blending is closely related to the Stoic theory of pneuma.[60] I believe this to be essentially correct, but since the thesis has not been worked out in detail in the light of all relevant texts it remains a rather vague intuition. In the following reconstruction I shall attempt to ground it more securely in the evidence, and fully explore its implications.

The theory of pneuma is the linchpin of Stoic physical theory as it was formulated by Chrysippus.[61] In recent years it has been subject to some elaborate interpretations through analogies with modern physical theories by Sambursky and Christensen.[62] I shall not try to evaluate these, nor attempt a complete interpretation of the doctrine.[63] I shall merely seek to establish its general character and then look closely at those aspects of it that are particularly relevant to the theory of total blending.[64]

The concept of pneuma must be seen in the context of the Stoic theory of principles (ἀρχαί). This held that an active principle totally pervaded a passive principle as the passage of body through body;[65] this relation was non-reciprocal, for the passive principle

---

[60] The point is made in very general terms by P. Duhem (*Système de Monde*, I, 305), E. Bréhier (*Chrysippe*, pp. 128-129), G. Verbeke (*La Doctrine du Pneuma*, pp. 70-71), and S. Sambursky (*Physics of the Stoics*, pp. 15-16). Cf. also n. 88 below.

[61] Chrysippus' authorship seems clear from *de mixt.* 3. 216. 14ff., and Ar. Did. fr. 28 (Stob. *Ecl.* I. xvii. 4 W.); see pp. 52-65 on these texts. He probably developed it from the Zenonian concept of the soul as "innate heat" (ἔμφυτον πνεῦμα); see Solmsen, *Meded. Nederl. Akad.* (1961) 286, with n. 79.

[62] S. Sambursky, *Physics of the Stoics*, ch. 2 *passim*; J. Christensen, *Essay on the Unity of Stoic Philosophy*, pp. 33-35.

[63] See Comm. on *de mixt.* 10, a major source for our knowledge of this doctrine, for further discussion.

[64] The supporting evidence for this account cannot therefore be scrutinised in detail: I shall simply quote the major source for my statements from *SVF* with the caveat that none of this evidence is primary and that it all deserves the close attention that I only give to the texts dealing with total blending in the course of this essay. I hope that my comments at sect. i, n. 9 above are sufficient justification for the use of this evidence at all.

[65] For the basic account see Diog. Laert. VII. 134 (*SVF* II 300); other material is gathered at *SVF* II 299-328. Cf. also the Comm. on *de mixt.* 224. 34, 225. 3 and 225. 5-18. In the latter note and elsewhere in the commentary

could not interact with the active but merely served as its medium.[66] It is thus often described also as substance (οὐσία) or qualityless matter (ὕλη) from which bodies evolve.[67] The result of this physical relation between the principles is a rational cosmos, for the active principle can also be regarded as Reason (the λόγος) or God.[68] The actual physical process by which the active principle pervades matter is explained in terms of the four elements. Probably by developing an Aristotelian distinction[69] Chrysippus identified Hot and Cold as active qualities, and Wet and Dry as passive.[70] In a monoqualitative theory of elements this suggests that the passive principle could be identified with Earth and Water, and the active with Fire and Air.[71] The latter identification seems to

---

on chs. 10 and 11 I survey some of the major ancient criticisms of this doctrine. A more positive characterisation of the Stoic principles, which I do not attempt to offer here, would have to be based on an analysis of such material.

[66] That is, ποιεῖν and πάσχειν ("Action" and "Suffering") did not describe physical *interaction* as they did for Aristotle, though at *de mixt.* 11. 226. 30-33a (+ 227. 23-25) Alexander attempts to impose this alien meaning on the Stoics—see Comm. *ad loc.*

[67] E.g. Diog. Laert. VII. 134 (*SVF* II 300): τὸ μὲν οὖν πάσχον εἶναι τὴν ἄποιον οὐσίαν, τὴν ὕλην. Also cf. Diog. Laert. VII. 150 for matter as that from which things come to be. At Diog. Laert. VII. 134 (*SVF* II 299) there is a perplexing and unparalleled report that the Stoics distinguished between principles and elements in that the former were incorporeal and shapeless (ἀσωμάτους . . . ἀμόρφους). The reading ἀσωμάτους is however suspect since it is preserved only in the Suidas and not the manuscript tradition. If this were correct, it would be difficult to make sense of the theory of pneuma in which both principles are given physical properties. It is best, I believe, to follow the manuscript reading and accept, as Alexander for example does (*de mixt.* 11. 224. 31-225. 5) or Aristocles (ap. Euseb. *P.E.* XIV. 816d), that the principles are bodies; this view is defended with closer reference to the literature by H.-J. Krämer, *Platonismus und Hellenistische Philosophie*, p. 108 n. 3. To argue this, it should be noted, is not *eo ipso* to offer an interpretation of the basic doctrine of principles; cf. n. 65.

[68] E.g. Diog. Laert. VII. 134, Alex. *de mixt.* 11. 224. 34.

[69] See *de gen. et corr.* B2. 329b 25-33, and Joachim *ad loc.* Elsewhere he brings out well the extent to which this can be regarded as an anticipation of the Stoic theory: "The Hot and the Cold in combination 'work on' the Moist and Dry with an 'immanent action' like that of Form on Matter in the seed of a plant or animal. It is this 'immanent' operation which plays the chief part in the production of the μιχθέν . . . ." (*J.P.* 29 [1904] 84).

[70] E.g. *SVF* II 444 and cf. *de mixt.* 218. 2-6 with Comm. *ad loc.*

[71] See *SVF* II 413 which describes the evolution of the elements from Fire. As Gould says (*Philosophy of Chrysippus*, p. 102) there is no explicit identification of pneuma with the active principle. But at the very least we can term the relation of pneuma to the passive elements an alternative description of the relation between the active and passive principles, which it very

be crystallised in the doctrine of pneuma which was a compound formed of these two elements [72] and so a body that pervaded matter which possessed the passive qualities necessary to receive it. The result of this process was the formation of a structured cosmos.[73] Thus the theory as a whole provided an explanation in qualitative physical terms of the unity and continuity of the cosmos.

The central feature of the theory of pneuma, then, is the motion of a body through a body. This in itself is not directly proved, although it is described in elaborate and suggestive terms. For example, pneuma is said to extend through matter (or the passive elements) [74] with a motion described as "tensional" (ἡ τονικὴ κίνησις).[75] This seems to be because it has its own "tension" (τόνος) or elasticity as a compound body [76] which gives it the capacity to move within matter, and to endow it and its individual parts with a similar tension.[77] In this way it creates a structured cosmos composed of individually structured parts, "structure" here meaning a continuity which it creates by (literally) "holding together" (συνέχειν) bodies and by creating a sympathy between the parts of the cosmos.[78] In this process individual bodies acquire a "tension" that makes them continuous (συνεκτικὸς τόνος),[79] or are said to be endowed with a "state" (ἕξις), or a principle of their struc-

obviously complements. On the relation between the active and passive elements and the four elements involved in cosmogony see Comm. on *de mixt.* 10. 224. 14-22, and 11. 225. 5-18.

[72] *SVF* II 444. Alexander's argument against the compound character of pneuma (*de mixt.* 224. 14-22 with Comm. *ad loc.*—cf. 225. 5-10) is a good indication that the Stoics claimed that it did have a compound character; cf. also Galen's criticism of it being a self-sustaining compound at *SVF* II 440.

[73] See *SVF* II 448 (Alex. *Mant.* 131. 5-10) which refers to pneuma's "holding together" (συνέχειν) the cosmos and bodies in it.

[74] Hereafter I shall generally refer to the medium of pneuma's motion as "matter," leaving it understood that it was qualified as Wet and Dry. Indeed at *SVF* II 439 (Galen) Earth and Water are referred to as ὑλικὴ οὐσία ("material substance").

[75] E.g. *SVF* II 448 (Alex. *Mant.* 131. 5-10).

[76] See *SVF* II 447 for the expression "pneumatic tension" (πνευματικὸς τόνος). See also Comm. on *de mixt.* 10. 223. 34-36 on pneuma's transmission of tension.

[77] At *Mant.* 115. 9-10 Alexander describes tension as a "particular form" (εἶδος ἴδιον).

[78] See *de mixt.* 216. 15-16 (= 223. 8-9). See Comm. *ad loc.*

[79] See *SVF* II 407 (Plutarch) in a discussion of the elements: also cf. *SVF* II 440 (p. 144. 31, 35) where Galen uses the expression συνεκτικὴ αἰτία ("cause that binds together").

ture.[80] This very basic description leaves aside many difficult philological and philosophical problems;[81] for the moment I shall only raise the limited issue of the relation between total blending and certain aspects of this intricate physical theory.

There are at the very least obvious parallels between the phenomenon of pneuma pervading matter and the blending of a wine drop with the sea. In each case one body totally pervades another: blending, as we saw, is not complete unless the wine drop becomes equal to the whole sea, and the cosmos—or an individual body—is not structured unless pneuma completely pervades matter.[82] Again verbs of extension are used to describe the process in both cases,[83] and here we may draw the parallels much closer, for as we have seen [84] the verb "to be extended by, or in line with" (ἀντιπαρεκτείνεσθαι) is used to describe the expansion of a wine drop through the sea so as to become equal to and thus totally blended with it.

Now pneuma is extended through matter because of its "tensional motion" (τονικὴ κίνησις) which is characterised as a form of oscillation in which it "moves itself from itself and into itself" or "backwards and forwards," [85] a process that some sources report as a simultaneous motion in opposite directions.[86] With this con-

---

[80] Cf. the standard definition of ἕξις as "pneuma containing body" (πνεῦμα σώματος συνεκτικόν) at *SVF* II 368. Cf. pp. 40-41 below.

[81] Apart from the discussions of this doctrine by Sambursky and Christensen (n. 62 above) see G. Verbeke, *La Doctrine du Pneuma*, pp. 61-90, and J. Gould, *The Philosophy of Chrysippus*, pp. 99-102.

[82] διὰ πάντων διῆκον is an almost formulaic description of pneuma's motion —e.g. see *SVF* II, p. 137. 30, p. 307. 8, and cf. *de mixt.* 216. 15.

[83] Mostly simple verbs of motion are used to characterise pneuma's movement—e.g. χωρεῖν (*SVF* II, p. 307. 8), or διέρχεσθαι (*SVF* II, p. 116. 12-13); but verbs of extension do describe the motion of sensory pneumata—e.g. διατείνειν (*SVF* II, p. 226. 8-9), ἐκτείνεσθαι (*SVF* II, p. 227. 25-26). On the relation between this form of pneumatic motion and the cosmic pneuma see my article "Συνέντασις and the Stoic theory of Perception," *Grazer Beiträge* II (1974) 251-261. In view of the nature of the "tensional motion" it is reasonable to assume that such language was employed to describe all aspects of the theory of pneuma.

[84] Pp. 32-33 above.

[85] Stob. *Ecl.* I. 154. 1-2 W (= *SVF* II 471); also cf. note on the text of *de mixt.* 224. 24-25.

[86] Alex. *de mixt.* 10. 224. 25 (cf. *Mant.* 131. 10, 16, 19-20). Both these texts are polemical and the notion of simultaneous motion may have been an accretion to the description that we find in the doxography (cf. p. 52). It is only reported elsewhere by Nemesius (at *SVF* II 451).

Sem. Griekse Letterkunde
Blandijnberg 2
Gent

tinuous expansion and contraction pneuma maintains the continuity of matter and the structure of individual bodies.[87] There would seem to be a very close parallel between this expansive stage of its motion and the passage of the wine drop through the sea which the term ἀντιπαρέκτασις (extension-through) is particularly apposite to describe. Apart from its general linguistic affinities to "tensional motion," it specifically describes the motion of a small body through a larger which must also be the character of pneuma's expansion "from itself" through matter. Transposed from the case of blending to that of pneumatic motion, ἀντιπαρέκτασις thus becomes associated with a physical theory in which the process that it describes is, if not logically demonstrated, at least made plausible by a set of specific physical concepts; whereas when applied to the case of a wine drop's motion through the sea, the term has a purely descriptive content with no theoretical significance.

Clearly if such a relation holds then the connection between total blending and the theory of pneuma is more than a parallel. Scholars who have attempted to be precise about such an interrelation have taken the view that total blending is either in some sense a support for the theory of pneuma, or that it is causally dependent on pneuma.[88] The debate, if it can be called that, seems sterile in that close attention has not been paid to *de mixtione* 4, the one text which presents us with a series of proofs for the extension (ἀντιπαρέκτασις) of one body through another.

The proofs (or suasions, πίστεις) listed by Alexander in this chapter (at 217. 13-26) are an example of a Stoic argument for total blending on the basis of the common notions.[89] They begin

[87] At *SVF* II 451, 452 pneuma's motion within a body is described as an extension to and from its limits, the former motion creating its unity, the latter its qualities. If this is an accurate report it would not apply to pneuma's motion through matter as a whole, but a species of this motion.

[88] K. Reinhardt was the clearest exponent of the latter view: "Chrysipp entwickelt seine Lehre von der Mischungen im Hinblick auf die Grundthese seiner Physik, im Hinblick auf den Satz von der Durchdringung der Materie durch das Pneuma" (*Kosmos und Sympathie*, p. 80; cf. *RE*, XXII-i [1953] col. 654). Sambursky's view (*Physics of the Stoics*, pp. 15-16) is along similar lines. Rex (*Chrysipps Mischungslehre*, p. 50) however argues that total blending is dependent on the passage of pneuma through matter, but, as far as I can see, this is (like Reinhardt's view) based on an interpretation of the relation between pneuma and the tripartite classification of mixtures in *de mixtione* 3. I argue against all such interpretations at pp. 60-65 below.

[89] πίστεις ("suasions," 217. 13) picks up πιστοῦσθαι ("to persuade," 217. 2), the verb used to describe proofs from the common notions for the differences

by offering a list of commonplace examples of bodies preserving their qualities when expanded (217. 14-17), and, more importantly, of bodies that are "helped" by other bodies to extend themselves in various ways (217. 17-26). This latter causal relation is clearly that classified elsewhere by the Stoics as a case of "cooperant causes" (συνεργὰ αἴτια) where one thing can easily contribute to produce an effect, as for example an additional hand can help men to lift a weight.[90] The inference that Alexander (at 217. 26-32) reports was drawn from these examples is as follows:

> Since this is the case, they say that there is nothing remarkable in the fact that certain bodies when helped by one another are in this way united together in their entirety so that being preserved along with their own qualities they have a complete mutual coextension through one another (ἀντιπαρεκτείνεσθαι ἀλλήλοις δι' ὅλων ὅλα), even if some of them are rather small in bulk and in themselves unable both to be spread to such an extent, and to preserve their own qualities; for in this way also the cup of wine is mixed with a large amount of water and helped by it to such a great extension.

The first clause presents a general inference that if bodies can be subject to cooperative causation, then they can extend through one another although one may be smaller than the other. The second clause on which I shall concentrate here[91] refers to the specific case of a wine drop being extended through water which we have seen described by Diogenes Laertius and Plutarch.[92] It is now presented as being entailed by the various commonplace

between mixtures (cf. Comm. on *de mixt.* 217. 2-13). For further discussion of this passage see my article at *Symb. Os.* 48 (1973) at 48-55.

[90] Cf. the definition at Sext. Emp. *P.H.* III. 15, which refers to this particular case, with 217. 23-25, "... and along with others we bear certain weights of which, when left alone, we could not bear the part that befalls us." Further on cooperant causes cf. Cicero *de fato* 41 (*SVF* II 974), Clement *Strom.* VIII. ix (at *SVF* II 346), and cf. Sext. Emp. *A.M.* IX. 243 (on which see O. Rieth, *Grundbegriffe der Stoischen Ethik*, p. 153).

[91] Clearly the idea of mutual coextension is incompatible with the preceding description of cooperative causation. For there bodies were helped *by others* (ὑπ' ἄλλων βοηθούμενα, 217. 18) from which it does not follow that they are helped *by one another* (βοηθούμενα ὑπ' ἀλλήλων, 217. 27). Only the case of the extension of wine through water, where it is helped by its medium (βοηθούμενον ὑπ' αὐτοῦ, 217. 32), is parallel to cooperative causation. On the reasons for this conflict see below, pp. 66-69.

[92] Above, pp. 31-33.

examples of cooperation, so that the large volume of water is the aid which causes the wine to extend through it. If this were just a theory of the mixture of liquids this argument would be nothing more than a rather weak analogy supporting an idiosyncratic theory. There are however clear indications that there is a further stage to this argument.

We have already seen that the extension (ἀντιπαρέκτασις) of the wine drop through water could be a precise description of the motion of pneuma through the passive elements. Now immediately following the passage we have quoted Alexander lists (at 217. 32-218. 9) some "clear evidence" (ἐναργῆ μαρτυρία) that one body extends through another.

1. They employ as clear evidence [ἐναργῆ μαρτυρία] that this is the case the fact that the soul which has its own substantiality, just like the body that receives it, pervades [διήκειν] the whole of the body while preserving its own substantiality in the mixture with it (for there is nothing in the body possessing the soul that does not partake of the soul); and the same holds for the Nature [φύσις] of plants, as also the State [ἕξις] in bodies held together by their State. (217. 33-218. 1)
2. Also they say that fire passes [χωρεῖν] completely through iron with each of them preserving its own substance. (218. 1-2)
3. And they say that two of the four elements, Fire and Air, being rare, light, and having tension, completely pervade [διαπεφοιτηκέναι] Earth and Water which are tense, heavy, and lack tension; and that each pair preserves its own nature and continuity. (218. 2-6)
4. They think that drugs that are deleterious, and all such odours, are blended with the bodies affected by them in a total juxtaposition. (218. 6-8)
5. Chrysippus also thinks that light is mixed with air. (218. 8-9)

The list is slightly confused,[93] but its principal items are the

[93] Pts. 2 and 4 scarcely belong in the same group; neither convincingly illustrates the blend of a small volume with a large one. Their relation to total blending must be as some looser form of empirical proof. 4 is unprecedented, but 2 is cited by Ar. Did. fr. 28 (*SVF* II, p. 153. 8) as an example of the total intermingling of bodies. Because of the terminological peculiarities of this doxographical text, on which Alexander was probably dependent, it may have been shifted to its present place in the exposition; see n. 141 below.

pervasion of the passive elements by the active (218. 2-6), and the motion of soul through body, nature through plants, and ἕξις (State) through the inanimate things that it holds together (217. 33-218. 1). The latter triad is a classification of the intracosmic manifestations of pneuma's action,[94] and the former dichotomy an alternative description of pneuma's motion through matter.[95] Now in itself this text is simply described as an additional set of proofs logically unrelated to the preceding proof for the extension of the wine drop through water. Yet since both are said to involve the motion, if not extension, of body through body [96] can we not regard the references to pneuma's motion in this list as in fact an inference from the preceding analogical argument?

That is, rather than taking items 1 and 3 as "proofs" for the blending of the wine drop with water we should, I suggest, take the latter as the premise from which they could be in turn analogically inferred. There is no discernible reason why the Stoics should use the theory of pneuma to demonstrate an idiosyncratic account of the blending of liquids, but a clear reason why such blending, itself derived by analogy, might then be used as a basis from which the relation between pneuma and matter could be derived. Also the term which describes the relation involved—ἀντιπαρέκτασις—is, as we have seen, particularly apposite to describe the case of pneuma's motion and, in that context, assumes a theoretically more significant role.[97] Indeed, in *de mixtione* 12 Alexander claims that the related doctrines of the unification of the cosmos, sympathy, fate, providence, the active and passive principles, and that of the soul are all dependent on total blending as body going through body, a clear statement of the principle which his own exposition in chapter four somewhat obscures.[98]

Such a reading of *de mixtione* 4 has important implications for the interpretation of total blending; these can be seen more clearly if we set out our reconstruction in a schematic form.[99]

---

[94] *SVF* II 366, 714-716, 1013, and III 160. This valuable evidence from Alexander is largely overlooked in discussions of this classification; it places the doctrine squarely in the context of orthodox Chrysippean physical theory (cf. p. 69, n. 189 below).

[95] See above, p. 35.

[96] These are interchangeable terms—cf. n. 83.

[97] See above, pp. 37-38.

[98] 227. 5-10; see Comm. on 12. 227. 9-10.

[99] At *Symb. Os.* 48 (1973) 53-54 I argue that the relation of (1) to (2) is

(1) cases of bodies aiding other bodies to produce effects—i.e. cooperative causes (συνεργὰ αἴτια) (*de mixtione* 4. 217. 13-26)

by analogy

(2) the extension of body through body, as in the extension of a wine drop through a large volume of water—the water being the cooperative cause (*de mixtione* 4. 217. 26-32)

by analogy

(3) the extension of pneuma through the passive elements.

The inference of (2) from (1) is described in the text of *de mixtione* 4, while that of (3) from (2) is, I hope to have shown, a plausible interpretation of the text. If this pattern of reasoning represents Chrysippus' thought,[100] then the theory of total blending is simply a stage in an argument and not an independent theory of mixture. It is an analogically formed conception of a causal relation from

---

that of presentations (φαντασίαι) to preconceptions (προλήψεις) or natural notions (φυσικαὶ ἔννοιαι), an interchangeable term. I then suggest that the common notions (αἱ κοιναὶ ἔννοιαι) be regarded as the general notions formed from the example in (1)—note the generalisations at 217. 14-15, 17-19—and that from these an independent theory of total blending (here expressed in [2]) could be derived. This I suggest, in very general terms, would serve as the physical basis for the theory of pneuma (p. 33 with n. 54). This earlier paper was an attempt to establish the interrelation of these epistemological terms, and to provide an interpretation of the doctrine of common notions, rather than to interpret the theory of total blending. Among other things (also cf. Comm. on *de mixt.* 3. 217. 2-13) I did not then recognise the case of cooperative causation expressed at (1) and (2), and so did not argue for as close a chain of reasoning between the specific examples at 217. 13-26, the conception of wine blending with water, and the related motion of pneuma through matter. However I still think that the generalisations of the cases grouped in (1) are specimens of the common notions; it is clearly not even necessary that there be any intermediary stage between (1) and (3), and so the common notion of cooperative causation could be the basis from which, by analogy, the motion of pneuma through matter could be directly inferred. Stage (2)—total blending—makes this a much clearer argument, but it is entirely dependent on the notion formed in (1), and thus this common notion, as the name suggests, would be the more fundamental of the two.

[100] At 217. 13-14 the whole argument in *de mixt.* 4 is simply attributed to proponents of the doctrine of total blending (οἱ προϊστάμενοι τῆσδε τῆς δόξης). But the presence of the case of the wine drop blending with water, which we know to be Chrysippean, at the core of the argument, and the fact that after chapter four Alexander always refers to the Stoics collectively, make it reasonable for us to attempt to integrate this text with Chrysippus' thought.

which, by a further analogy, the central relation of the Stoic physical system, that of pneuma to matter, can be inferred.

It should be stressed here that in each case what is inferred is a certain relation between bodies and nothing about their internal properties. So at (3) only the motion of pneuma through matter is implied—the τονικὴ κίνησις—and not the elaborate qualitative physics on which this concept depends. The argument addresses itself to the physical principle involved in this motion rather than the whole physical theory of which it is a part: that would have to be demonstrated on other more elaborate grounds.[101]

We can now also speculate that the relation expressed at (3) was like those at (1) and (2) regarded as a case of cooperative causation, with matter *qua* the passive elements being, like the water, the agent which makes it possible for pneuma to pervade it as the wine can extend through the water. The concept of cooperation is a way of expressing the limited dependency of the pervading (or blending) body on its medium; although the passive elements do not interact with pneuma [102] they are still a necessary condition for its action, as presumably is the water which does not extend through the wine as the wine extends through it. The relation of extension or motion-through that holds between pneuma (or the wine drop) and the passive elements (or the water) is not reciprocal, but merely involves one body cooperating with another so that this other body can produce an effect on it.

The overall effect of the argument as reconstructed here is to show that a causal relation embodied in a physical theory can be derived from sense experience. That is, the same relation of "coop-

[101] So, for example, in addition to establishing the relation between pneuma and the elements, sketched in above (p. 35; cf. *Comm.* on 10. 224. 14-22), we would need to determine its relation to the theory of infinite division, which is itself the foundation of the Stoic continuum (I have discussed some aspects of this relation at *Apeiron* VII-i [1973] 21-29). For by its motion pneuma creates a material continuum, whereas such a continuum is a theoretical consequence of infinite division. The suggestions that have been made about the relation between infinite division and total blending (cf. n. 116 below) may therefore be relevant only to the doctrine of pneuma which we argue it supports. This might also be true of the Stoic theory of surfaces reported by Plutarch *de comm. not.* 40; the references to total blending there (1080F-1081A) seem to be polemical (cf. the conditional at p. 110. 18 Teubner ed.) and not grounds for suggesting that the doctrine of surfaces is the theoretical basis of total blending *pace* Sambursky (*Physics of the Stoics*, p. 96—cf. also E. Bréhier, *Théorie des Incorporels*, p. 40).

[102] Above, p. 35, n. 66.

eration" is exemplified by commonplace phenomena, and can be analogically transposed into a single clear conception, from which a theory also embodying this relation can be similarly derived. But since, as we have noted,[103] the theory of pneuma would have to be demonstrable independently of this argument, this chain of reasoning would appear to be circular. However this would be true only if it followed a pattern in which premise entailed conclusion by necessity, rather than by analogy; for analogies merely suggest [104] a conclusion already independently established rather than leading to one as yet undemonstrated.[105]

The importance of such reasoning for the Stoics must have been that since, for them, nature (in the sense of the empirical world) was a result of Nature (in the sense of the world order, which pneuma's motion through matter serves to establish), then the character of the latter would have to be demonstrable in terms of experience: the fundamental laws of nature had to be manifested in and discernible through experience. Such discernment did not take the form of inductive reasoning, but as this argument shows, is achieved through the recognition of a pattern of similar causality linking ordinary experience and fundamental physical theory. The simple pattern of analogical argument that expresses this is therefore an *a posteriori* confirmation of the world order.[106]

In the light of this interpretation the texts from Diogenes Laertius and Plutarch that were examined earlier must be regarded as incomplete representations of total blending. They report only

---

[103] N. 101.

[104] Hence, even if it is Alexander's own term, the examples introduced at 217. 14-26 are well described as "suasions" (πίστεις); cf. Comm. on 217. 13.

[105] For this reason in particular I would reject O. Rieth's attempt (*Grundbegriffe der Stoischen Ethik*, pp. 188-189) to ground this argument (at least to 217. 14-32) in the Stoic theory of "demonstrative sign" (ἐνδεικτικὸν σημεῖον) as reported e.g. at Sext. Emp. *P.H.* II. 97-103. Rieth's interpretation is developed without attention to the overall context of Alexander's, and I shall not attempt to refute it in detail.

[106] At *Symb. Os.* 48 (1973) 58-60 I develop the thesis of this paragraph at greater length, and suggest that by the common notions (κοιναὶ ἔννοιαι) the Stoics meant to establish simply conceptions that would demonstrate the nature of the world-order. On some aspects of this thesis see Comm. on *de mixt.* 3. 217. 2-13. This whole form of argumentation is comparable to the allegorical interpretation of poetry that was widely practised by Stoics, where for example poetic descriptions of relations between gods were interpreted as relations between natural phenomena; see the useful account given by P. de Lacy, *AJP* 69 (1948) at 256-263.

stage (2) of our reconstructed argument as though it were an independent account of mixture, rather than as an intermediary stage in an analogical argument. If the conception of a small volume of liquid extending through a much larger one had referred to an actual process, critics like Plutarch could have justifiably extracted it from its context and ridiculed it as a highly paradoxical notion. However it seems clear that this process should be thought of as having only conceptual status; that is, it is a fiction, and not a conception representing an actual state of affairs.

To support this interpretation we can invoke a text which the Stoics are said to have allowed for the formation of mental conceptions (νοούμενα) by analogies such as that contained in the transition from stage (1) to stage (2) of our reconstructed argument.[107] The cases listed in this text cover conceptions formed of either something not immediately perceptible (e.g. Socrates from his picture); or something contingently imperceptible (e.g. the centre of the earth by analogy with smaller spheres); or complete fictions (e.g. the Cyclops, formed "by analogy through a process of increase"). The case of the conception of a wine drop pervading water could be grouped in this last class; it is in principle imperceptible, and thus not seen to happen. Unless the Stoics wanted to offend common sense they would then best be understood to have regarded this conception as a fiction which clarifies the relation of cooperative causation that occurs between bodies but is not itself a direct example of it. As such it links the actual cases of cooperative causation listed at stage (1) of our argument with the theory of pneuma's motion through matter—stage (3)—without itself being an actual physical process.

Such an interpretation is not solely designed to free the Stoics from apparent embarrassment; it makes greater sense of the analogical argument which we have reconstructed by establishing the case of total blending as an intermediary stage between actual physical processes available to sense experience and an analogically related physical theory, rather than as a paradoxical duplication of perfectly normal physical processes entirely unsupported by any

[107] Diocles Magnes ap. Diog. Laert VII. 52 (= *SVF* II 87). See *Symb. Os.* 48 (1973) 51-53 where I accept Sandbach's association of this passage with the account of concept-formation at Aet. IV. 11 (= *SVF* II 83) (see *CQ* 24 (1930) 46) and argue that these conceptions are what the Stoics described as natural notions (φυσικαὶ ἔννοιαι) or preconceptions (προλήψεις).

physical theory. The importance of the economy thus introduced into our picture of the argument reconstructed from *de mixtione* 4 can be seen if we examine some traditional interpretations of total blending.

Ancient critics, for example, took the idea of total blending to mean that two bodies occupied the same place, a reasonable construction perhaps to put on the isolated case of a wine drop pervading the sea.[108] However if this phenomenon is an analogically formed conception it cannot be taken literally and subjected to analysis as a physical theory. Only the theory of pneuma's motion through matter which it serves to illustrate can be treated in this way;[109] for here it is correct to claim that body pervades body, and that two bodies occupy the same place, because this is explained, as we have seen, by an elaborate qualitative physics.[110] On analysis, we might find this an unsatisfactory physical theory,[111] but it is not inherently paradoxical as is the bald claim attributed to the Stoics by their ancient critics that a small volume of liquid moves through and occupies the place of a larger one. This theory is not based on an account of the intrinsic properties of these liquids which might explain this motion; such an account is only given in the case of pneuma's motion through matter. Should not the latter then, as I have suggested, be regarded as illustrated by the case of the blend of unequal volumes to which it is so precise a parallel? This would relieve the Stoics of the absurdity of having claimed that initially unequal volumes of liquid could occupy the same place, when they meant such a case simply as a conception crystallising a relation explicable in terms of a more elaborate physical theory.

The same point can be sustained if we consider suggestions made by modern scholars that total blending, as it applies to liquids, meant that the Stoics employed a special definition of body as peculiarly fluid.[112] Certainly such a definition is required and is

---

[108] On this whole pattern of criticism see further below, sect. III *passim*.

[109] Although in the structure of our reconstructed argument the motion of pneuma through matter is derived from the blending of the wine drop with water, it can also be said to be "illustrated" by it inasmuch as this reasoning is *a posteriori*.

[110] See above, pp. 34-37.

[111] Cf. Comm. on *de mixt.* I. 213. 2-13.

[112] E.g. E. Oegereau, *Essai sur le Système Stoïcien*, p. 44 and G. Rodier, *Études de Philosophie Grecque*, p. 258. Cf. next note.

given for the case of pneuma's motion through the passive elements; but for the case of a wine drop blending with the sea no such definition is necessary since this would not be a description of the behaviour of physical bodies, but a mental conception analogically formed from relations between such bodies. Thus in terms of our reconstructed argument, there would be two operative definitions of body: at stage (1) in the commonplace examples of cooperative causation body meets the standard definition of a three-dimensional solid, and the Stoic endorsement of this [113] could then be taken to apply at this level of sense experience. At stage (3), where pneuma's motion through matter is involved, a totally different definition of body is required, and is given, in the different qualitative features assigned to these bodies.[114]

These two definitions correspond to the two aspects of Nature—

---

[113] It is attributed to Apollodorus at Diog. Laert. VII. 135 (cf. also *SVF* II 357, 501). Baeumker suggested that such solidity was only a relative concept that could accommodate the relation that occurs in total blending (*Das Problem der Materie*, p. 335). However E. Weil (*Mélanges A. Koyré*, II, 565) has argued that this definition was only designed as a contrast with the definition required to explain the relation between the active and passive principles described at Diog. Laert. VII. 134, which he describes as "*corps-energie*" (p. 565 n. 23). This view is, I believe, essentially correct, if we will also allow that the Stoics recognised an area in which the ordinary definition of body applied, represented here by stage (1) of our argument. Also cf. Appollodorus' view reported at Diog. Laert. VII. 150 where he defines substance (οὐσία) as a body that is acted-on (παθητός), which supports the view that there was this dual conception of body. (I have drawn here on M. Reesor's suggestions at *AJP* 75 [1954] 56-57.) The case of fire pervading iron (cf. n. 93 above) perhaps appealed to the Stoics as a good empirical illustration of body pervading body, where one body, at least, could not be conventionally described as three-dimensional.

We might also contrast the Stoic theory of tensional motion with their general doctrine of motion as simple change of place (e.g. *SVF* II 492); the former characterises the motion of pneuma, while the latter, as far as we can judge from the scanty evidence, can only apply to the behaviour of intracosmic bodies. This distinction might be derived from the two senses of body that I have attempted to distinguish.

[114] Thus scholars who claim that the Stoics did hold that two bodies occupied the same place (e.g. Sambursky, *Physics of the Stoics*, pp. 15-16, L. Edelstein, *The Meaning of Stoicism*, pp. 23-24) ought, I believe, to make this distinction more firmly. Otherwise the extent to which the ancient critics misrepresented the Stoics by taking the case of total blending as a description of the behaviour of random three-dimensional bodies can hardly be appreciated (see further below, pp. 72, 83). Needless to say, the two senses of "body" that I advocate here do not constitute a complete interpretation of the Stoic concept of body; that too would require a separate analysis of the sources.

as the intracosmic world confronted in sense experience, and as the constitutive processes of the natural order. The former depends on the latter, and reflects its character, as our reconstructed argument shows, but does not directly correspond to it. If it did there would be no need to argue that body pervades body, for this would be exemplified in our experience. The case of the wine drop blending with water links the two parts of this dichotomy as a conception which crystallises the essential features of a causal relation evident to sense experience, and thereby provides a bridge to the same kind of causal relation occurring at the level of basic physical theory. It does not offend common sense because as a mental conception it does not involve a relation between three-dimensional solids; nor on the other hand does it assert a relation between peculiarly fluid bodies, because it is not part of a physical theory. The Stoics' opponents in antiquity could then be said to have erred in failing to see that total blending is the kind of relation possible only between pneuma and matter, while insisting that it occur between three-dimensional solids, hence forcing on the Stoics the unsupported claim that two bodies can occupy the same place.[115]

Finally, if total blending has this very circumscribed role as a mental conception in a chain of reasoning linking two levels of physical reality, there is no point in treating it as a theory which can be explained in terms of other physical theories. So, for example, attempts to ground total blending in the Stoic theory of infinite division,[116] or in their theory of the continuum,[117] are irrelevant to

---

[115] See sect. iii below *passim*, and Comm. on *de mixt.* 6. 219. 11-12.

[116] Both Pohlenz (*Die Stoa*, I. 73) and V. Goldschmidt (*Le Système Stoïcien et l'Idée du Temps*, p. 45) seem to regard the totality of a blend as an integration of bodies at an infinite number of points to form thereby a continuous whole (cf. also H. Dörrie, *Porphyrios' "Symmikta Zetemata"*, p. 25 for a closely related view). This ignores, erroneously in my opinion, the fact that total blending specifically occurs between a small and a large body and that its totality is the extension of the former through the latter (cf. p. 32 above). Also see H. Wolfson (*Philosophy of the Church Fathers*, I² 382) who finds a connection between total blending and infinite division at Nemesius *de nat. hom.* 3. 128 Matth. although the text only refers to blending escaping notice "because of the fineness of the bodies that are blended" (διὰ τὸ λεπτομερὲς τῶν κεκραμένων). Alexander's own discussion in *de mixt.* 8 of infinite division in relation to total blending is, as the Commentary will show, a scholastic exercise and not a reflection of Stoic doctrine.

[117] See Sambursky, *Physics of the Stoics*, p. 14, where total blending is said to be a "consequence of their idea of continuity." Since continuity was grounded in the theory of infinite divisibility (as Sambursky, p. 98, allows)

the kind of conception that we have argued it is. These suggestions have perhaps some validity if restricted to the case of pneuma's motion through matter [118] which total blending serves to illustrate, but none if directly applied to the case of blending itself.

These revisionary implications of our interpretation of total blending cannot be fully justified until a further set of texts is carefully examined. So far a reconstruction has been established on the view, grounded in our primary sources, that total blending was solely a phenomenon occurring between a small and a large volume of liquid. We must now confront texts in which this process is presented as the interpenetration of unspecified volumes, is classified along with other types of mixture, and, as such, appears to be placed in a relation of causal dependency on the theory of pneuma.

The Secondary Sources—The Classification of Mixtures

Aristotle distinguished two principal types of mixture (μῖξις)—the juxtaposition of corpuscles (σύνθεσις "composition"), and blending (κρᾶσις).[119] The crucial difference between them was the degree of unification attained: juxtaposed constituents were not really mixed, though in some cases they might appear to form a unity,[120] but when bodies were blended they were unified to form a homogeneous compound over and above their individual qualities.[121] At the same time these constituents were not destroyed but preserved in potentiality, so that it was possible for them to be separated out again.[122] The additional case of constituents being so mixed that they formed a new compound from which they were

---

this suggestion can be criticised along the same lines as the views discussed in the preceding note. In general it would have been helpful if Sambursky had clarified somewhere the relation between his separate claims that total blending (1) gives "a firm foundation for [the] conception of hexis" (*Physics of the Stoics*, p. 16), (2) is explained by the Stoic theory of surfaces (p. 96, cf. n. 101 above), and (3) follows from their theory of continuity. All these suggestions are probably relevant to a full interpretation of the theory of pneuma, but only if we admit from the outset that there was some internal structure to the Stoic physical system.

[118] Cf. note 101 above.

[119] *de gen. et corr.* A10 327b 31-328a 18. Alexander *de mixt.* 13. 228. 25-229. 3 has a more elaborate taxonomy.

[120] *de gen. et corr.* A10 328a 12-15; cf. Comm. on *de mixt.* 2. 214. 18-28.

[121] See above, p. 33 n. 58.

[122] *de gen. et corr.* A10 327b 23-28. Cf. Comm. on *de mixt.* 1. 213. 7-8, and 15. 231. 12-29.

irretrievable was recognised in theory by Aristotle but not given any significant place in his discussion of mixture.[123]

So far we have seen one text (Diogenes Laertius VII. 151) in which Chrysippus is said to have distinguished total blends (κράσεις δι' ὅλων) from surface juxtaposition. This appeared to be simply a contrast designed to clarify the nature of a complete blend as the passage of one body through another. In other texts, however, this contrast is embodied in a tripartite classification of mixtures attributed to the Stoics: total blends are grouped along with juxtapositions (παραθέσεις) and a third limiting case termed "fusion" (σύγχυσις) or "joint-destruction" (σύμφθαρσις), in which bodies completely and inseparably mixed to form a new compound. Total blending is then defined as a process from which the constituents can be separated, thus distinguishing it from fusion. This also entails that the bodies that are blended are actually present in the blend and that it is not a homogeneous unity as it was for Aristotle.

We find this trichotomy[124] in Philo who employs it quite independently of any other aspect of Stoic physical theory in a passage in which he is attempting to define fusion (σύγχυσις) as applied to the biblical confusion of tongues.[125]

> Now mixture is held to occur among dry substances, blending among moist.
>
> Mixture [μῖξις] is the juxtaposition [παράθεσις] of different bodies in no orderly fashion, as if one were to make a heap of barley, wheat, and vetch and other kinds of grain and pile them together.
>
> Blending [κρᾶσις] is not juxtaposition but the complete mutual coextension [δι' ὅλων ἀντιπαρέκτασις] of dissimilar parts interpenetrating one another, though their qualities can still be separated by some device as they say happens in the case of wine and water. When these substances come together they produce a blend, while the product of the blend can nonetheless

---

[123] *de gen. et corr.* A10 327b 4-6; cf. 327b 30.

[124] The tripartite classification attributed to Plato at Anonymus Londiniensis XIV. 15-26 Diels of σύνφθαρσις (or σύγχυσις), μῖξις (or παράθεσις), and διάκρασις is also to be regarded as Stoic rather than an original part of Menon's Ἰατρικά reported in this papyrus. See W. H. S. Jones, *The Medical Writings of Anonymus Londiniensis*, pp. 3 and 61.

[125] *de conf. ling.* 184, II, p. 264 Wendland (= *SVF* II 472).

be resolved into the qualities from which it was composed, for the water is absorbed by an oil-filled sponge while the wine is left behind; probably because the sponge is produced from water it tends to absorb water, the substance kindred to it, while the alien substance, wine, is left as a residue.

Fusion [σύγχυσις] is the destruction of the original qualities which are coextended [ἀντιπαρεκτεινομένων] with all the parts [of the constituents] to produce a single different quality, as occurs with the quadruple drug in medicine; for wax and tallow and pitch and resin, I believe, come together to produce this, and when it is compounded it is impossible for the qualities from which it was composed to be separated, but each of them is annihilated and the destruction of them all produces a single distinct peculiar quality.

The illustration of juxtaposition in this text goes back to Aristotle [126] although the term for it—παράθεσις—is new. The case of fusion is illustrated here, and in many other places, by the combination of substances to form the quadruple drug.[127] Most noticeable in the light of our earlier discussion is the use of terms for "extension" (ἀντιπαρέκτασις, ἀντιπαρεκτείνεσθαι) to describe the *interrelation* of constituents in blending,[128] for as that discussion showed these terms were uniquely fitted to describe the relation of a smaller entity to a larger one and the process by which it attained its magnitude.[129] Indeed they are only used to mean a process of mutual coextension—or a reciprocal interpenetration—in this passage from Philo and in other secondary reports of the tripartite classification of mixture.[130]

On the face of it, then, we are presented with two conflicting descriptions of total blending, and the conflict is heightened by the fact that two texts describe the three types of mixture classified

---

[126] *de gen. et corr.* A10 328a 2-3.

[127] Celsus V. 19. 9, Anon. Lond. XIV. 19 Diels, Galen *de const. art. med.* I. 242 K, *In Hippocr. de nat. hom.* II. 18 K; cf. *de el. sec. Hippocr.* I. 452 K.

[128] Its use to describe the case of fusion in this passage must be a misunderstanding—see below, n. 143.

[129] See above, pp. 32-33.

[130] Apart from those in the two texts quoted below see Hippolytus (at *SVF* II 469) if Diels' conjecture (*Dox. Graec.*, p. 571. 24) is correct; for παρεκτείνεσθαι in this sense see Alexander *Mant.* 140. 9, 141. 22-23 (cf. *de mixt.* 218. 6, 220. 5), and Sextus Empiricus *P.H.* III. 60, and ἐκτείνεσθαι at *Mant.* 141. 16, and ἐπιτείνεσθαι at *Mant.* 141. 24.

by Philo as causally dependent on the motion of pneuma through matter. Since our preliminary interpretation established a quite specific relation between total blending and the process of pneumatic motion it can only be sustained if this additional evidence can be shown to be erroneous or in some sense complementary to that earlier thesis. The two relevant texts are closely related and need to be examined in detail; I quote them in parallel.

Stobaeus *Ecl.* I. xvii. 4. 153. 24-155. 14 Wachsmuth (= Arius Didymus *Fr. Phys.* 28 = *SVF* II 471)	Alexander *de mixt.* 3. 216. 14-217. 2 (= *SVF* II 473)
1a. Χρύσιππος δὲ τοιοῦτόν τι διεβεβαιοῦτο· εἶναι τὸ ὂν πνεῦμα κινοῦν ἑαυτὸ πρὸς ἑαυτὸ καὶ ἐξ αὑτοῦ, ἢ πνεῦμα ἑαυτὸ κινοῦν πρόσω καὶ ὀπίσω· 1b. πνεῦμα δὲ εἴληπται διὰ τὸ λέγεσθαι αὐτὸ ἀέρα εἶναι κινούμενον· ἀνάλογον δὲ γίνεσθαι κἀπὶ τοῦ αἰθέρος, ὥστε καὶ εἰς κοινὸν λόγον πεσεῖν αὐτά. 1c. ἡ τοιαύτη δὲ κίνησις κατὰ μόνους γίνεται τοὺς νομίζοντας τὴν οὐσίαν πᾶσαν μεταβολὴν ἐπιδέχεσθαι καὶ σύγχυσιν καὶ σύστασιν καὶ σύμμιξιν καὶ σύμφυσιν καὶ τὰ τούτοις παραπλήσια.	1. ἔστι δὲ ἡ Χρυσίππου δόξα περὶ κράσεως ἥδε· ἡνῶσθαι μὲν ὑποτίθεται τὴν σύμπασαν οὐσίαν, πνεύματός τινος διὰ πάσης αὐτῆς διήκοντος, ὑφ' οὗ συνέχεταί τε καὶ συμμένει καὶ σύμπαθές ἐστιν αὐτῷ τὸ πᾶν,
2a. Διαφέρειν γὰρ ἀρέσκει τοῖς ἀπὸ τῆς Στωικῆς αἱρέσεως παράθεσιν, μῖξιν, κρᾶσιν, σύγχυσιν.	2a. τῶν δὲ μιγνυμένων ἐν αὐτῇ σωμάτων,
2b. παράθεσιν μὲν γὰρ εἶναι σωμάτων συναφὴν κατὰ τὰς ἐπιφανείας, ὡς ἐπὶ τῶν σωρῶν ὁρῶμεν, ἐν οἷς πυροί τε καὶ κριθαὶ καὶ φακοὶ καὶ εἴ τινα τούτοις ἄλλα παραπλήσια περιέχεται καὶ τῶν ἐπὶ τῶν αἰγιαλῶν ψήφων καὶ ἄμμων.	2b. τὰς μὲν παραθέσει μίξεις γίνεσθαι λέγει, δύο τινῶν ἢ καὶ πλείονων οὐσιῶν εἰς ταὐτὸν συντεθείμενων καὶ παρατιθεμένων ἀλλήλαις, ὥς φησιν, καθ' ἁρμήν, σωζούσης ἑκάστης αὐτῶν ἐν τῇ τοιαύτῃ παραθέσει κατὰ τὴν περιγραφὴν τὴν οἰκείαν οὐσίαν τε καὶ ποιότητα, ὡς ἐπὶ κυάμων φέρε εἰπεῖν καὶ πυρῶν ἐν τῇ παρ' ἀλλήλους θέσει· (216. 17-22)
2c. μῖξιν δ' εἶναι δύο ἢ καὶ πλειόνων σωμάτων ἀντιπαρέκτασιν δι' ὅλων, ὑπομενουσῶν τῶν συμφυῶν περὶ αὐτὰ ποιοτήτων, ὡς ἐπὶ τοῦ πυρὸς ἔχει καὶ τοῦ πεπυρακτωμένου σιδήρου, ἐπὶ τούτων γὰρ <δι'> ὅλων γίγνεσθαι τῶν σωμάτων τὴν ἀντιπαρέκτασιν. ὁμοίως δὲ κἀπὶ τῶν ἐν ἡμῖν ψυχῶν ἔχειν· δι' ὅλων γὰρ τῶν σωμάτων ἡμῶν ἀντιπαρεκτείνουσιν, ἀρέσκει γὰρ αὐτοῖς	2c. τὰς δέ τινας γίνεσθαι μίξεις λέγει, δι' ὅλων τινῶν οὐσιῶν τε καὶ τῶν τούτων ποιοτήτων ἀντιπαρεκτεινομένων ἀλλήλαις μετὰ τοῦ τὰς ἐξ ἀρχῆς οὐσίας τε καὶ ποιότητας σῴζειν ἐν τῇ μίξει τῇ τοιᾷδε, ἥντινα τῶν μίξεων κρᾶσιν ἰδίως εἶναι λέγει. τὴν γὰρ δύο ἢ καὶ πλείονων τινῶν σωμάτων ὅλων δι' ὅλων ἀντιπαρέκτασιν ἀλλήλοις οὕτως, ὡς σώζειν ἕκαστον αὐτῶν ἐν τῇ

σῶμα διὰ σώματος ἀντιπαρήκειν. κρᾶσιν δὲ εἶναι λέγουσι δύο ἢ καὶ πλειόνων σωμάτων ὑγρῶν δι' ὅλων ἀντιπαρέκτασιν τῶν περὶ αὐτὰ ποιοτήτων ὑπομενουσῶν· [omitting 154. 24-155. 3] συνεκφαίνεσθαι γὰρ ἐκ τῆς κράσεως τὴν ἑκάστου τῶν συγκραθέντων ὑγρῶν ποιότητα, οἷον οἴνου, μέλιτος, ὕδατος, ὄξους, τῶν παραπλησίων. ὅτι δ' ἐπὶ τοιούτων κράσεων διαμένουσιν αἱ ποιότητες τῶν συγκραθέντων, πρόδηλον ἐκ τοῦ πολλάκις ἐξ ἐπιμηχανήσεως ἀποχωρίζεσθαι ταῦτα ἀπ' ἀλλήλων. ἐὰν γοῦν σπόγγον ἠλαιωμένον καθῇ τις εἰς οἶνον ὕδατι κεκραμένον, ἀποχωρίσει τὸ ὕδωρ τοῦ οἴνου, ἀναδραμόντος τοῦ ὕδατος εἰς τὸν σπόγγον.

μίζει τῇ τοιαύτῃ τήν τε οἰκείαν οὐσίαν καὶ τὰς ἐν αὐτῇ ποιότητας, λέγει κρᾶσιν εἶναι μόνην τῶν μίζεων· εἶναι γὰρ ἴδιον τῶν κεκραμένων τὸ δύνασθαι χωρίζεσθαι πάλιν ἀπ' ἀλλήλων, ὃ μόνως γίνεσθαι τῷ σῴζειν ἐν τῇ μίζει τὰ κεκραμένα τὰς αὐτῶν φύσεις.
(216. 25-217. 2)

2d. τὴν δὲ σύγχυσιν δύο <ἢ> καὶ πλειόνων ποιοτήτων περὶ τὰ σώματα μεταβολὴν εἰς ἑτέρας διαφερούσης τούτων ποιότητος γένεσιν, ὡς ἐπὶ τῆς συνθέσεως τῶν μύρων καὶ τῶν ἰατρικῶν φαρμάκων.

2d. γίνεσθαι τὰς δέ τινας συγχύσει δι' ὅλων τῶν τε οὐσιῶν αὐτῶν καὶ τῶν ἐν αὐταῖς ποιοτήτων συμφθειρομένων ἀλλήλαις, ὡς γίνεσθαί φησιν ἐπὶ τῶν ἰατρικῶν φαρμάκων κατὰ σύμφθαρσιν τῶν μιγνυμένων, ἄλλου τινὸς ἐξ αὐτῶν γεννωμένου σώματος. (216. 22-25)

Translation

(1a) Chrysippus emphasised the following sort of view: that there is that which is pneuma moving itself to itself and from itself, or pneuma moving itself backwards and forwards;

(1b) it is conceived of as pneuma because it is described as being air in motion; this occurs similarly also with the aether, so that it even shares its definition.

(1c) Such motion occurs only according to those who think that the whole of substance receives change and fusion and combination and comixture and cohesion and similar effects.

(1) Chrysippus' theory of blending is as follows: he holds that while the whole of substance is unified because it is totally pervaded by a pneuma through which the whole is held together, is stable, and is sympathetic with itself,

(2a) For in the opinion of members of the Stoic sect juxtaposition mixture blending and fusion differ.

(2b) For juxtaposition is the contact of bodies at their surfaces as

(2a) yet some of the mixtures of bodies mixed in this substance

(2b) occur by juxtaposition, through two or more substances

we see with heaps in which corn and barley and figs and other such things are combined and with heaps of stones along the shores and sand-grains.

being composed into the same mass and juxtaposed with one another "by juncture" as he says, and with each of them preserving the surface of their own substance and quality in such a juxtaposition, as, one will grant, happens with beans and wheat-grains in their juxtaposition;

(2c) Mixture is the complete co-extension of two or more bodies while their inherent qualities remain stable, as in the case of fire and heated iron, for here complete coextension of bodies occurs. Similarly with our souls; for they extend through the whole of our bodies—for it is their opinion that body extends through body.

Blending they describe as the complete coextension of two or even more moist bodies and their qualities [154. 24-155. 3] for they say that the quality of each of the moist bodies blended together appears together from the blend, as with wine, honey, water, vinegar, and the like. It is clear that with such blends the qualities of the bodies blended together remain stable from the fact that they can often be separated from one another by an artificial device. Certainly if one places an oil-drenched sponge in wine blended with water it will separate the water from the wine as the water retreats into the sponge.

(2c) the third type of mixture he says occurs through certain substances and their qualities being mutually coextended in their entirety and preserving their original substance and qualities in such a mixture: this mixture is blending in the strict sense of the term. The mutual coextension of some two or even more bodies in their entirety with one another so that each of them preserves their own substance and its qualities in such a mixture—this, he says, alone of the mixtures is blending; for it is a peculiarity of bodies that have been blended that they can be separated again from one another, and this only occurs through the blended bodies preserving their own natures in the mixture.

(2d) Fusion is the alteration of two or even more qualities that belong to bodies to create another quality differing from these, as in the composition of perfumes and medical drugs.

(2d) Other mixtures occur by total fusion with both the substances and their qualities being destroyed together, as he says happens with medical drugs in the joint-destruction of the constituents and the production of some other body from them.

Both these texts are clearly the product of a long tradition. They each paraphrase some source; 1b of the doxographical passage, for example, is the rationalisation of pneuma's motion rather than a direct description,[131] while 2c of Alexander's report excludes the

[131] Cf. Alexander *de mixt.* 10. 223. 34-224. 6 where another kind of attempt to rationalise the motion of pneuma through body is made. The words

illustration of mechanical separation found in the parallel text only for its author to employ it independently in his account of the Aristotelian theory of blending later in the *de mixtione*.[132] Now both passages could be dependent on a similar source which they have used in their own way; on the other hand, it is plausible to suggest in principle that Alexander could have relied on a doxographic report such as that preserved by Stobaeus. We know of his acquaintance with doxographical material in Aetius [133] and, of course, of his use of the Peripatetic source of the doxography, Theophrastus' *Opinions of the Physicists*.[134] By examining separately the details of the two texts quoted here I shall try to show that all the significant differences between Alexander's report and the fragment of Arius Didymus can be explained as independent systematisations by the Peripatetic exegete, so that it is entirely possible that he employed this text as his source. More importantly this analysis will draw attention to the numerous difficulties involved in regarding either text as direct evidence for our knowledge of the theory of total blending, and so prepare the way for an attempt to integrate this evidence with the interpretation of total blending already reached on the basis of the primary sources.

(1) The classification of mixtures (2b, 2c, 2d).

Both texts are very similar in their descriptions of juxtaposition and fusion.[135] The only important difference is Alexander's use of the expression "by juncture" (καθ' ἁρμήν) to describe juxtaposition. This is unparalleled and may emanate from contemporary

---

τοιοῦτον τι at 1-a of the doxographical passage also suggest that only an approximate paraphrase is being offered.

[132] 15. 232. 2-3.

[133] Cf. *de an.* 83. 2-8 with Aetius IV. 11. 1-2 (= *SVF* II 83) and see above, p. 28, n. 35; and *Mant.* 130. 14-17 with Aetius IV. 15. 3 (= *SVF* II 866).

[134] See Diels, *Dox. Graec.*, 481. 9, 482. 5, 483. 8, 485. 4, 492. 2, 494. 4, and cf. E. Howald, *Hermes* 55 (1920) 86-87. A detailed investigation of Alexander's reports on the Presocratics would be helpful. Cf. also the doxographic presentation of theories of matter at *de mixt.* 1. 213. 18-214. 6 (with Comm. *ad loc.*), and on 213. 23-214. 2 see Diels, *Dox. Graec.*, p. 174.

[135] Neither seems aware of the illustration of this by the quadruple drug (cf. n. 127). Cf. also Pseudo-Galen *de qual. incorp.* XIX. 471 K (= p. 7. 20-25 Westenberger) who similarly draws a very general connection between fusion and the formation of drugs from blends. Alexander's ignorance here is probably also reflected at *de an.* 24. 24-25.

Stoicism.[136] There is however a major difference in the accounts of blending (κρᾶσις): for Alexander, as for Philo, this is simply the mutual coextension of moist bodies, but the doxographical report (at 2c) attempts to distinguish it from "mixture" (μῖξις) as the complete coextension of dry bodies. The restriction of this term to dry bodies is found in Philo's classification, but there it is more naturally associated with juxtaposition, a usage that Alexander himself slips into at one point,[137] though he normally employs "mixture" in a generic sense. The doxographer's examples of "mixture" as the passage of iron through fire is a standard Stoic illustration of total blending,[138] while the case of soul pervading body is often presented as a case of body going through body.[139] But the isolation of these cases as the total mixture of dry bodies, and the consequently unusual use of the term "mixture," must be regarded as doxographic pedantry.[140] If Alexander was dependent on this source it would be natural for him to eliminate this case and to group its examples as cases of total blending, as seems to have happened in *de mixtione* 4;[141] the term "mixture" could then be used generically, in an Aristotelian fashion.[142]

Both texts use the term "extension through" (ἀντιπαρέκτασις) to describe the mutual coextension of constituents of blends; hence the relative initial volumes being blended are not specified, and, as we have seen, this case differs radically from the form of total blending described in our primary sources. It is left entirely unclear why such coextension should not also occur in fusion.

---

136 The expression is discussed in detail by Rex, *Chrysipps Mischungslehre*, pp. 31-33.

137 *de mixt.* 7. 221. 7.

138 Cf. Comm. on *de mixt.* 4. 218. 1-2.

139 This is clear even from polemics against the Stoic theory of corporeal soul; cf. the texts cited at p. 84, n. 243 below

140 It recurs only at Aetius I. 24. 3; cf. pp. 61-62 below.

141 At 218. 1-2 the case of fire pervading iron obtrudes into the group of illustrations of total blending (cf. p. 40, n. 93); it may owe its presence here to Alexander's revision of his doxographic source. The case of soul pervading body is dealt with at 217. 33-36 in relation to the cases of φύσις (Nature) and ἕξις (State).

142 Cf. also the use of τὸ πᾶν (literally "The All") at *de mixt.* 3. 216. 15-16 (cf. 10. 223. 8) to describe the medium of pneuma's action. Conventionally (*SVF* II 522-525) Stoics distinguished this from τὸ ὅλον (literally "the whole") as comprising τὸ ὅλον (the material universe) along with the infinite external void. Alexander, I suspect, has simply fallen back on the Aristotelian usage of τὸ πᾶν to mean the universe; cf. Bekker, *Ind. Ar.* 571b 54-59.

There would seem to be nothing in the intrinsic properties of wine and water to explain their extending through one another and occupying the same volume while certain chemicals blend without such interpenetration.[143] Fusion therefore only provides a contrast with blending insofar as it is irreversible, and the difference between the two processes is indicated, but not explained, by the use of ἀντιπαρεκτείνεσθαι to describe blending. It would be purely speculative in this context to try to explain this term by reference to the motion of pneuma as described in part 1 of both passages. It is clear from 2a that this motion is supposed to cause *all three* kinds of mixture; but if, as we shall argue,[144] this is an inaccurate report there can be no reason for segregating total blending from the tripartite classification in this way.

Both texts also see blending as a reversible process—the constituents are separable because they are actually present together in the blend. In the doxographical passage this is described as their "being manifest together" (συνεκφαίνεσθαι)[145] while Alexander terms it their "preserving their own substances and qualities," or their own "natures," or elsewhere, their own "surfaces." This language is unique to Alexander and, as I argue in the Commentary, can be regarded as his own essentially polemical redescription of this aspect of total blending.[146] It makes explicit the requirement that constituents be actually present in the blend which, as we saw, was implicit in the description of the blending of a wine drop with the sea in the primary sources.[147] More generally, this would seem to be required by the Stoic avoidance, if not rejection, of the concept of potential being, the state which the constituents of a blend attained in Aristotle's theory.[148]

---

143 This may be the rationale behind Philo's use of ἀντιπαρεκτείνεσθαι ("extension-through") to describe the relation of constituents in fusion (cf. n. 128). In a less technical presentation these two cases could easily be assimilated.

144 Pp. 60-65 below.

145 The closest parallel is Alexander *de an.* 19. 32-33, referring to the general principle that constituents are preserved in a blend: εἴ γε δεῖ μὲν ἐμφαίνεσθαι τῷ ἐκ τῆς μίξεώς τινι ὄντι πάσας τὰς δυνάμεις τὰς ἐκείνων, ἐξ ὧν μέμικται.

146 Cf. on *de mixt.* 1. 213. 2-13.

147 See pp. 32-33 above. Such actual copresence of constituents is also, of course, required by the theory of pneuma's motion through matter which such blending, we have argued, illustrated.

148 It is not explicitly employed by the Stoics, and its absence is criticised

As a whole the tripartite classification of mixtures is entirely descriptive. While the cases of juxtaposition and fusion are exemplified by commonplace examples, the process of mutual coextension involved in total blending is left mysteriously unexplained. The fact that blending is neither of the other two limiting cases does nothing to demonstrate exactly what sort of physical process it is. We shall shortly try to assess the significance of this clearly rather circumscribed account.

(2) The identification of the constituents (2b, 2c, 2d).

In Alexander's report of the tripartite classification the constituents of the various mixtures are characterised as substances with qualities. A looser version of this description is also given in Philo's report, while the doxographic account conceives of constituents as bodies with qualities. It might seem necessary to link these descriptions with the Stoic theory of substance and quality, which held that the whole of substance was qualified in its parts, so that the substance of all bodies broke down, as it were, into particular substances with their individual qualities.[149] I believe that this would be a mistaken systematisation of Stoic thinking. As the examples of blending and fusion show, it is not really substances that interact but unspecified masses. Since a doctrine of substance is normally directed towards an account of the individuality of a body it is doubtful that such masses would qualify as substances.[150] Certainly the Stoic doctrine of the ἰδίως ποιόν ("particularising quality") is an attempt to establish a theory of uniquely individuating qualities for bodies that could hardly include random volumes of water and wine.[151] Another point is that although the Stoic account of substance and quality allowed for qualitative

by Plotinus *Enn.* VI. 1. 26 (= *SVF* II 315). Cf. my comments at *Apeiron* VII-i (1973) 23, n. 14 on the supposed potentiality of the Stoic infinite.

[149] Diog. Laert. VII. 150 (= *SVF* II 316).

[150] At *Met.* Z16. 1040b 5-10 Aristotle suggests that the elements—the primary constituents of blends—are not strictly speaking substances, but "like a heap" (οἷον σωρός). This makes it difficult to see in his case too in what strict metaphysical sense substances could blend: cf. R. Sokolowski, *JHP* 7 (1970) 271-272 on this and Comm. on *de mixt.* 13. 229. 18-30.

[151] Presumably this doctrine deals only with what the Stoics termed "unified" bodies (ἡνωμένα), or bodies that can be specified as having a ἕξις (*SVF* II 366, 368). On the whole doctrine of categories see Rist, *Stoic Philosophy*, pp. 152-172 where the major literature is discussed.

change, there is no evidence that it allowed for the interaction of substance with substance, as in blending, or the loss by substances of their qualities and identity, as in fusion.[152] In Aristotelian terms, the Stoics do not seem to have had a theory of "simple change" (γένεσις ἁπλῆ) or change in the category of substance; their metaphysics concentrated on an explanation of the relation between individual substances and substance as a whole (the part-whole relation) rather than the interrelations between these quasi-independent substances.[153]

For these reasons the manner in which our texts characterise the constituents of mixtures is not, I believe, indicative of an underlying metaphysical theory. Alexander's description in particular can almost certainly be traced to a standard report of total blending in which the Peripatetics and Stoics are contrasted, the former said to hold that only qualities are totally blended, while for the latter both substances and qualities are involved. Found explicitly in Galen [154] and in a vestigial form in Sextus Empiricus and Plotinus,[155] this report seems to be a piece of syncretistic doxography. The basic categories of substance and quality are applied to two schools whose doctrine in this area differed considerably, and are in all probability derived from Aristotelian sources.[156] Then the Stoic concept of total blending is again applied to both schools, and so entirely misrepresents Aristotle who denied that qualities could blend by themselves and insisted that bodies (if not substances) be the constituents of blends.[157] This inherently untrustworthy

---

152 I disagree with Rist. (*Stoic Philosophy*, pp. 158-159) that *SVF* I 92 (Galen *In Hippocr. de hum.* XVI. 32 K, cf. *Nat. Fac.* I. 2, II. 5K) shows that Zeno held that substances and qualities were capable of change. The texts attributed to Zeno the thesis that total blends occur between such entities. As I have argued, this is implausible (see above, p. 30, n. 44; cf. n. 154 below); in the rest of this section I shall argue that the thesis itself is a misleading systematisation of Stoic doctrine.

153 The evolution of elements from Fire (*SVF* II 579-581) and the combination of elements to form compounds (*SVF* II 413) do not constitute an exception. The process here is not strictly coming-to-be but the manifestation of a single physical principle in a variety of roles.

154 *In Hippocr. de nat. hom.* I XV. 32 K (= *SVF* II 463); cf. *de el. sec. Hippocr.* I. 9, I. 489 K (= *SVF* II 464).

155 Respectively *P.H.* III. 57-62 and *Enn.* II. 7. 1. 8-12.

156 We know that Galen was prejudiced towards the "Peripatetic" view that blends occurred between qualities (cf. IV. 762 K and X. 16 K).

157 *de gen. et corr.* A 10 327b 15-22. The term "substances" (οὐσίαι) is not used, but only the expression "separables" (χωριστά). It is left to Alexander

doxa is, I suggest, the source of Alexander's report in 2c where total blending is defined as the complete coextension of substances and qualities, and of his general characterisation of the constituents of all three types of mixtures as substances. Such a description would provide a clear coordination between the preliminary description of pneuma pervading the whole of substance (πάσῃ οὐσίᾳ) and a causally dependent trichotomy of mixtures between substances. As we shall see, it was in Alexander's interests to establish such a relation, and I suspect that his identification of constituents as substances with qualities was a stage in its attainment, rather than a genuine Stoic attempt to establish a relation between the whole of substance and its parts.[158]

(3) Pneuma's motion through substance: its relation to the tripartite classification (1, 2a).

On a superficial reading both these texts could be interpreted to mean that pneuma is the cause of the three kinds of mixture.[159] I shall argue that on close analysis neither yield such a meaning, and that the presence of a description of pneuma's motion through matter along with the tripartite classification of mixtures does not indicate any relation of physical causation.

Part 1 of the doxographic report first describes Chrysippus' doctrine of tensional motion in standard terms (1a), then provides a rationalisation of this by suggesting the physical properties pneuma must possess (1b),[160] and finally says that such motion can only occur if substance (or matter) can be altered by entering into various relations of compounding (1c).[161] The three types of

(*de mixt.* 13. 228. 13-25) to emphasise that substances are the constituents of blends, though whether or not without difficulty is questionable; see Comm. on 13. 228. 10-25.

[158] There is no other text in Alexander that shows any awareness of the Stoic doctrine of substance.

[159] H. Leisegang, Art. "Physik," *RE* XX-i (1960), col. 1052 flatly describes, *SVF* II 471ff. as showing "pneuma als Ursache der παράθεσις, μῖξις, κρᾶσις, σύγχυσις . . ."

[160] Cf. above, p. 36, and for Alexander's similar attempt to equate pneuma with the Aristotelian aether (presumably what the doxographer has in mind here) see *de mixt.* 10. 223. 9-17, 30-34.

[161] The list of συν-compounds here is unparalleled. The only comparable text seems to be Arius Didymus fr. 20 (= Stobaeus *Ecl.* I. xi. 5[a], p. 133 W) where Chrysippus is said to hold that matter (ὕλη) can receive "division and fusion in its parts" (διαίρεσιν δὲ καὶ σύγχυσιν . . . κατὰ μέρη) and fusion is said to be that kind of change by which destruction (φθορᾶς) occurs.

mixture are then said to differ because of this situation (2a). The initial difficulty is to see whether it is all or part of the preceding paragraph (1) that is regarded as explaining these processes. If we assume that the whole phenomenon of pneuma's motion through matter explains the three mixtures, the causal connection is very unclear. Are these mixtures processes that occur between pneuma and matter, as the presence of σύγχυσις (fusion) in both the description of matter's flexibility and the tripartite classification might suggest?[162] If so this would be an unprecedented account of pneuma's motion, and highly implausible as far as the case of juxtaposition is concerned. On the other hand, could these mixtures occur after pneuma has pervaded matter and so be processes for which it provides the physical precondition? This, as we shall see, is Alexander's interpretation, but it is still unsatisfactory in that there is no reason why these particular processes should necessarily occur as a result of pneuma's motion: as we have noted (pp. 56-57 above), no special connection is drawn in these texts between the extension of bodies through bodies in blending and the pneumatic motion.

It is, I believe, more informative to look at the relation between the third sentence of part one of the doxographic report (1c) and the opening sentence of part two. Here a description of the affectability of matter is immediately followed by an account of the three types of mixture with the claim that the latter depends on the former. We find an almost identical relation expressed in another doxographic text at Aetius I. 24. 3.[163]

> Πυθαγόρας καὶ πάντες ὅσοι παθητὴν τὴν ὕλην ὑποτίθενται, κυρίως γενέσεις καὶ φθορὰς γίνεσθαι· ἐκ γὰρ ἀλλοιώσεως τῶν στοιχείων καὶ τροπῆς καὶ ἀναλύσεως[164] [γενέσεως καὶ φθορᾶς][165] παράθεσιν καὶ κρᾶσίν τε καὶ σύγχυσιν γίγνεσθαι.

> (Pythagoras and all who posit matter as affectable [say] that coming to be and passing away occur in a strict sense; for from the qualitative change of the elements and their modification and dissolution juxtaposition and mixture, and blending and fusion occur.)

---

[162] On the various contexts in which fusion occurs cf. n. 186 below.

[163] Diels, *Dox. Graec.*, p. 320. 29-35.

[164] From here to the end I cite Stobaeus *Ecl.* I. xx. 1d^a^, p. 170 24-25 W; Pseudo-Plutarch *Epit.* I. 24 has only the words ταῦτα γίνεσθαι after ἀναλύσεως.

[165] Heeren regarded this as a gloss, and it is so recorded by Diels.

Here the three types of mixture are said to be directly dependent on the "affectability" of matter once the evolution of the elements has occurred; that is, because matter has the capacity to undergo change, then certain specific types of change can occur. If we isolate 1c and 2a of the text from Stobaeus a similar type of reasoning can be detected; that because matter can receive various changes, then the specific changes described in the tripartite classification can occur. Only the reference to the elements is omitted.[166]

In the text from Aetius the connection drawn between matter's affectability and the various mixtures is attributed to a wide range of thinkers,[167] while in the text from Stobaeus it is narrowed down to the Stoics. But we can still detect the doxographical organisation of this evidence, for 1c is a very general description of matter's capacity for change, and the Chrysippean theory of pneumatic motion is said to justify including the Stoics among "those who hold" (οἱ νομίζοντες) this theory. That is, a specific doctrine of Chrysippus (1a) leads to the association of the Stoics with a general doxographical description of matter's affectability (1c) which we have seen has elsewhere (at Aetius I. 24. 3) a causal relation to the view that change occurs by the specific processes of juxtaposition, blending and fusion. Although the classification of these mixtures may have been Stoic in origin (but perhaps not Chrysippean) [168] it is enmeshed in both texts in a doxographic thought-pattern that seems to have a validity independent of the specific description of Stoic doctrine.

This criticism removes the possibility that Arius Didymus fr. 28 describes a causal relation between pneuma and the tripartite classification of mixtures; for the latter is now seen to be dependent on a description of matter (1c) couched in the general terms of a doxographer's explanation of pneuma's motion, rather than on an

---

[166] This need not be significant. In the doxography the active and passive principles were grouped along with the four elements as the principles (ἀρχαί) of the Stoic system; cf. Comm. on *de mixt.* 11. 225. 5-18.

[167] Cf. Aetius *Plac.* I. 9. 2 (= *SVF* II 324) where the Stoics are grouped with Thales, Pythagoras, and Heraclitus with respect to their belief in the affectability of matter (τρεπτὴν καὶ ἀλλοιωτὴν καὶ μεταβλητὴν καὶ ῥευστὴν ὅλην δι' ὅλης τὴν ὕλην); also cf. *SVF* II 305.

[168] At 2a it is only attributed to the school collectively, and its apparent attribution to Chrysippus by Alexander is questionable (cf. n. 175). It may, therefore, have been developed by later Stoics (see p. 69 below).

integral part of Chrysippus' account. In this context it would be wrong to expect that there was a precise physical theory in Stoicism prescribing the relation between pneuma and the three types of mixture. The chain of reasoning in the first four sentences of the text preserved by Stobaeus (1, 2a) is that of a doxographer, not a philosopher. Taken separately each sentence expresses a certain truth about Stoic physics, but in their interrelation they express reasoning established by the doxography and not emanating from Chrysippean physical theory.

Much the same criticism can be made of Alexander, despite a superficially more satisfactory account. He characterises pneuma's motion in very general terms and concentrates, unlike the doxographer, on describing its effects. These are that substance is unified, and this unity involves continuity, stability, and sympathy with itself. This formulaic description is an accurate account of pneuma's activity,[169] but it is not clear what relevance it has to the tripartite classification. As Alexander reports it, the constituents of mixtures (τὰ μιγνύμενα) are bodies that are "in" the whole of substance that has been unified. Being in such a physical context should mean that they are all interrelated,[170] yet what we find described are a series of physical interactions that seem to have no connection at all with this basic situation.[171] There is simply the assertion that a unified substance is a necessary condition of there being mixtures of bodies, but no indication is given of why these particular mixtures should occur. As in the doxographic report neither the identification of the constituents nor their interaction is related to pneuma's motion. We might for example expect that if these three mixtures are somehow caused by pneuma, then constituents, as inanimate bodies, might be identified by their possession of a ἕξις (a state),[172] or that the extension of body through body (2c) might be grounded in pneuma's motion. As it is, the sole indication of a relation between pneuma and the tripartite classification is the very vague description in 2a.

Since Alexander's report is part of a treatise on mixture we can,

[169] Cf. Comm. on *de mixt.* 3. 216. 15-16.

[170] As Alexander himself assumes in one of his arguments against the theory of pneuma, at *de mixt.* 10. 223. 27-29.

[171] At pp. 58-60 above we have ruled out any correlation in the Stoic theory between the individual substances that Alexander describes as being mixed, and the whole of substance (πάσῃ οὐσίᾳ) in which this occurs.

[172] On this see above, p. 37.

I think, appreciate its character by setting it in context. In the opening chapter of the *de mixtione* Alexander asserts that theories of mixture depend on theories of matter, and proceeds to divide the latter into the corpuscular theory and the theory that matter is unified.[173] At the outset of chapter three he announces his return to proponents of the latter theory after a discussion of the Atomists and Epicurus in chapter two.[174] We can therefore, I think, take the description of the whole of substance being unified by pneuma (216. 14-15) as Alexander's deliberate description of the theory of pneuma to fit his general programme in the *de mixtione*. This could also explain his deviation from a source like the doxography which offers a much more diffuse account of pneuma's motion. Then after classifying the Stoics with those who believe that matter is unified Alexander has to show that their theory of mixture is derived from this theory of matter. This he does rather vaguely, as we have seen, by saying that the constituents of mixtures are bodies interacting within the unified material matrix created by pneuma's motion. But the vagueness of this derivation, its absence from any other source, and the manner in which it fulfills the broader programme of the *de mixtione* prove, I believe, that at 216. 17 it is Alexander himself who has provided a link between the motion of pneuma through matter and the tripartite classification of mixtures.[175]

There are other considerations supporting this conclusion. In chapter two of the *de mixtione*—as the Commentary will show in detail—Alexander provides a similar correlation between Atomist and Epicurean theories of mixture and theories of matter only by the totally ahistorical approach of arbitrarily attributing to them theories of blending derived from Aristotle's discussion in *de generatione et corruptione* A10.[176] Though the correlation has a greater inherent plausibility than that asserted for the case of Chrysippus, it is still formulated solely to fulfill the general organisational principle of the treatise that theories of mixture be

---

[173] *de mixt.* 1. 213. 15-18, 214. 6-10.

[174] *de mixt.* 3. 216. 1-2 (μετέλθωμεν δὲ ἐπὶ τοὺς κοινῶς ἡνῶσθαι τὴν ὕλην λέγοντας . . .). On this whole conception see Comm. on 3. 216. 1-4.

[175] In doing so he could also have followed the doxographer in attributing the authorship of the tripartite classification to Chrysippus rather than to the school collectively: cf. the grammar of parts 1 and 2a of the parallel passages. Further on the Chrysippean authorship of the doctrines described in *de mixt.* 3 and 4 see Comm. on 3. 217. 2-13, and 4. 217. 14 and 218. 8-9.

[176] See especially Comm. on 214. 18-28 and 214. 28-215. 8.

derivable from theories of matter. We have also seen in section (I) that Alexander is frequently capable of arbitrarily assimilating Stoic ideas and terminology for his own philosophical purposes. Here, I believe, he is simply carrying out the same kind of restructuring of his opponents' ideas, much in the manner of Aristotle, to conform with his methods of writing the history of philosophy.

From this final discussion we can conclude that the two sources separately analysed here are indeed the product of a complex process of transmission. They place a description of pneuma's motion through matter in a causal relation to a tripartite classification of mixtures when these are clearly independent doctrines. At the same time the details of each of our two sources are affected both by the doxographical tradition, and, in Alexander's case, by an apparent attempt to systematise arbitrarily a doxographic source. The value of these texts as evidence for the Stoic theory of total blending is therefore severely circumscribed.

As for their interrelation, enough evidence has, I think, been produced to claim confidently that Alexander is at least dependent on the classification of mixtures at Arius Didymus fr. 28. Whether his attempt to systematise the relation between pneuma's motion through matter and this classification (Alexander, 1, 2a) in order to meet the general framework of the *de mixtione* was a reformulation of the more diffuse account in the same source is less certain. There are other cases where he radically systematises doxographic material to fit his needs, and so in the present instance it is entirely possible that he followed the same practice.

A Final Reconstruction

On turning back to the preliminary interpretation of the theory of total blending, the results of the preceding discussion seem at first sight unpromising. In our major secondary sources total blending is a process of mutual coextension, contrasted with two other kinds of mixture, and yet causally unrelated to the theory of pneuma. Earlier, on the basis of the primary sources, we agreed that it was the extension of a small volume through a large one, but such only as a conception analogically formed from empirical data and thence the basis for a further analogical derivation of pneuma's motion through matter. Yet because the latter account is challenged by texts in which we have found certain inherent deficiencies as a result of a doxographic tradition we are justified in attempting to retain it in the face of such clearly unreliable secondary evidence. This, I believe, can be done principally by examining carefully the structure of *de mixtione* 3 and 4, for this is the one text in which both these conflicting accounts of total blending can be found in an unreconciled form.

In chapter four, when presenting the analogical argument for the conception of a small volume of liquid extending through a larger one, Alexander makes this process a special case of the general relation of the mutual coextension of bodies (217. 26-32). Yet the various phenomena illustrating cooperative causation (at 217. 13-26), from which this conception is analogically formed, can only be reformulated as a non-reciprocal relation between bodies; that is, in this context, if A is helped by B to extend itself, it does not follow that B is helped by A; so the water helps the wine drop to extend through it, but not vice versa.[177] The attempt that Alexander makes to import into this context the case of the *mutual coextension* of bodies would seem to be a result of the description of total blending in these terms in the tripartite classification in *de mixtione* 3 (our 2c).[178]

One text which does suggest an entirely compatible relation between these apparently conflicting conceptions of blending is chapter 37 of Plutarch's *de communibus notitiis*. This is almost entirely a polemic against total blending, but with primary emphasis on the paradoxicality of a small body pervading or being blended with a larger one. It is however striking that after an argument dealing with this specific relation [179] Plutarch makes the following claim:

---

[177] See p. 39 n. 91 above.

[178] Cf. also *de mixt.* 3. 217. 2-9 where Alexander reports that Chrysippus claimed that the difference between the three types of mixture was recognisable by means of the common notions. In the Comm. on 3. 217. 2-13. I argue that this too is a rationalisation of the tripartite classification of mixture in the face of the argument in chapter four which shows that the common notions were only used to prove that a body could extend through a body. This indicates that Alexander is attempting to coordinate the doxographic account of the three mixtures with a proof for total blending probably derived from contemporary Stoic sources (see Comm. on *de mixt.* 4. 217. 36).

[179] 1078A (Teubner, p. 104. 8-20); this involves assigning unit value to unequal volumes. I discuss it further at pp. 86-87 below.

> And this is the wonderful result for those who force bodies into a body, and this the inconceivability of their containing one another. For it is necessary that when bodies go into one another [εἰς ἄλληλα χωρούντων] in blending, one does not surround the other while the other is surrounded, and one receive while the other inheres (for in this way there will not be blending but the contact and touching of surfaces, with the inner surface enclosed while the outer surface surrounds, while the other parts are differentiated as unmixed and pure); but it is necessary that when the blending that they uphold occurs, the constituents should come to be in one another [ἐν ἀλλήλοις . . . γίνεσθαι] and that the same body should, in a single mass, be surrounded by its inherence [in the other body] and surround the other body by receiving it, and neither of them should again be capable of existing [independently], when blending forces both to go through one another [δι' ἀλλήλων διιέναι], and absolutely nothing be left behind but both to be completely filled up with another.[180]

Plutarch then goes on to describe Arcesilaus' argument against a small body blending with a larger one,[181] and to quote Chrysippus' description of the wine drop blending with the sea that we discussed earlier,[182] but with no sense of incompatibility between this very general attack on intermingling, where the relative volumes of the constituents are unimportant, and a set of arguments directed exclusively against the paradox of a small volume extending through a much larger one.[183] In fact the argument that we have quoted seems to be based on the same kind of general distinction between total blending and juxtaposition made by Diogenes Laertius at VII. 151 before he also describes the case of

---

[180] 1078B-C (Teubner, pp. 104. 20-105. 5). Cf. *de mixt.* 7. 221. 7-20 for a more definite polemic against the idea that constituents of a total blend lose their identity; cf. Comm. *ad loc.*

[181] 1078C-D (Teubner, p. 105. 5-23).

[182] 1078E (Teubner, p. 165. 23-28 = *SVF* II 480); see above, pp. 31-33.

[183] Cf. also *de comm. not.* 37. 1078C (Teubner, p. 105. 7) where the premise of Arcesilaus' argument against a small body blending with a larger is εἰ γάρ εἰσιν αἱ κράσεις δι' ὅλων, again an apparently general characterisation of intermingling; and 37. 1077E-F (Teubner, pp. 103. 29-104. 3) where a description of the extension of a small volume through a large one is said to be propounded by "those who blend bodies with bodies as wholes with wholes" (τοὺς τὰ σώματα τοῖς σώμασιν ὅλοις ὅλα κεραννύντας).

a wine drop extending through the sea. Even Alexander draws a similar contrast at the end of *de mixtione* 3 before he proceeds in chapter four to describe the argument which primarily demonstrates the blend of a small volume with a larger.[184] Yet while both Plutarch and Diogenes supplement this general contrast with a description of total blending as *uniquely* the blend of a small volume with a larger, Alexander offers a tripartite classification of mixtures and, as we have just seen, attempts to subordinate the case of the blend of a small with a large volume to the case of mutual coextension described in that classification. There are, I believe, reasons why his evidence takes this form, and these yield a reading of *de mixtione* 3 and 4 that makes it a basically consistent account of total blending which Alexander himself has cast into a conflicting form.

As we saw earlier,[185] the tripartite classification of mixtures seemed to be designed to clarify the unique status of total blending by contrasting it with juxtaposition and fusion. As such it addressed itself only to the general characteristics of a total blend—the complete coextension of constituents and their separability—and not to the nature of the process by which such a blend was achieved. It resembled the simple contrast between juxtaposition and fusion found in Plutarch and Diogenes Laertius, with only the addition of fusion which serves to illustrate Stoic physical theories in several other doxographical texts.[186] Now just as the description of blending in that bipartite contrast did not prevent its authors giving an account of total blending as uniquely the blend of a small with a large volume, neither need the tripartite classification conflict with such an account. Blending as mutual coextension is entirely neutral as to the relative initial volumes of constituents of a blend, and can easily complement a description of the blend of unequals or a proof of such a process as offered by Alexander in *de mixtione* 4. In the tripartite classification total blending is a case of mutual coextension because it is presented in general terms: in *de mixtione* 4 it is the extension of a small volume

---

[184] *de mixt.* 217. 9-13.

[185] Pp. 55-58.

[186] At *de aet. mund.* 16. Philo lists fusion as one of four types of destruction (τρόποι φθορᾶς) apparently invoked by the Stoics in discussing the conflagration of the universe. Cf. also Arius Didymus fr. 20 where it is said to be one of the effects that matter can receive, and cf. fr. 27 where it is listed as one of the types of change recognised by Posidonius.

through a larger one because it occurs in an argument establishing its precise relation to central doctrines of Stoic physics. There are not therefore two conflicting senses of the verb "to extend through" (ἀντιπαρεκτείνειν); rather there are two interdependent and complementing descriptions of total blending that this verb is called upon to characterise.

The description of blending in the tripartite classification should then be regarded as part of a systematic clarification of the more complex doctrine which we have argued underlay the Stoic theory of pneuma. Conceivably it was developed in Stoic scholasticism as a pedagogic effort to elucidate its complexities after Chrysippus had formulated the theory that we reconstructed earlier.[187] Certainly Plutarch who knew this Stoic's works seems to have had no knowledge of a special sense of fusion (σύγχυσις) and uses the term synonymously with blending (κρᾶσις).[188]

This whole classification of mixtures is indeed closely comparable to another Stoic classification in which bodies are said to be either unified (ἡνωμένα), divided (ἐκ διεστώτων), or "interconnected" (ἐκ συναπτομένων).[189] This is designed to demonstrate, through a simple contrast, the nature of the class of unified bodies which are held together by pneuma,[190] without however specifying the details of that physical theory. Equally the classification of mixtures simply clarifies the nature of a physical relation ultimately central to Stoic physics but without proving how it occurs.[191] Both accounts

---

187 Since Chrysippus' authorship of this classification is anyway in some doubt (cf. n. 175 above), it is possible that it was formulated by a later generation of Stoic scholastics. Such triadic systematisations of doctrines are certainly very common in Stoicism: see M. Fuhrmann, *Das Systematische Lehrbuch*, pp. 147-148, especially the references at p. 148 n. 1.

188 *de comm. not.* 40. 1081A (Teubner, p. 110. 27). Cf. 37. 1078A (Teubner, p. 104. 17) for the use of διάχυσις = κρᾶσις. See note 48 above for Diogenes Laertius' probable use of συμφθείρεσθαι to mean "blending."

189 *SVF* II 366 (Plutarch), II 368 (Achilles), and cf. II 1013 and III 160. This rather simple classification has been subjected to a great deal of elaborate speculation. I share Graeser's view (*Plotinus and the Stoics*, p. 73, where earlier literature is summarised) that *SVF* II 367 clearly shows that in authorship it is Chrysippean. On its possible relation to the tripartite classification of mixtures, and the triad State/Soul/Nature see H. Dörrie's speculations (they are no more) at *Porphyrios' "Symmikta Zetemata"*, pp. 27-28.

190 Made clear by Achilles (*SVF* II 368) where unified bodies are described as those controlled by a ἕξις (State) which is πνεῦμα συνεκτικὸν σώματος ("pneuma capable of containing a body").

191 Thus E. Bréhier's point (*Théorie des Incorporels*, p. 42) that juxtaposition is "une concession aux apparences" is correct. The classification

are illustrative descriptions parasitic upon independently demonstrable theories.

There is therefore no reason to find an inherent incompatibility between the account of blending as mutual coextension in *de mixtione* 3, or its doxographical source, and its description as the extension of a small volume through a large one conveyed by our primary sources, and by *de mixtione* 4. Since we have also shown that the tripartite classification of mixtures is not causally dependent on the motion of pneuma through matter we can now restate our preliminary interpretation of total blending in the light of the analysis of sources undertaken in this section. The relevant texts have contained three main components:

I. The motion of pneuma through matter.
II. The tripartite classification of mixtures in which total blending is the mutual coextension of bodies.
III. The proof by analogy that total blending is the extension of a small volume through a large one.

Two of our sources (Arius Didymus fr. 28 and Alexander *de mixtione* 3. 216. 16-217. 2) describe I and II, and, as we have argued, do not show, despite appearances, that I is the cause of II. In addition we have just seen that II can be regarded as an independent illustration of total blending rather than a comprehensive classification of mixtures that occur in the empirical world. III is reported by Alexander in *de mixtione* 4 and accords with the conception of total blending expressed in our primary sources, Diogenes Laertius and Plutarch. As we have argued,[192] the case of a small volume extending through a larger one should be taken as a conception from which the relation of pneuma to matter is analogically derived. This, in effect, means that I is to be derived from III, leaving II, as we have suggested on independent grounds, as an adjunct to the whole theory illustrating only the general characteristics of the process described in III.

We can now suggest an account of the genesis of Alexander's

---

of mixtures is not a complete account of the material continuum, in which juxtaposition is impossible; rather it is an attempt to clarify that relation (total blending) which helps to explain the formation of the continuum by the motion of pneuma through matter by contrasting it with juxtaposition.

[192] Above, pp. 38-45.

evidence in *de mixtione* 3 and 4.[193] As we have argued, his attempt in chapter three to establish a causal relation between I and II is a systematisation, probably based on his doxographic source, and certainly formulated to meet the general programme of the *de mixtione*. However in chapter four he presents us with III. This places him in the difficult position of asserting two quite different conceptions of total blending which he attempts to reconcile only with some strain. At the same time he introduces the case of pneuma's motion through the passive elements as a "clear proof" that a wine drop extends through water.[194] Thus in a single exposition of the Stoic theory of total blending he asserts two different relations between pneuma and total blending, one in which it is the cause of a mutual coextension of bodies (the relation between I and II in chapter three) and another in which it somehow illustrates the extension of a small volume of wine through a large volume of water (the relation between I and III in chapter four). I have suggested a number of reasons why the first of these is Alexander's own creation, probably based on his doxographic source, and not a part of Stoic physics; the second I argued earlier should be taken as the derivation of pneuma's motion through matter from the analogically formed conception of a wine drop's extension through water.

If these conclusions are correct then as a whole *de mixtione* 3 and 4 is an attempt on the part of its author to coordinate the first of these relations with an argument that partly conceals the second. As evidence it does not therefore challenge the thesis argued earlier that the second of these relations represents a correct account of the role of total blending in Stoic physical theory.

To sum up, the overall interpretation advanced in this essay is that the so-called Stoic theory of total blending had a limited but important role in their physical system. It was an analogically

---

193 More tentatively we can also speculate on the genesis of the doxographical report at Arius Didymus fr. 28. II, as a systematic description, would gain a place in the doxography more readily than the complex argument in III. At the same time, its relation to I had to be acknowledged, and so I and II were juxtaposed and, as we have argued, placed in a causal relation by a doxographic rationale. III would probably be preserved only by Stoics, and it is from such sources that Alexander may have directly derived it; cf. Comm. on *de mixt.* 4. 217. 36.

194 *de mixt.* 4. 218. 2-6.

formed conception of the blending of a small volume with a large one designed, not as a representation of physical reality, but as a purely mental conception which could serve as the illustrative prefigurement of the theory of pneuma's motion through matter. As such it was only part of an analogical demonstration of this physical relation and could not have been the sole proof of this intricate physical theory. Its importance was that it made a complex central doctrine of Stoic physics readily comprehensible, and this is probably why it is quite widely reported, extensively criticised, and often influential as the model for the description of other relations.

If this interpretation is correct, it is perhaps not surprising that the ancient critics of Stoicism failed to appreciate its complex and circumscribed character. They saw total blending simply as a factual claim that two bodies went through one another and occupied the same place, and none of them discussed the doctrine in the context of the Stoic physical system.[195] Their polemics in effect amount to a detachment of the case of total blending, whether as the extension of a small volume through a larger one or as the coextension of two random volumes, either from the place that we have argued it occupied in the proof for the motion of pneuma through matter, or from its place in the tripartite classification of mixtures. Thus isolated it could be easily ridiculed as the theory that two bodies occupied the same place *simpliciter*, or that a random body could somehow expand to occupy the place of another.[196]

This segregation of Stoic doctrines from the relevant part of the system which gave them meaning is fairly typical among ancient authors whether for the purposes of criticism, assimilation or description.[197] Were our sources more complete, or works by early Stoics extant, this phenomenon would be little more than a minor

---

[195] The references to body going through body at *de mixt.* 10. 223. 18 and 11. 225. 4-5 in the discussion of pneuma and the Stoic deity do not constitute an exception—see introductory note to chapters ten and eleven.

[196] In Alexander's case in the *de mixtione* the description in the tripartite classification influences his conception of blending as a reciprocal process of "body going through body" (σῶμα διὰ σώματος χωρεῖν). He only refers to the extension of a small volume through a larger one once in the *de mixtione* at 220. 13-23: cf. Comm. *ad loc.* Note also Calcidius *In Tim.* 221 where total blending (*concretio*) is said to involve two bodies being in the same place.

[197] See my discussion at *Symb. Os.* 48 (1973) 60-63 of the assimilation of the terminology of the language of the Stoic theory of common notions.

episode in the history of ancient philosophical scholasticism. But when the authors responsible for this maltreatment of Stoic doctrine are also among our major sources their methods must be appreciated in all their detail. Only in this way can interpretation avoid mistaking criticism for direct reports of doctrines, or hope to fully grasp the significance of given doctrines in the overall context of the Stoic system. For this reason the next section of this essay presents a detailed analysis of the form and sources of some of the major arguments brought against the theory of total blending by ancient critics.

III. "BODY GOING THROUGH BODY"

Alexander and the Criticism of the Stoic Theory of Total Blending (apropos de mixtione 5 and 6) [198]

The earliest criticism of the notion of total blending was an attack delivered, according to Plutarch, by Chrysippus' former teacher,[199] the sceptical Platonist Arcesilaus. The mixture of a small body with a much larger one meant, he claimed, that a leg could be cut off, thrown into the sea and expand to the point where Xerxes and the Greeks could fight a sea battle in it.[200] Although

---

[198] The general topic of ancient criticisms of total blending has been dealt with by H. Dörrie, *Porphyrios' "Symmikta Zetemata"*, pp. 24-35, but mainly with reference to the Neoplatonic tradition. His discussion is also not accompanied by a full interpretation of the theory of total blending, nor is the Peripatetic tradition—the importance of which he is aware (cf. p. 29 n. 1)—examined in detail. The argument of this section, if correct, will necessitate some revision in his views of at least the background to the later critique of total blending; see nn. 200, 211 below. I have also discussed the topic of "body going through body" in the Peripatetic and exegetical tradition in a more limited context in an article on Pseudo-Alexander *Quaest.* II. 12 at *Philologus* 116 (1972) 293-305.

[199] See Diogenes Laertius VII. 184. It is plausible on chronological grounds: Arcesilaus' dates are 316/15 to 214/40 (v. Von Arnim, *RE* II, col. 1164), and Chrysippus' 281/277 to 208/204 (*idem, RE* III, col. 2502). The tradition that made Arcesilaus an opponent of Zeno (Cicero *Acad.* I. 12. 44) is dubious, and the text from Plutarch referred to in the next note cannot be taken as evidence that Zeno propounded the theory of total blending, as some of the older literature claimed (see K. Troost, *Zenonis Citiensis de rebus physicis doctrinae*, p. 52 n. 2).

[200] Plutarch *de comm. not.* 37. 1078C-D. There is no evidence of Carneades' critique of total blending, and though Dörrie (*Porphyrios' "Symmikta Zetemata"*, p. 29 n. 1) may be right to speculate that he did deal with it, he was certainly not "der Ahnherr dieser Kritik." Conceivably Sextus Empiricus *P.H.* III. 56-62 is derived from him; cf. n. 251 below.

the tone of ridicule was to be maintained, later criticism was somewhat more intricate, mainly stressing the paradoxical idea that one body could go completely through another, expressed in the phrase σῶμα διὰ σώματος χωρεῖν. In some texts this appears as an actual description of total blending [201] and perhaps for this reason Von Arnim employed it as a subtitle to his collection of texts on the topic.[202]

It is not however used by Stoics in any primary sources and it is most unlikely that it could ever have served them as a genuinely informative description of total blending.[203] The verb "to go" (χωρεῖν) conceals the complexity of the relation of extension (ἀντιπαρέκτασις) which is central to this process; for it suggests that it was reciprocal when, if we have reasoned correctly, it could only have involved the extension of a small volume through a large one.[204] Again, apart from the fact that body going through body is principally treated as a paradoxical relation occurring between three-dimensional solids by critics of Stoicism,[205] there are a number of texts in which it is discussed and elaborated without any reference whatsoever to Stoic physical theory. All of these occur in the works of the Aristotelian commentators and yet none of this material has ever been recognised or investigated in relation to the critique of the Stoic theory of blending. I hope to establish

---

[201] Hippolytus *Philos.* 21 (at *SVF* II 469), Themistius *Phys.* 104. 14-18 (*SVF* II 468), both very general accounts. On Plutarch *de comm. not.* 37. 1077E (*SVF* II 465) cf. the *Philologus* article cited above (n. 198) at p. 296, and n. 220 below. Oddly Dörrie (*Porphyrios' "Symmikta Zetemata"*, p. 25 nn. 5 and 7) attributes the doctrine of body going through body to Zeno and Chrysippus on the basis respectively of Arius Didymus fr. 38 (*Dox. Graec.*, p. 270. 4) and fr. 28 (*Dox. Graec.*, p. 463. 20ff. = *SVF* II 471; quoted at pp. 52-53 above). Yet neither of these texts does more than use verbs of motion to describe the blending of the elements in the first case, and of soul and body in the second, without even employing the expression σῶμα διὰ σώματος χωρεῖν.

[202] *SVF* II, p. 151. 7.

[203] Decisive on this score is *de mixt.* 12. 227. 2 where Alexander says that "body going through body" was not appreciated by the Stoics themselves as the fundamental notion in their doctrines of God, fate, the soul, etc. Cf. Comm. *ad loc.*

[204] Cf. pp. 32-33, 65-71 above. In *de mixt.* 5 and 6 it is clearly irrelevant which of the two bodies goes through which, and even if a one-way process is described in a given text it is clearly reversible. This is made quite explicit in other arguments against the extension of a small body through a large one which critics explicitly regarded as reversible; cf. pp. 87-88 below.

[205] Comm. on *de mixt.* 6. 219. 11-12; cf. p. 83 below.

here that this evidence is so extensive and so closely dependent on Aristotelian texts as to leave no doubt that it was this exegetical tradition that developed most of the arguments and paradoxes centred around body going through body that are employed by the ancient critics of total blending, and moreover that it is this tradition that Alexander has inherited in his critique of this theory in the *de mixtione* and elsewhere.

Peripatetic Exegesis and the Origins of the Criticism

After the detailed exposition of the Stoic theory of mixture in *de mixtione* 3 and 4, the very general argument inaugurated in chapter 5 seems rather abrupt. Instead of attacking the details of the theory just described Alexander isolates the general principle that body goes through body,[206] and sets out to show that this is impossible either between solids (in chapter 6) [207] or if there are void pores in bodies (chapter 5). Now this whole disjunction seems unnecessary when the Stoics would never have posited the latter theory [208], and to discover its roots we shall need to analyse texts from Aristotelian commentators for whom, as we shall show, it became in the process of exegesis a natural disjunction. Alexander himself will not be represented, except indirectly, since none of his commentaries on the relevant Aristotelian works are extant; nonetheless as these commentators themselves indicate they inherit a tradition detectable in his works.

The exegetical technique that we shall observe in this section is a special case of the methodology described earlier [209] whereby a commentator can exceed the Aristotelian text through various forms of interpretative paraphrase. We shall see an issue of minor significance in Aristotle converted in the exegetical tradition into a philosophical principle by the synthesis of some loosely-related texts. Since these texts served Aristotle in his criticism of various doctrines, once it can be shown that any part of the principle developed from them is entailed by a Stoic doctrine, that doctrine can be accordingly redescribed and criticised. The basic framework is the result of reflection on the Aristotelian text, its use against the Stoics a subsequent development.

---

[206] 219. 11-220. 6.
[207] 218. 24-219. 9.
[208] As Alexander implies in this context at *de mixt.* 5. 218. 25-26.
[209] See above, Pt. I, p. 15 and cf. Pt. II pp. 25-26 on the attribution of general theses to opponents.

Sem. Griekse Letterkunde
Blandijnberg 2
Gent

In *de mixtione* 5 it is clear that Alexander uses the expression "body going through body" (σῶμα διὰ σώματος χωρεῖν) simply to mean that two bodies are in the same place.[210] This redescription is not to be found in Aristotle's numerous references to the paradox of two bodies being in the same place (δύο σώματα ἐν ταὐτῷ), but these texts form the natural starting place for our enquiry. For Aristotle this paradox is held to be entailed by a variety of views and is used to justify their summary dismissal. These are (1) that place is body (*Phys.* 209a 4-7), (2) that growth occurs by body (*Phys.* 214b 6, *de gen. et corr.* 321a 5-10), (3) that light is a body (*de an.* 418b 13-18),[211] (4) that soul is a body (*de an.* 409b 3), and (5) that the elements are generated from body (*de caelo* 305a 19-20). In each case Aristotle's argument is entirely *ad hominem*, and he makes no attempt to systematise his dismissals of these various doctrines, though taken together they might loosely be said to constitute a thesis of anti-materialism;[212] for in each of the first four cases Aristotle's own doctrine involved the use of a non-material concept.[213] His commentators however did recognise similarities between these arguments.

It is clear that the argument that if place is a body then two bodies will occupy the same place is fundamental to this pattern, and that the other four are simply special cases of it. The commentators' treatment of the text from *Physics* Δ1 in which this occurs (209a 4-7) therefore provides some guide to their treatment of the other arguments.

In the most extensive comment on this text Simplicius[214] presents a capsule of the argument of *de mixtione* 5 and 6. He describes the paradox of two bodies in the same place as "body

---

[210] Cf. 218. 15-16 with 218. 18-21.

[211] This is the only one of these closely related passages noted by Dörrie (*Porphyrios' "Symmikta Zetemata"*, p. 25 n. 6) who, erroneously I believe (cf. n. 198), regards "body going through body" as the Stoic contradiction of this Aristotelian principle. This is a good example of speculative reasoning establishing historical relations without close attention to the quality of the sources.

[212] It might be added to the "presuppositions of Aristotle's physics" collected by G Boas, *AJP* 57 (1936) 24-36.

[213] Although Aristotle does not use this language, clearly place as the limit of surrounding body is incorporeal (*Phys.* Δ 4), as is soul *qua* form (*de an.* B1-2), the form and not matter in which growth occurs (*de gen. et corr.* A5), and light as the actualisation of the transparent medium (*de an.* B7).

[214] Cf. also the article cited at n. 198, at p. 296f.

going through body" (σῶμα διὰ σώματος χωρεῖν),[215] points out that it is entailed by Stoic materialism, and finally refers to Alexander's arguments against this paradox "in a separate treatise and in his commentaries." [216] He then quotes a long passage from Alexander, presumably from the latter's note on *Physics* Δ1 209a 4-7, in which an argument against bodily interpenetration through pores is included.[217] This would seem to confirm that the roots of the argument in *de mixtione* 5 and 6 do rest in the exegesis of this Aristotelian argument. Elsewhere among the commentators Themistius [218] refers explicitly to the anti-Stoicism of the *de mixtione* in his comment on this same text, and Philoponus [219] uses the expression "body going through body" in the same context. Finally, passing to a pre-Alexandrian author, we find Plutarch identifying bodily interpenetration with the theory that body is place, an equation that he may have derived from a Peripatetic source.[220]

Before these connections can be elaborated any further we need to establish the Aristotelian roots of the whole disjunctive argument in *de mixtione* 5 and 6. For although the alternative of bodily interpenetration through pores is broached in some of the texts just reviewed its source is still not clear.

---

[215] *Phys.* 530. 9-30: lines 9-14 = *SVF* II 467.

[216] *Phys.* 530. 15-16 (δι' ἀφωρισμένου συγγράμματος . . . καὶ ἐν τοῖς ὑπομνήμασιν).

[217] *Phys.* 530. 21-22. This corresponds to Alexander *Mant.* 139. 33-34 which is part of a general critique of bodily interpenetration. There are loose parallels between Simplicius *Phys.* 530. 16-19 and *Mant.* 140. 6-8, *Phys.* 530. 24-26 and *Mant.* 140. 4-5, and *Phys.* 530. 26-30 and *Mant.* 140. 10-20. This suggests that the whole of this Alexandrian text (*Mant.* 139. 29-141. 28) is an abstract of Alexander's note on *Phys.* 209a 4-7. Cf. *Mant.* 127. 27-130. 12, a critique of vision through light-rays, which has many parallels with texts in Alexander's commentary on the *de sensu*; see Bruns's *apparatus criticus* where they are collected.

[218] *Phys.* 104. 18-22 (= *SVF* II 468). Cf. also p. 30, n. 44 above.

[219] *Phys.* 505. 22-23.

[220] *de comm. not.* 37. 1077E: παρὰ τὴν ἔννοιάν ἐστι σῶμα σώματος εἶναι τόπον καὶ σῶμα χωρεῖν διὰ σώματος, κενὸν μηδετέρου περιέχοντος—ἀλλὰ τοῦ πλήρους εἰς τὸ πλῆρες ἐνδυομένου καὶ δεχομένου τὸ ἐπιμιγνύμενον τοῦ διάστασιν οὐκ ἔχοντος οὐδὲ χώραν ἐν αὑτῷ διὰ τὴν συνέχειαν. ("It is contrary to the general conception for body to be the place of body and body to go through body with neither containing void but solid sinking into solid and receiving the body admixed with it as one without any interval or space in itself because of its continuity.") Cf. Alexander *de mixt.* 5. 218. 15-18 and 21-24, both of which resemble the above and seem to confirm that Plutarch is offering a polemical redescription of the Stoic theory rather than (as M. Pohlenz, *Hermes* 74 [1939] 30 implies) a direct report.

Here we need first to examine the second case in which Aristotle held that a theory entailed that two bodies were in the same place, that of growth occurring by body. At *Physics* Δ7 214b 5-9 he points out that the void is not necessary for the explanation of growth; for either two bodies will be in the same place, or the whole of the body that grows will be void, if indeed growth is through the body and through void (διὰ κενοῦ).[221] Then at *de generatione et corruptione* A5 321a 6-9 he argues that if growth is either by body or by the incorporeal, then in the first case there will be two bodies in the same place, and in the second there will be a separable void.[222] These texts are, I believe, another source of the general disjunction that we find in *de mixtione* 5 and 6. In Philoponus' note on the passage from *de generatione et corruptione* the argument from *Physics* Δ7 214b 5-9 is used although it is not in the Aristotelian text,[223] and again, both here and in Alexander's discussion of growth in *de mixtione* 16, the void through which growth might occur is referred to as a pore or a set of void spaces (κενά).[224] We can therefore see how by interpreting these Aristotelian texts, quite naturally, as posing the alternative between growth through bodily interpenetration or through interstitial pores,[225] com-

[221] "For either bodies are not increased at all, or not by body, or it is possible for two bodies to be in the same place (then they claim to solve a general difficulty, but do not show that void exists) or the whole body must be void if it is increased in every part and increased by void." This answers the earlier argument of proponents of the void that growth occurs through void as in the case of ash absorbing water (*Phys.* 213b 18-22), otherwise two bodies will be in the same place. Cf. *de mixt.* 220. 10 where Alexander, in a critique of body going through body, refers to the case of ash as one which does not prove this thesis.

[222] "If [it grows] by the incorporeal there will be separable void; but it is impossible that there be separable matter for a magnitude, as was said earlier." The reference is to 320b 17-25.

[223] Cf. *Phys.* 214b 6-9 ("Either two bodies can be in the same place . . . or the whole body must be void, if it grows throughout and grows through void.") with Philoponus *de gen. et corr.* 90. 12-15 (". . . the whole body that grows [must be] assumed to be a pore, or body [must] pervade body."). We find the same basic alternative in Alexander's discussion of growth at *de mixt.* 16. 234. 23-32.

[224] See preceding note, and for the expression "void spaces" (κενά) see *de mixt.* 16. 234. 29. I have discussed elsewhere the extent to which the discussion of pores both here, and especially in dealing with the theory that light is a body (see note 246 below), might also have been provoked by Strato of Lampsacus' theory of interstitial κενά; see the Appendix to the article on *Quaest.* II. 12 cited above, n. 198.

[225] Joachim—presumably relying on the commentators—actually uses this

mentators might be led to generalise the alternatives, and to use them in other contexts. In *de mixtione* 5 and 6 they are transferred to a context in which bodily interpenetration is criticised in completely general terms, and one which, as we saw, has other roots in the Aristotelian attack on the theory that body is place.[226]

There is one more Aristotelian text to be reviewed before we can understand fully the origins of the notion of bodily interpenetration through pores. This is the argument against Empedocles and the Atomists in *de generatione et corruptione* A8 (at 325b 5-9 and 326b 6-14)[227] that physical interaction (ποιεῖν and πάσχειν) is impossible through void pores. This complements the alternative of interpenetration through the void posed briefly in the two Aristotelian texts on growth discussed above by providing arguments against the general notion of bodily interaction through void spaces. These can be very easily transferred to a context in which simple motion (χωρεῖν) of body through body is under consideration. Not surprisingly then we find that the arguments against bodily interpenetration through pores in *de mixtione* 5 are clearly dependent on those from *de generatione et corruptione* A8, and that in the one extant commentary on this work Philoponus restates Aristotle's arguments in this text in terms of a disjunction resembling that of *de mixtione* 5 between bodily interpenetration through solids or through pores.[228]

So we see the formation of a scholastic synthesis. Aristotle occasionally uses the paradox of two bodies occupying the same place in *ad hominem* and unsystematically related arguments. Then in one case, because he is arguing against a specific aspect of

---

alternative (on *de gen. et corr.* 321a 5-9). Perhaps Aristotle did imply a theory of interstitial void in this argument, but of an entirely undefined sort.

[226] That such a concept of place is implicit in these texts is clear from *de mixt.* 5. 218. 18-21; 6. 219. 22-27.

[227] These are, briefly, (a) that pores must be spaces in indivisible bodies (325b 5-7), since (b) if bodies were completely divisible then there would only be pores and no intervening solids (325b 7-9); (c) if pores are filled then they are superfluous (326b 6-10), whereas if they are empty they must contain body since pores so small as to be unable to receive body are inconceivable. (a) and (b) equate Empedocles with Atomism, while (c) is directed just against the former.

[228] This can be shown by comparing (1) *de mixt.* 218. 33-219. 1 with 326b 6-10; (2) 219. 1-3 with 325b 5-7; and (3) 219. 3-5 with 325b 7-9. Also cf. the texts in (3) with *Phys.* 214b 8-9 and Philoponus *de gen. et corr.* 178. 5-20 (on 326b 6) which is relevant although the commentator does not particularly emphasise here that the pores are void.

the theory of the void, he states the alternative of growth through body or through the void. In the commentators this becomes a disjunction between a phenomenon described, in non-Aristotelian terms, as "body going through body" (σῶμα διὰ σώματος χωρεῖν), and its occurrence through void pores. This can then be used in paraphrasing an Aristotelian text, even if Aristotle himself does not pose the alternative explicitly, as in his discussion of growth,[229] or even hint at it, as in his arguments against interaction (ποιεῖν/πάσχειν) through void pores, or in his argument against the view that light, or soul, is a body.[230] In this way the commentators build on an Aristotelian foundation to develop a synthesis which allows for the elaborate paraphrase of Aristotelian texts without serious distortion. We see, in effect, one technique by which a faithful exegete can express his originality, even if in a very limited form.

Finally, for future reference, we need to note a further elaboration that the commentators brought to the paradox of bodily interpenetration. This involved two minor additions both based very probably on the same Aristotelian text, *Physics* Δ6 213b 5-12. This text cites an argument by proponents of the void that if there were motion without void then (1) two bodies, or any number of bodies, could occupy the same place, and (2) a very large body could occupy the place of a smaller one by being divided into parts equal to it and, by the first paradox, occupying their place. In view of the technique of textual synthesis that we have observed it is not surprising that the commentators use both of these arguments, and particularly the second, to further emphasise the paradoxicality of two bodies being in the same place.[231] To illustrate the second paradox they regularly invoke the image of "heaven being in a millet-seed" that Aristotle used (at *Physics* Δ12 221a 22) in a somewhat different argument in his discussion of time, where

---

[229] See nn. 221 and 222 above.

[230] On light, cf. Philoponus *de an.* 344. 4-8 and several passages from the *Mantissa* (see below, n. 246). This might have been derived from *de gen. et corr.* A8 326b 10-14 which transposes the arguments at 326b 5-10 (cf. n. 227 above) to the case of vision; cf. Philoponus *ad loc.* (*de gen. et corr.* 179. 15-18) where the expression "body going through body" (σῶμα διὰ σώματος χωρεῖν) is used. See also Philoponus *de an.* 173. 21-26 (on the soul), or Simplicius *de caelo* 269. 3 (à propos Aristotle on the motion of the elements) where interpenetration through void is mentioned.

[231] See Simplicius *Phys.* 530. 22-24, *de caelo* 629. 29-30; Philoponus *Phys.* 505. 23-27, *de an.* 343. 34-344. 4. Cf. Alexander *de an.* 20. 14-15.

he claimed that two things existing contemporaneously are not "in" one another, otherwise everything would be in anything. This was a sufficiently similar point for this vivid illustration to be used of bodies of unequal size being in one another.[232] Its potential relevance to the Stoic idea that unequal volumes could blend is very clear and will be demonstrated in detail below.

I hope for the moment to have established that the arguments of *de mixtione* 5 and 6 must be regarded as an outgrowth of the Peripatetic exegetical tradition. Since the same disjunction between body going through body or through void is to be found in Alexander's predecessors Plutarch and Sextus Empiricus,[233] we can assume that this tradition had been established prior to Alexander. Whether it originated after Andronicus' "edition," or in an earlier period, cannot be determined. Alexander's contribution, at any rate, must be considered the scholastic achievement of maintaining this tradition rather than an independent achievement in Aristotelian scholarship.

Alexander and "Body going through body" (σῶμα διὰ σώματος χωρεῖν)

Simplicius, as we saw, reported (at *Physics* 530. 9-30) that Alexander dealt independently with the topic of "body going through body" and considered Stoic materialism in terms of this paradox. Since all the relevant commentaries of Alexander on the Aristotelian works in which the paradox of two bodies being in the same place occurs are lost, his contribution to its elaboration must be gleaned from his treatises and schooltexts, principally the *Mantissa*.[234] The following Alexandrian texts all contain a similar

---

232 On this and the conjoint use of Stoic imagery in such texts see below, pp. 87-88.

233 Plutarch *de comm. not.* 37. 1077E (cf. n. 220 above); Sextus Empiricus *A.M.* IX 256-257 (on his date v. Part I, n. 93). Cf. also Pseudo-Galen *de qual. incorp.* XIX. 474 K (= p. 10 Westenberger), a work of uncertain date, though in view of its general similarities to texts by Alexander and Albinus (cf. n. 247 below) probably to be dated to the second or early third century A.D. Its authorship by Albinus, on the basis of a Syrian testimonium, is argued by E. Orth, *Ant. Class.* 16 (1947) 113-114.

234 There are problems about the authenticity of this collection as a whole (cf. above, Pt. I, pp. 18-19). I shall draw evidence from four texts all of which attack various materialistic theses: (1) "that soul is incorporeal" (113. 25-118. 4); (2) "that qualities are not bodies" (122. 16-125. 4); (3) "that light is not a body" (138. 4-139. 28); (4) "that body cannot pervade body" (ὅτι σῶμα διὰ σώματος ἀδύνατον διήκειν, 139. 9-141. 28). Of these (4) is probably derived from Alexander's lost commentary on the *Physics* (above,

set of arguments or assertions employing either the notion that body goes through body, or that two bodies are in the same place.

(1) *de mixtione* 5 and 6, *Mantissa* 139. 29-141. 28, and *Quaestiones* II. 12 attack the general notion of body going through body.
(2) *de mixtione* 16. 234. 24-32 attacks the view that growth occurs by body, and can thus be considered a gloss on Aristotle *Physics* Δ7 214b 3-9, and *de generatione et corruptione* A5 321a 5-10.
(3) *Mantissa* 115. 35-116. 1 and *de anima* 20. 6-19 (cf. *Topics* 173. 14-16) deal with the theory that soul is body (cf. Aristotle *de anima* 409b 3).
(4) *Mantissa* 138. 4-139. 28 attacks the view that light is a body (cf. *de anima* 418b 13-18).
(5) *Mantissa* 124. 13-20 (cf. 123. 12-13) attacks the doctrine that qualities are bodies, a specifically Stoic theory (cf. *SVF* II 377, 380).

In most of these texts we find the same pattern of argument that we traced in the preceding section: a given theory is held to entail that body goes through body, with, in some cases, the added rider that this interpenetration is impossible through pores. In the texts previously examined commentators were elucidating Aristotelian texts; here however Alexander is dealing with what appear to be Stoic doctrines cast in the same mould as the materialistic doctrines Aristotle attacked. We need to measure the effect of a critique so obviously rooted in different philosophical assumptions on the presentation of Stoic doctrine.

(1) As we have shown, the arguments in *de mixtione* 5 and 6 and some sections of the texts from the *Mantissa* are clearly derived from a tradition of Aristotelian scholarship. The argument at *Quaestiones* II. 12 against body going through body by compression (συστέλλεσθαι) can similarly be shown to be derived from Aristotelian sources, with the Stoic theory of total blending entirely contained by its Peripatetic redescription.[235] In none of these texts is

n. 217); (3) has parallels in Philoponus *de an.*, suggesting a source in Alexander's commentary on this work (below, n. 246), and (1) could be related to the same commentary. (2) is perhaps the most questionable, but it has parallels within the *Mantissa* (cf. for example 124. 9-7 with 141. 1-4). Given the evident unity of theme in these four texts and the above evidence, we can confidently employ them pending further research on the *Mantissa*.

[235] See my article cited in n. 198 above for the full argument.

there any attempt to argue that the Stoic theory must be reducible to a case of body going through body; it is simply assumed that this is its essential thesis. As we noted at the end of the preceding section the basic trend of the ancient criticism of total blending is to detach it from its place in the Stoic physical system and treat it as an isolated assertion of a physical paradox. Thus the expression "body going through body" retains very little of the dynamic character of the relation between bodies conveyed by the verb "to extend through" (ἀντιπαρεκτείνειν) and related terms, when transposed from the blend of liquids to the case of pneuma's motion through matter.[236] Among the Aristotelian commentators "body going through body" means nothing more than that there are two bodies in the same place, while in Alexander we can find the verb of going (χωρεῖν) used interchangeably with such verbs as "to receive" (δέχεσθαι) or "to come to be in" (γίνεσθαι ἐν) [237] that convey only a minimal sense of motion.

Again, the conception of body on which these polemics rely is the traditional one of a body as a three-dimensional solid;[238] if our interpretation is correct, then for the Stoics the total blending of liquids as an analogically formed conception and not an actual case of mixture was independent of any theory of body. They could therefore have invoked the traditional definition of body in characterising relations between intracosmic bodies, while explaining the motion of body through body in the case of pneuma's motion through matter in terms of a set of qualitative physical concepts.[239]

(2) Alexander's arguments against the view that growth is by body occur in *de mixtione* 16. Since this treatise is predominantly anti-Stoic these might also be considered part of the polemic. However as they follow the standard pattern of a disjunction between bodily interpenetration or motion through pores as a gloss on *de generatione et corruptione* 321a 6,[240] and since *de mixtione* 16

---

[236] In the course of *de mixt.* 5 and 6, at 218. 18 and 219. 23 for example. Cf. also the use of verbs of motion and extension interchangeably at *de mixt.* 5. 218. 15-16; see Comm. *ad loc.*

[237] See Comm. on *de mixt.* 5. 218. 17-18 on δέχεσθαι and for γίνεσθαι ἐν Comm. on 6. 219. 23.

[238] See *de mixt.* 6. 219. 16-19 and Comm. on 6. 219. 11-12. This assumption is of course implicit in the Aristotelian commentators given the nature of the texts being discussed.

[239] Cf. above pp. 47-48.

[240] See Comm. on *de mixt.* 16. 234. 23-32.

as a whole is a restatement of the Aristotelian theory of growth, it is unlikely that this is so. This chapter is directed against "those who explain growth by bodily interpenetration" (*de mixtione* 16. 238. 23), but there is clear evidence to indicate that these could not have been the Stoics.[241] What we have is a restatement of the Aristotelian theory of growth in the face of an assumed materialistic theory, just as *de mixtione* 13-15 is a restatement of the Aristotelian theory of mixture in the face of a similar materialist account of blending. The fact that the latter is clearly Stoic does not entail that the former is also; for the basic arguments against bodily interpenetration in *de mixtione* 5 and 6 are in entirely general terms, and are reproduced more briefly in similar terms in chapter 16. We have to accept that the *de mixtione* is a generalised attack on the notion of "body going through body," one case of which (total blending) is Stoic, while another (growth by body) is not. This would be entirely consistent with Alexander's tendency, much clearer in the ostensibly anti-Stoic *de fato*, to attack general theses rather than engage in detailed criticism of the specific doctrines of an opponent.[242]

In the three other cases, (3), (4), and (5) above, where Alexander introduces the paradox of "body going through body" the Stoics are clearly the main quarry of his arguments.[243] However there

---

[241] Plutarch (*de comm. not.* 44. 1083C-D = *SVF* II 762) describes Chrysippus' solution of the problem of how the identity of an individual is maintained throughout growth and diminution—that these processes occur in a stable quality, while matter is continually unstable. This is comparable to the Aristotelian doctrine (*de gen. et corr.* A5) that growth occurs in form and not matter expounded in *de mixt.* 16. A. C. Pearson, *JP* 30 (1907) 211-214, who has an excellent discussion of this whole passage, compares the Alexandrian *Quaestiones* I. 5, a basic statement of the Aristotelian position (cf. introductory note to *de mixt.* 16).

[242] On this see also Comm. on *de mixt.* 16, introductory note.

[243] Cf. *Mant.* 117. 9-10, 21-22 with *SVF* I 518 and II 790 respectively (on corporeal soul). *Mant.* 139. 10-11 (on corporeal light) refers to the Stoic rejection of intra-cosmic void (cf. *SVF* II 522-524). The attack on the doctrine of a corporeal soul in terms of the paradox of two bodies being in the same place, or with reference to the classification of mixtures, or both, can be widely documented from the Neoplatonist tradition; cf. Calcidius *In Tim.* 221 (= pp. 234-5 Waszink), with Waszink's note on 234. 5-6 where references are gathered and the importance of Alexander's contribution noted. All this does not mean that total blending was primarily designed as an explanation of the relation of soul to body (*pace* Gould, *Philosophy of Chrysippus*, p. 112 n. 4), only that that relation was generally regarded as the philosophically

are indications of the independent character of his polemic. For example, the claim that if soul is a body then two bodies will be in the same place is reproduced and elaborated at *de anima* 20. 6-19. But this is certainly no more than a restatement of the point that Alexander would have made in discussing *de anima* A5 409b 3, where Aristotle introduces this paradox. The fact that it could be used against the Stoics does not make it anti-Stoic in this context. Much clearer in this regard are the cases where Alexander derives the paradox of body going through body from the thesis that light is a body. In one text of the *Mantissa* this is obviously anti-Stoic,[244] but in another he applies it to the manifestly non-Stoic theory of vision by light-rays.[245] The fact that Philoponus reproduces several of these Alexandrian arguments as his commentary on *de anima* B7 (where Aristotle argues that light is not body), without any anti-Stoic emphasis, is further indication of their very general character.[246]

Finally (5), the attack on the theory that bodies are qualities gives us a good insight into the scholastic nature of Alexander's anti-Stoicism. This text is essentially a Peripatetic version of a standard critique of the Stoics, other traces of which can be found in the Pseudo-Galen *de qualitatibus incorporeis* and Albinus. In all three the paradox of several bodies being in the same place is introduced in basically the same way.[247] The autonomy of this

---

most challenging aspect of the Stoic doctrine of pneuma's motion through the whole of matter.

244 That explicitly entitled "that light is not body" at *Mant.* 138. 3-139. 28, part of which is rightly quoted by Von Arnim at *SVF* II 432.

245 See *Mant.* 128. 34-129. 1 and 129. 24-32.

246 Two parallels are worth noting: (1) the argument that if light travels through void pores then there will be a separable void (*Mant.* 129. 25-26, 139. 9-11, and Philoponus *de an.* 344. 7-8); (2) if light moves through pores then there will be a discontinuous pattern of light and dark (*Mant.* 129. 29-32, 139. 14-17 and Philoponus *de an.* 326. 20-26). Both are particular applications of general arguments against void pores: cf. (1) with *de mixt.* 5. 218. 24-26 and (2) with *de mixt.* 5. 219. 1-3, where interpretation through pores is said to result in juxtaposition (παράθεσις), i.e. a discontinuous series of bodies.

247 See *Mant.* 123. 12-13 and Albinus *Didask.* 11, p. 65 Louis; cf. Pseudo-Galen *de qual. incorp.* XIX. 470-471 K (= pp. 7-8 Westenberger). The latter passage examines the problem of the copresence of several corporeal qualities in a single body in terms of the tripartite classification of mixtures (juxtaposition, fusion, blending—cf. pp. 49-65 above), and concludes that they must be fused together in the same place. This seems to be an entirely polem-

notion and its mechanical application to a Stoic doctrine for polemical purposes is very clear; for neither this theory nor the theories that soul and light are bodies, could have been formulated in this completely generalised fashion. Such a presentation overlooks their systematic relation to other central doctrines of Stoic physics, principally the doctrine of pneuma. Soul and qualities were produced by the action of pneuma on passive matter,[248] while light was almost certainly a case of tensional motion in air, probably similar to the wave motions that accompanies vision and hearing.[249] It may not be unreasonable for a critic to reduce these theories to their general characteristics, given the Peripatetic background we have sketched in, and if, as is quite likely, Stoics themselves had argued for some of them in highly general terms.[250] But we must also appreciate how this reduction facilitates the preordained arguments of their Peripatetic critic, and those of critics outside the Peripatetic school; and these criticisms must not be confused with direct reports.

To put this more generally: we need to recognise that just as the doxographical tradition in ancient philosophy has distorted theories by cataloguing them under simplified headings, so many of the ancient opponents of Stoicism (who are unfortunately so often our sources) employ what we may term a method of hostile doxography by which simplified versions of Stoic doctrines are made the object of polemics. It is one example of the operation of this method that this section has uncovered.

Finally both in Alexander and other critics the reduction of total blending to the general notion of bodily interpenetration had

---

ical exercise, and would not sustain Pohlenz' general claim (*Die Stoa*, I, 72) that the particular purpose of the theory of total blending was to explain how many qualities would be in one body. All three of the texts cited here deserve close analysis in the context of a reconstruction of the Stoic doctrine of quality; the general doctrine "that qualities are bodies," against which they argue, may well have emerged, like "body going through body," solely within an anti-Stoic tradition.

[248] Cf. above, p. 41.

[249] A vestigium from Plutarch (*SVF* II 433) refers to air being illuminated κατὰ νύξιν ἢ ψαῦσιν ("by agitation or contact"). νύξις ("agitation") recalls the use of νύττεσθαι to describe the agitation of the air in vision (*SVF* II 866, p. 233. 29-30).

[250] Cf. for example the very general arguments for the corporeality of the soul at *SVF* II 790, 791, and I. 518. There is, however, no clear evidence of such arguments for the case of body going through body.

its effect on polemics against the view that mixture could involve the extension of a small volume through a very large one; for if mixture is regarded as interpenetration, then the paradox of this process would be its reversibility. It is exactly this argument that we find used against the Stoics by Alexander in the *Mantissa*, where he reiterates an argument found also in Sextus, and partly echoed by Plutarch, that ridicules this reciprocal mixture of unequal quantities by giving them a unit value, so that a small quantity can be said to be equal to a larger, and vice-versa.[251] The Aristotelian roots of this argument would seem to be evident from the commentators' use of the argument (attributed to proponents of the void at *Physics* Δ6 213b 5-12) that if solids interpenetrate then a very large body will occupy the place of a very small one.[252] This could easily be adapted to meet the Stoic claim that mixture was in fact the opposite process, and converted into a new counter-claim that Stoic mixture involved *both* processes. Later commentators, probably copying Alexander, certainly used the example of the mixture of liquids to illustrate the mixture of a very large with a very small body, taking the image of wine pervading the sea used by the Stoics and making it part of their own general arguments against body going through body.[253] Philoponus actually employs this originally anti-Stoic argument that a reciprocal mixture of liquids will occur between unequal quantities if

---

[251] Cf. *Mant.* 141. 9-16 with Sextus Empiricus *P.H.* III. 60-61 and Plutarch *de comm. not.* 37. 1078A-B. The latter deals only with the blending of a small with a large volume, although at 1078E-F he refers in general terms to the opposite process, the compression of many bodies into one. Alexander's critique also shows signs of being influenced by more abstract arguments against the contact of part with whole: cf. *Mant.* 141. 16-19 with Sextus Empiricus *A.M.* IX. 261. *Mant.* 140. 10-24 and 141. 19-22 are more general arguments against "reciprocal" blending. These particular arguments could well have originated in the sceptical tradition (cf. n. 200 above) and have been grafted onto the Peripatetic arguments against "body going through body."

[252] Discussed at pp. 80-81 above.

[253] At *Mant.* 140. 22-23 the image of a cup of wine mixing with the sea is used. In the commentators this image is so common that the expression "the sea in a cup" (ἡ θάλασσα ἐν κυάθῳ) is used along with that of "heaven in a millet seed" (ὁ οὐρανὸς ἐν τῇ κέγχρῳ), derived from Aristotle *Phys.* Δ12 221a 22 (cf. above, p. 80); see Simplicius *Phys.* 530. 24, *de caelo* 629. 30, Philoponus *Phys.* 505. 24-25, *de an.* 344. 4; cf. Pseudo-Galen *de qual. incorp.* XIX. 473 K (p. 9. 14-15 Westenberger), and Calcidius *In Tim.* 221 (p. 235. 3-4 Waszink).

body goes through body,[254] but presents it as a support for the Aristotelian text (*Physics* 209a 4-7) in which the paradox of two bodies being in the same place is said to be entailed by the theory that place is body. The Stoics have been forgotten, but a polemic against them continues latent in Aristotelian exegesis.[255]

---

[254] *Phys.* 506. 2-14. That is, it employs the standard arguments referred to in n. 251 above.

[255] Another example of this is the reappearance of Alexander's general critique of the Stoic theory of presentation (φαντασία) at *de an.* 68. 12-18 in Byzantine commentaries—cf. Michael *Parv. Nat.* (*CAG* XXII-1) 3. 7-16, and Sophonias (Ps.-Themistius) *Parv. Nat.* (*CAG* V-6) 4. 25-31.

PART THREE

THE *DE MIXTIONE*

INTRODUCTION

The *de mixtione* is undoubtedly an authentic work by Alexander of Aphrodisias. Themistius refers to it by title in his commentary on Aristotle's *Physics* as a work which attacks the notion of body going through body (σῶμα διὰ σώματος χωρεῖν),[1] and Simplicius, as we have seen, must also have been referring to it in mentioning a "separate treatise" in which Alexander dealt with this topic.[2] Alexander himself may have had it in mind when in a text in Simplicius' commentary on the *de caelo* he refers to a more extensive critique of the Stoic doctrine of pneuma, possibly *de mixtione* 10.[3] Detailed evidence of its relation to other Alexandrian works, both in its content and in the use of certain terms and expressions, will emerge in the commentary.

In its superficial structure our treatise is similar to those Aristotelian works or portions of works in which predecessors' views are described and criticised before a theory is stated. Thus after an introductory chapter in which the anti-Stoicism of the work is announced and theories of mixture are related to theories of matter, it passes to an exposition and critique of the Atomist and Epicurean theories of blending (2), a description of the Stoic theory and its supporting proof (3-4), a lengthy critique of this theory (5-12) including a digression on its ramifications in the theories of pneuma and God (10-11), and finally an exposition of the Aristotelian theory of blending (13-15). The final chapter sixteen on the Aristotelian theory of organic growth is by way of an appendix.

Commentary. The underlying complexity of this structure will be examined in the commentary; the outline preceding the translation is only intended to help the reader follow the general argument of the treatise. The other main object of the commentary will be to provide a complement to the two preliminary essays. In the case of chapters three to twelve this means reinforcing the analysis of Alexander's presentation and critique of the Stoic theory of total blending offered in sections ii and iii of Part II by examining

[1] *Phys.* 104. 18-22.
[2] *Phys.* 530. 14-16; see above, p. 77.
[3] *de caelo* 286. 23.

the numerous detailed issues irrelevant to the main themes of that essay. In particular I have tried to set this text in the perspectives of the Aristotelian works that were Alexander's ultimate authorities, of the contribution of other authors—especially critics of Stoicism—who dealt with the same topics, and finally of other works in the Alexandrian corpus. This is designed to be a contribution more to our knowledge of the source material for Stoic physics than to the full interpretation of the doctrines discussed in these chapters. This is particularly true of the commentary on chapters ten and eleven which would anyway not be the place to reconstruct such complex doctrines as the Stoic theories of principles, elements, God, and pneuma. With a few exceptions it does not elaborate the reconstruction of the Stoic theory of mixture offered in Part II. Finally, in dealing with chapters thirteen to sixteen I try to indicate the methods by which Alexander's scholasticism operated in the exposition of Aristotelian doctrine: this will provide some detailed support for the general picture of his scholastic achievement offered in Part I.

In preparing this part of the work I have been greatly assisted by Dr. Rex's dissertation, the first part of which analyses the *de mixtione* in detail. However, since he explicitly disavows any attempt to set this treatise in the context of other source material for Stoic physics,[4] and neglects other Alexandrian material, I have not found it helpful to refer to it extensively in a commentary founded on such different principles.

Translation. This is, I believe, the first complete translation of the *de mixtione* into English; Sambursky (*Physics of the Stoics*, pp. 120-122) translated only a few passages. Part of chapter fourteen was put into French by Moraux (*Alexandre d'Aphrodise*, pp. 40-41), and the first complete translation into a modern language was into German by Friedemann Rex in his Frankfurt dissertation of 1966. There were three translations into Latin in the sixteenth century. The most widely circulated was that by Angelus Caninius, first published at Venice in 1546; I have consulted a copy of the third edition of 1555.[5] All these translations

[4] *Chrysipps Mischungslehre*, p. 22.

[5] See F. E. Cranz, *Catalogus Translationum*, I, 113-114 on these translations. He lists one other published translation by Jacob Schegck (Tübingen, 1540), which was also known to Bruns, and one available only in manuscript

appear to have been based on the Aldine edition of 1527 or a closely related manuscript.[6]

It would be too much to hope that a translation of a work such as the *de mixtione* could be entirely readable, but I have tried, as far as possible, to achieve an accuracy that is not directly literal. The major constraint that I have placed on myself has been to translate the considerable amount of technical or specialised terminology in the treatise in a consistent form. The glossary of terms indicates the equivalents that, with a few unavoidable exceptions, have been adopted.

Texts and Editions. The following manuscripts contain the complete text of the *de mixtione*; I cite them with Bruns's sigla.

A = Venetus Marcianus Gr. 257 Zanetti, 13th-14th century
B = Taurinensis C. I. 15, 16th century
C = Vaticanus Gr. 1302, 16th century
L = Escoraliensis X. I. 11, 16th century
P = Parisinus Gr. 2028, 16th century
R = Parisinus Gr. 1848, 16th century
S = Parisinus Gr. 2540, 16th century

Before Bruns' critical edition of 1892 there had been two printed editions of the *de mixtione*. The Aldine edition of 1527 (abbreviated "a" in the apparatus) appears to have been based on the same source as R,[7] and it in turn served as the sole basis of J. Ideler's text of 1836 [8] which did, however, include a number of valuable emendations. For his critical edition Bruns personally collated only A, B, P and C, and depended on A. Elter's reports of R and S. He also had a report of the beginning of L. His brief analysis of this material led him to posit A as the archetype although he did

---

form by Gaspar Gabriel of Padua made between 1541 and 1543. Also cf. Cranz's interesting discussion of the *weltanschauung* of some of these Renaissance translators at *Amer. Philos. Soc.* 102 (1958) 510-546. It should be noted that Schegck's translation occurs as a complement to an essay entitled *de causa continente* in which Alexander's critique of the Stoics is discussed with reference to Galen's treatment of their theory of causes. It is perhaps the earliest piece of scholarship on the *de mixtione*.

6 See Bruns, *SA* II-ii, *Praef.* p. xxxviii and Montanari, *Atti dell'Accademia Toscana* 36 (1971) 29-30.

7) See Montanari, *op. cit.*, pp. 33-35, 50.

8) In *Aristotelis Meteorologica* Vol. II, Leipzig 1836, at pp. 589-624.

not develop a detailed stemma.[9] In particular he did not define clearly the relation between A and R; while recognising that the latter showed signs of independence he still thought that it was probably a copy of A.[10] Only recently have all the manuscripts been described and thoroughly collated by Elio Montanari who has severely criticised Bruns' analysis and has formulated a stemma in which A represents a subordinate part of the tradition.[11] Briefly, Montanari shows that both A and L on the one hand, and R and a on the other, are separately derived from different sources each of which are in turn separately derived from the common source of all our manuscripts. Montanari also introduces for the first time a fourteenth century manuscript containing most of the last four chapters of the *de mixtione*, 229. 13-238. 23, Riccardianus Gr. 63 (his F). This was first described by H. Vitelli after the publication of Bruns' edition,[12] and is of particular importance as the only direct copy of the common source of our manuscripts.

The new critical edition to which Montanari's paper looks forward will not be published for some time and my translation and commentary must therefore be founded on Bruns' edition. His text naturally followed A for the most part; I have counted only about thirty places where he accepts the reading in R or other manuscripts. On the basis of Bruns' admittedly second-hand reports I disagree with his adjudications at 233. 17[13] and 235. 24. He also incorporated over eighty emendations in his text and in preparing my translation and commentary I have found it necessary to increase this number to over two hundred. In the circumstances I have decided that the most useful apparatus to accompany my text would be one listing all these changes; this will enable the reader to see where I have deviated from what is still the standard text, and to see where the text as a whole deviates from readings in the manuscripts.

Apart then from the places where I have altered Bruns' text in the light of Montanari's reports of the previously neglected F, and

---

[9] Bruns, *Praef.* pp. xxxvi-xxxviii.

[10] Bruns, *Praef.* p. xxxviii.

[11] *Op. cit.* note 6 above; see especially pp. 50 and 55-58.

[12] See Montanari, *op. cit.* note 6 above, pp. 26-28, and Vitelli at *Studi Italiani di Filologia Classica* 3 (1895) 379-381.

[13] See note on the text *ad loc.* For examples of the deficiencies in Bruns' knowledge of R see Montanari, *op. cit.* p. 44.

a few other special cases, my apparatus will contain the emendations of various scholars. Apart from my own contributions I have, like Bruns, included some by Ideler, H. Diels, E. Schwartz, and notably O. Apelt.[14] In several places I have preferred conjectures made by Bruns himself or these other scholars that are reported in his apparatus but not included in the text; these are all specifically introduced by "coni." In addition I have drawn on subsequent discussions of the text by Rodier, who offered a number of corrections to Ideler's edition,[15] Von Arnim in *Stoicorum Veterum Fragmenta* Volume II,[16] A. Brinkmann,[17] and by Dr. Rex and Dr. Matthias Schramm in the former's recent dissertation. Unless otherwise indicated, it may be assumed that an emendation listed is included in Bruns's text. I have not, it should be noted, attempted to compile a complete repertorium of alternative emendations, except for most of the editor's own. Those changes that seemed to require a more detailed defence or explanation than could be provided in the apparatus are discussed in a separate set of notes; it is to these that "*v. infra*" in the apparatus will refer. Inevitably I have been selective in compiling these since it has not been my overall purpose to provide a full linguistic commentary.

All this is naturally no substitute for the new critical edition that Montanari has shown to be necessary. I hope nonetheless that this interim presentation of the text may be some contribution to the criticism of a text that abounds in minor difficulties. Many of my emendations are naturally an off-shoot of my commentary and the preliminary studies, and reveal the importance of reading Alexander's text in the light of his Aristotelian and other sources, of other works in his corpus, and of course of the internal characteristics of the *de mixtione*. No text of the *de mixtione* can fail to be illuminated from these sources and I shall be satisfied to have brought some of them to the forefront.[18]

---

[14] Most of Apelt's were published at *Philologus* 45 (1886) 82-99; those of the others were communicated privately to Bruns.

[15] *Revue de Philologie* 17 n.s. 1 (1893) 10-13.

[16] See the list of quotations of the *de mixtione* in *SVF* at p. 253 below.

[17] *RM* 57 (1902) 488-491. See the apparatus at 226. 33a-34.

[18] For emendations based on Aristotelian precedents see 219. 30, 222. 9, 234. 20 (F's reading), 235. 30, 237. 1-2, 11; for others supported by parallels in the Alexandrian corpus see 215. 22, 218. 15, 219. 10-11, 224. 25, 226. 13, 235. 30; and for some derived from internal considerations see 213. 4, 216. 3, 218. 28, 222. 9, 228. 3, 230. 16, 234. 24, 237. 24.

Finally I would note the following abbreviations used in the apparatus: "codd." will refer only to those manuscripts that Bruns used, his "libri," and thus excludes both L and F; "A^1" used in a few places refers to A as corrected by the original scribe, and "A^2" to this manuscript as corrected by a later hand; "Br." abbreviates Bruns himself. It should also be noted that as a result of emendation my page and line numbers do not correspond exactly to those of Bruns in some places; see 213. 14-15, 222. 18-19, 224. 14-27, 226. 33-34, 229. 21-22, 231. 28-232. 15, 233. 16-234. 1, and 234. 20-24.

ANALYTICAL OUTLINE

1. *Introduction: The problem of blending*

213. 2-13. Total blending involves bodies preserving their own character and being separable; this is much more paradoxical than saying that bodies are separable but not totally mixed.

213. 13-214. 10. Hence the notion of body going through body is as problematic as any theory of blending. For not only is there disagreement on this point between those who hold that matter is unified and those who believe that it is composed of separate bodies (the latter's views are summarised [213. 18-214. 6]) but also among proponents of each theory.

214. 10-16. This is because of the inherent difficulty in formulating a satisfactory explanation of the simple fact of mixture.

2. *Democritus' and Epicurus' theory of blending*

214. 16-18. The review of the more significant theories of mixture begins.

214. 18-28. Democritus, like those who make the uniform bodies principles, thinks that mixture is really the juxtaposition of corpuscles; only through the weakness of our perception does it seem to be a homogeneous blend.

214. 28-215. 8. Epicurus avoided this by holding that mixture is the juxtaposition of the atoms of which compounds are formed, thus conflating mixture with passing-away (of bodies into atoms) and coming-to-be (of atoms into the blend).

215. 8-22. These theories offend the two main assumptions about mixture—homogeneity, and the preservation of their character by the constituents. Juxtaposition does not create a homogeneous blend, and a reduction of constituents to qualityless atoms means that they lack an identifiable character.

215. 22-23. Those who hold that bodies are composed of planes follow the Epicurean theory of blending.

215. 23-32. The worthlessness of theories of mixture as juxtaposition makes further examination of their various forms unnecessary. The object of this work is not a detailed disquisition, but an attempt to discover what theory of mixture conforms with our "perceptions and common preconceptions."

3. *The Stoic theory of total blending*

216. 1-4. The theories of mixture held by those who believe that matter is unified are taken up.

216. 5-14. Of these, the Stoics are the principal theorists. Despite doctrinal differences as a result of Aristotle's influence, Chrysippus' theory of mixture is the most widely accepted.

216. 14-217. 2. Chrysippus' theory is that pneuma unifies substance by totally pervading it, and that in it mixtures can be classified as juxtaposition, fusion, or blending.

217. 2-9. This classification depends on the common notions which are the natural standards of truth. The product of each mixture has a different sense-presentation which indicates that they result from a different process.

217. 9-13. Blending is the complete mutual coextension of constituents, not their juxtaposition.

4. *Proofs for the Stoic theory of total blending*

217. 13-32. The theory of total blending relies on commonplace observations that through a cooperative agency bodies may behave in a manner that would be impossible for them when alone. This supports the theory that bodies of disparate magnitudes may be totally mixed while preserving their own qualities, and the phenomenon of the measure of wine mixing with a large amount of water.

217. 32-218. 10. Clear evidence of this can be seen in the soul-body relation, and parallel relations among plants and inanimate bodies; in the relation between elements of differing density; and in the presence of fire in iron, light in air, and drugs in bodies.

5. *Total blending and "body going through body":*

(i) *interpenetration through pores*

218. 10-15. Though the Stoics claim to base their theories on general conceptions they neglect them completely, particularly in the case of their theory of total mixture.

218. 15-24. That theory offends the natural conception that one solid cannot receive another in it. Since this could only occur if there were place and a solid has no place in it, it is ludicrous.

218. 24-219. 9. Nor can this penetration be through pores or empty spaces. If they are filled with another body either (1) it will be displaced, in which case penetration will not occur (218. 26-29); or (2) it will be an immovable body unable to receive another body in it (218. 29-33); but (3) if penetration occurred through filled pores, this would make them superfluous (218. 33-219. 1). In general mixture through pores does not produce a blend but only a juxtaposition of bodies, unless the whole body penetrated were a pore. But this is ludicrous, since a pore is a void interval in a continuous body, and body that is a total pore would be non-existent (219. 1-9).

6. *Total blending and "body going through body":*

(ii) *interpenetration through solids*

219. 9-22. Bodily interpenetration contravenes the principle that when two bodies are compounded an increase occurs.

219. 22-220. 11. (1) If two bodies interpenetrate then the place occupied by the penetrating body will be left void (219. 22-28). (2) The penetrating body is not even a body if it is not in its own place, nor in that of the receiving body (219. 28-32). (3) The receiving body is not a body because it can only receive a body in the spaces between divisions in it, and these are not corporeal (219. 32-220. 3). So the receptivity of body by body entails no increase in the size of the receptive body; but this never occurs in the case of the mixture of bodies. Where there is no increase on compounding, two bodies are not involved, as in the case of matter and form (e.g. soul and body), or quality and body (220. 3-11).

220. 11-13. Again, if body penetrates body, there is no longer need of mutual replacement.

220. 13-23. The total blending of bodies of unequal volume is absurd, and is not supported by the phenomenon of heated incense.

7. *Total blending as either juxtaposition or fusion*

220. 23-29. While bodily interpenetration offends general conceptions, even more so does the claim that in a mixture bodies preserve their original qualities while being totally mixed.

220. 29-35. This claim is the basis of the distinction between blending and fusion; on it depend the notions of total blending and of the separability of the constituents.

220. 35-221. 15. But there is no alternative between juxtaposition and fusion for the Stoic theory: either the constituents of a blend are preserved in their original condition and therefore juxtaposed (220. 25-221. 7); or they are totally destroyed, and thus blended and fused (221. 7-11). And if they are fused they cannot be separated from the blend (221. 11-15).

221. 15-25. So if total mixture is to result in a unified blend, the constituents cannot be preserved; for a unified body has a single State, and the constituents forming it must be inseparable; yet we see that constituents are separable, and so this consequence too is unacceptable.

8. *Total blending and the divisibility of bodies*

221. 25-222. 3. If blending occurs through the constituents dividing one another, then if any bodies are left over in the division they will not be totally blended but once again juxtaposed (221. 28-34), while if nothing is left undivided division will have terminated in incorporeal divisions from which the constituents cannot be composed (221. 34-222.3).

222. 4-6. But if they say that bodies are divided to infinity, then as a complete and actual division of a body this cannot explain total mixture through bodies being divided into actual sets of infinite parts.

222. 6-13. For if they claim that infinite division is simply an unending process this will not explain total mixture, since there will always be a residue of undivided bodies,

222. 13-17 but rather to explain total mixture bodies must be completely divided into actual infinite sets of parts,

222. 17-22 and if these parts have physical size each constituent will be infinitely large, since the sum of an infinite number of bodies is an infinite body—and each of these parts will also be infinitely large,

222. 22-26 unless the parts of the completely divided bodies are not bodies (they cannot be indivisible bodies), in which case the constituents will be bodies that are not composed of bodies.

9. *Total blending and the relation between matter and form*

222. 26-222. 35. Our preconception of mixture—that the constituents be separable as self-subsistent bodies—cannot be applied to relations such as those of soul and body, light and air, Nature and plants, State and inanimate bodies; for enmattered form cannot be separated from matter and bodies.

222. 35-223. 6. Fire is not mixed with iron or any material; for since primary matter is not mixed with form, only matter, *qua* material of the fire, is destroyed by it and loses its form, while primary matter is indestructible.

10. *The Stoic doctrine of pneuma*

223. 6-17. By failing to recognise the ordering power of the fifth body on the cosmos, the Stoics attribute its unity to the pervasion of it by a material force, pneuma.

223. 17-19. Now the doctrine of pneuma employs the doctrine of body going through body, but more importantly is false for other reasons:

223. 20-27. No actual pneuma (air) could be contained within bodies; if it were, it would soon be expelled.

223. 27-30. Pneuma only unifies, while bodies are both continuous and discontinuous.

Sem. Griekse Letterkunde Blandijnberg 2 Gent

223. 30-34. Pneuma's function of unifying individual bodies, and causing sympathy between them, is best divided between form on the one hand, and primary matter and the action of the heavenly bodies on the other.

223. 34-224. 6. What is pneuma's unifying tension, when it is itself so materially tenuous that it acquires force only by being acted on by another body?

224. 6-14. The notion of a single unifying force is ludicrous when some bodies can be decomposed and divided.

224. 14-22. If composed of fire and air, pneuma is neither a simple body, nor composite if its constituents depend for their existence on it.

224. 23-24. How can pneuma be a compound of the Hot and the Cold?

224. 24-27. The tensional motion of pneuma as a simultaneous motion in opposite directions is inconceivable.

11. *The Stoic doctrine of God*

224. 27-32. The Stoics do not distinguish form and matter, but conflate the roles of both in their doctrine of pneuma.

224. 32-225. 4. Also for the Stoics, God and matter are both material principles, the former totally mixing with the latter and structuring it. The following criticisms thus apply:

225. 4-5. This involves body going through body.

225. 5-18. God must either be a simple body, or compound, or a fifth body. If a simple body or a compound He would have to pervade matter before being created in order to create Himself. But as such He would also have to be dependent on matter; being so created God would only be eternal in name.

225. 19-226. 10. The Stoic argument that God pervades matter because natural bodies have an internal structure (225. 19-27) can be shown to be invalid by a review of the main examples of natural causation, where a creator is always external (225. 28-226. 10).

226. 10-16. If God is the form of matter, then matter cannot be formless.

226. 16-24. If God is the form of matter at the time of the conflagration then his form will change when fire changes into other bodies. And if this change is caused by God, he will be self-destructive, which is ludicrous.

226. 24-30. The picture of God as an active creator of the most insignificant bodies is unworthy of our preconception of him.

226. 30-33b (+ 227. 23-25). If bodies that are mixed are reciprocally active and passive, then God will be acted on by matter.

12. *"Body going through body": concluding criticisms*

226. 34-227. 10. The preceding critique dealt with a challenge to the Aristotelian doctrine of the fifth body. The Stoics do not understand that the origin of their most important doctrines is in the notion of body going through body; it underlies all the aspects of their doctrine of God—fate, providence, the active and passive principles, and the unity and interrelatedness of the cosmos.

227. 10-17. The theory of body going through body takes its support from an apparently obvious phenomenon like the heating of iron by fire, which the Stoics describe as the penetration of one body by another.

227. 17-228. 4. This explanation of the phenomenon can be criticised on the following grounds: (1) How can the fire be present in the iron if it has no support? (227. 17-22). (2) It cannot be supported by its own matter, because no ashes are found in the iron (227. 22-30). (3) It must be separated from its own matter, and supported by the moisture present in the iron itself, until that runs out (227. 30-228. 4).

13. *The Aristotelian theory of blending:*

(i) *constituents of blends, types of mixture, conditions for blending, the interaction of the elements*

228. 5-10. Only the Aristotelian theory of mixture can provide an adequate response to the illogical theories previously discussed.

228. 10-25. The constituents of any mixture must be substances, and for this reason they are separable. Qualities, and forms, are inseparable from substances, and therefore not mixed. These substances are physical bodies.

228. 25-229. 3. The terminology of mixture is applied according to the following *differentiae*:

Mixture	
Compounding	Blending
of unlikes; but the compounding of like with like is not mixture.	the unification of malleable, wet, and dissimilar bodies; the blending of like with like or of dissimilar incompatibles is not mixture.

229. 3-21. Blending only occurs among bodies that can act and be acted on by one another. This requires that they share in the same matter, and be mutual opposites.

229. 21-30. Opposites act and are acted on not only at their extremes, but at points between the extremes, for they have the same form as the extremes. In general, all such interaction of opposites depends on their sharing in the same matter, while differing only in forms and accidental qualities.

229. 30-230. 13. The differing qualities of the four elements make it possible for them to change into one another. A change occurs when one opposite dominates another, and its form supervenes on it.

14. *The Aristotelian theory of blending:*

(ii) *the difference between blending and coming-to-be and passing-away*

230. 14-20. Since blending is a form of dissolution different from coming-to-be and passing-away (for the constituents are preserved throughout the process), this difference needs to be explained.

230. 20-26. Though both occur with matter as a base, in blending there is not total change from one contrary to another, as in coming-to-be and passing-away.

230. 26-34. In blending neither constituent supervenes on the other, but there is a balance in their powers, which leads to the unification of their matter, and the creation of a single quality from these powers.

230. 34-231. 1. Blending only occurs among moist bodies, or bodies accompanied by moisture; these lose their shape and are easily unified.

231. 1-231. 10. Blending therefore requires bodies (1) to be unified in matter, (2) juxtaposed because of their moistness, and (3) have equally balanced powers. Under these conditions, neither constituent will supervene, and only blending will occur.

231. 10-12. Blending may be defined as the unification of juxtaposed bodies that are altered but not destroyed.

15. *The Aristotelian theory of blending:*

(iii) *the separability of constituents from a blend*

231. 12-29. Moist bodies are easily mixed, since they are easily divisible, and the blend is a single body with the constituents present in it in potentiality. Therefore the recovery of the constituents is an easily effected change, being neither absolute coming-to-be, nor the recovery of the original constituents (in that case they would only have been juxtaposed).

231. 29-232. 18. Such recovery of the constituents of a blend is illustrated as follows: As cheese and whey are recovered from the uniform compound milk when a heated stone is immersed in it, so is water from wine by a sponge dipped into the mixture; the constituents are not separate parts of the mixture, but are recovered from the whole of it (231. 29-232. 5). Similarly as, on fermentation, sugar can be decomposed into air and wine, though air could not be actually present in any liquid, so this sort of change occurs when constituents are separated from a blend (232. 5-13). Thus, as in these cases there is only a slight change when the constituents are separated from the blend with the help of some small factor, and not a change into an opposite state, so this is the type of change that occurs with the constituents of any blend (232. 13-18).

232. 18-31. The separation of constituents from a blend can be distinguished from both the separation of juxtaposed bodies and the separation that occurs when there is coming-to-be and passing-away, as when air is separated from water. It falls between these cases, for the constituents are in potentiality in a blend, and regain the actuality they lost when they cancelled each other out on being blended.

232. 31-233. 2. If this is true, and if the previous arguments conform with the facts and general assumptions about blending, Aristotle's theory will be the only correct one.

233. 2-14. The process of blending can be perceived when constituents differ in colour and in density.

16. *The Aristotelian theory of organic growth and "body going through body"*

233. 14-24. The Aristotelian doctrine of growth and nutrition needs to be described, since some have tried to explain growth by the theory of body going through body.

233. 24-30. The discussion will be confined to organisms with a nutritive faculty. Their characteristics must be described as a basis for the discussion.

233. 30-234. 15. The candidates for growth are bodies which absorb food from outside themselves proportionately in all their parts, and remain stable throughout the process. This rules out both mixture, and coming-to-be and destruction as examples of growth. The body that grows must be stable and must retain its nutritive faculty; thus the food that causes growth does not itself grow.

234. 15-32. Growth must either occur by a corporeal or incorporeal addition to a body. But both are impossible; the incorporeal, because growth cannot occur through something that does not occupy place; body, because this involves a complete bodily interpenetration. The latter cannot be saved by claiming that penetration occurs through empty spaces in the body that grows unless the whole of the body, which must grow in all its parts, is held to be empty.

234. 32-235. 14. The bodies that grow must be precisely defined: they are the uniform parts of bodies, and all growth in the non-uniform parts is dependent on their growth. Since growth only occurs in stable bodies, the uniform parts must be stable.

235. 14-236. 5. A uniform part is a compound of matter and form. If it grows to the extent that it is stable, it must grow with respect to its form, which remains numerically the same, while its matter is different on different occasions.

236. 6-236. 26. The actual process of growth involves nutriment being assimilated by pre-existent flesh, and preserving the body when it is less than, or equal to, the deteriorating matter of the flesh, but increasing it when it is greater than this.

236. 26-237. 5. The central factor in the process of growth is that the body receiving food preserves its shape intact throughout, even though the space occupied by the body is different as a result of the even distribution of nutriment through its parts.

237. 5-238. 10. Illustrations of this: As the earth preserves its spherical shape and central position after the addition of a larger body, because that body is distributed throughout its mass, so do uniform parts maintain their shape through the even distribution of nutriment (237. 5-20). The non-uniform parts have their shapes preserved in exactly the same way as those of the uniform parts (237. 20-25). This process can be compared with the case of liquid moving along a channel and preserving its shape, though its quantity may vary from time to time; it is the stable form (the shape of the channel) that can be said to increase and decrease, while the matter (the varying quantities of liquid) is different on different occasions (237. 25-238. 10).

238. 10-17. Nourishment is not the assimilation of food followed by its absorption: the two occur simultaneously.

238. 17-20. Nourishment is like water being made into wine as it is poured in, or like fire acting on combustible bodies.

238. 20-23. Thus the theory of growth outlined here does not support the absurd doctrine of body going through body.

213 ΑΛΕΞΑΝΔΡΟΥ ΑΦΡΟΔΙΣΙΕΩΣ ΠΕΡΙ ΚΡΑΣΕΩΣ ΚΑΙ ΑΥΞΗΣΕΩΣ

I * * * πῶς δ' ἂν παραδέξαιτό τις τὸ ἐν τῇ τοιαύτῃ κράσει τῶν σωμάτων ἕκαστον τῶν κιρναμένων φυλάττειν δύνασθαι τὴν οἰκείαν ἐπιφάνειαν, ὡς ὁμοῦ μὲν μηδὲ τὸ τυχὸν αὐτῶν μόριον εἶναί [τε] καθ' αὑτὸ κεχωρισμένον θατέρου, ὁμοῦ δὲ φυλάττειν ἕκαστον αὐτῶν τὴν ἐπιφάνειαν τὴν ἑαυτοῦ, ἣν εἶχε καὶ πρὸ τῆς κράσεως; τοῦτο γὰρ ὑπεραίρει καὶ τὰς ἐν τοῖς μύθοις παραδοξολογίας, ὃ τίθησι Χρύσιππος ὅσον ἐν τούτῳ [τὸ] δύνασθαι τὰ κεκραμένα χωρίζεσθαι πάλιν. τοῦ τε γὰρ λέγειν δύνασθαι χωρίζεσθαι τὰ κεκραμένα, ἀλλὰ καὶ τοῦ μηδ' ὅλως κιρνᾶσθαί τινα δύνασθαι λέγειν μακρῷ παραδοξότερον καὶ παρὰ τὰς ἁπάντων ἐννοίας τὸ λέγειν ὁμοῦ μὲν δι' ὅλων ἀλλήλων χωρεῖν τινα σώματα, ὡς μηδὲν αὐτῶν μόριον εἶναι καθ' αὑτό, ὁμοῦ δ' ἕκαστον αὐτῶν ὑπὸ τῆς οἰκείας ἐπιφανείας περιέχεσθαι, ὑφ' ἧς περιείχετο καὶ πρὸ τῆς μίξεως.

τό τε οὖν σῶμα διὰ σώματος <χωρεῖν> οὐδεμῶς ἐλάττω διαφορὰν ἔχειν συμβέβηκε τῶν κατὰ φιλοσοφίαν δογμάτων καὶ τῶν περὶ μίξεώς τε καὶ κράσεως λόγων. οὐ γὰρ μόνον διηνέχθησαν πρὸς ἀλλήλους περὶ τοῦδε τοῦ δόγματος οἱ μίαν ὕλην ὑποκεῖσθαι πᾶσι τοῖς ἐν γενέσει σώμασιν λέγοντες πρὸς τοὺς ἐκ διωρισμένων τε καὶ κεχωρισμένων σωμάτων ποιοῦντας αὐτήν

(ὧν οἱ μὲν ἄτομα σώματα ἄπειρα τῷ πλήθει, κατὰ σχῆμα καὶ μέγεθος μόνον τὴν πρὸς ἄλληλα διαφορὰν ἔχοντα, τὰς ἀρχὰς καὶ τὰ στοιχεῖά φασιν εἶναι, καὶ τῇ τούτων συνθέσει τε καὶ ποιᾷ περιπλοκῇ ἔτι τε τάξει καὶ θέσει τἆλλα γίνεσθαι· ἐφ' ἧς δόξης πρῶτοι μὲν Λεύκιππός τε καὶ Δημόκριτος γενέσθαι δοκοῦσιν, ὕστεροι δὲ Ἐπίκουρός τε καὶ οἱ τὴν αὐτὴν τούτῳ τραπέντες· οἱ δὲ αὐτῶν,

---

213. 1 περὶ κράσεως καὶ μίξεως B: περὶ μίξεως Ra: cf. Montanari p. 17 n. 2

213. 4 μηδὲ Br.: μήτε codd. αὐτῶν Rex, cf. 213. 5, 11, 12: αὐτοῦ codd. τε secl. Br.

213. 7 τὸ seclusi, v. infra

213. 14 χωρεῖν supplevi, cf. 213. 11 et passim

213. 14 οὐδεμῶς scripsi: οὐδεμιᾶς ABCPS, obelo not. Br.: μηδεμιᾶς Ra

TRANSLATION

213. 1 *Alexander of Aphrodisias on Blending and Growth*

213. 2 I ... how could one grant that in such a blend each of the constituents could maintain its own surface so that in a uniform body no random part of them is in itself separate from another, while each constituent maintains its own surface which it had even before the blend? This transcends even the marvels of the myths, inasmuch as Chrysippus claims that in this case bodies that have been blended can be separated again. For compared with saying that bodies that have been blended can be separated, but also saying that certain bodies cannot be completely blended, it is much more remarkable and against universally held notions to say that in a uniform body some bodies can go through one another completely so that no part of them exists independently, yet each of them is surrounded by its own surface by which it was surrounded even before the mixture.

13. 13 It follows, then, that the notion of body < going > [1] through body causes no less disagreement than do philosophical principles and theories of mixture and blending; for not only do those who hold that a single matter underlies all bodies that come to be disagree on this question with those who create matter from discrete and separate bodies

13. 18 (some of the latter say that the first-principles and elements are infinitely numerous indivisible bodies only distinguished from one another by size and shape, and that other things come to be by the composition and particular interlinking of these things as also by their position and order; Leucippus and Democritus seem to have been the first to take this view, and later Epicurus and those of the same persuasion; others [214] say that

[1] This indicates an addition to the Greek text; I have only noted the more significant of these. Square brackets carry my own explanatory addenda.

214 οὐκ ἀτόμους, ὁμοιομερῆ δέ τινά φασιν ἄπειρα εἶναι σώματα, ἐξ ὧν ἡ τῶν αἰσθητῶν γένεσις σωμάτων γινομένη κατὰ σύγκρισιν καὶ σύνθεσιν, <ἐφ'> ἧς δόξης Ἀναξαγόρας τε καὶ Ἀρχέλαος δοκοῦσι γεγονέναι· ἤδη δέ τινες καὶ ἀμερῆ τινα σώματα τὰς ἀρχὰς καὶ στοιχεῖα τῶν πάντων προήχθησαν εἰπεῖν· ἔστι δέ τις δόξα καὶ ἐξ ἐπιπέδων τὴν γένεσιν ποιοῦσα τῶν σωμάτων καὶ ἐξ ἀριθμῶν τις ἄλλη),

οὐ δὴ πρὸς ἀλλήλους μόνον, ὡς ἔφην, οὗτοι διηνέχθησαν κατὰ τὴν πρὸς τὰς ἀρχὰς διαφόρους οὔσας ἀκολουθίαν καὶ τὰς κράσεις διαφόρως γίνεσθαι λέγοντες, ἀλλὰ καὶ αὐτῶν τῶν ἡνωμένην τὴν ὕλην λεγόντων καὶ πάλιν αὖ τῶν διωρισμένην τε καὶ κεχωρισμένην ἐστί τις πρὸς ἀλλήλους διαφωνία.

αἰτία δ' αὐτοῖς τῆς διαστάσεως τῆς τοσῆσδε ἡ χαλεπότης τοῦ δόγματος. <τῷ> μὲν γὰρ ἐναργές εἶναι κιρνᾶσθαί τινα τῶν σωμάτων πρὸς ἄλληλα πάντες σχεδὸν οἱ περὶ φύσεώς τε καὶ τῶν γινομένων φύσει φιλοσοφοῦντες ἦλθον ἐπὶ τὸ ζητεῖν τὴν αἰτίαν αὐτοῦ, τῷ δ' εἶναι χαλεπήν τε αὐτοῦ τὴν εὕρεσιν καὶ ἑκάστῃ τῶν ἀποδιδομένων αἰτιῶν οἰκείας τινας ἕπεσθαι δυσχερείας ἄλλος ἄλλῃ ἀπετράπετο.

II Διὸ οὐ χεῖρον καὶ ἡμᾶς περὶ αὐτῆς διαλαβεῖν καὶ μηνῦσαι τὰ δοκοῦντα ἡμῖν εὐλογώτερα λέγεσθαι τῶν περὶ κράσεως εἰρημένων τοῦ δοκεῖν οὕτως ἔχειν παρεχόμενους τὰς αἰτίας. Δημόκριτος μὲν οὖν ἡγούμενος τὴν λεγομένην κρᾶσιν γίνεσθαι κατὰ παράθεσιν σωμάτων, διαιρουμένων τῶν κιρναμένων εἰς μικρὰ καὶ τῇ παρ' ἄλληλα θέσει τὴν μῖξιν ποιουμένων, οὐδὲ τὴν ἀρχήν φησιν εἶναι πρὸς ἀλήθειάν τινα κεκραμένα, ἀλλ' εἶναι τὴν δοκοῦσαν κρᾶσιν παράθεσιν σωμάτων ἀλλήλοις κατὰ μικρὰ σωζόντων αὐτῶν ἑκάστου τὴν οἰκείαν φύσιν, ἣν εἶχον καὶ πρὸ τῆς μίξεως· δοκεῖν δ' αὐτὰ κεκρᾶσθαι τῷ τὴν αἴσθησιν διὰ μικρότητα τῶν παρακειμένων μηδενὸς αὐτῶν αἰσθάνεσθαι δύνασθαι μόνου. φιλαλήθως τε καὶ φιλοσόφως οὐκ ὤκνησεν εἰπεῖν τὸ ἑπόμενον τοῖς οὕτω τὰς κράσεις λέγουσι γίνεσθαι. παραθέσει δὲ τοιαύτῃ τὰς κράσεις ἀνάπτουσι καὶ οἱ τὰς ὁμοιομερείας ὕλην εἶναι τῶν γινομένων λέγοντες.

---

214. 1 οὐκ ἀτόμους Diels Apelt: οὐ κατ' ἄλλους codd.
214. 2 <ἐφ'> coni. Br., cf. 213. 21
214. 9 αὖ τῶν Diels: αὐτῶν codd.
214. 11 <τῷ> μὲν Br.: ὅτι μὲν BPS: μὲν ARa
214. 14 χαλεπήν τε Diels: χαλεπὴν (χαλεπὸν SP) δὲ ABSCP: χαλεπὴν Ra
214. 19-20 τῶν κιρναμένων secl. Kranz, *DK* II, 100. 28 ad loc.
214. 26 ὤκνησεν Schwartz: ὤκνησαν codd.

there are not atoms but certain uniform and infinitely numerous bodies from which the coming-to-be of perceptible bodies occurs by compounding and composition, a theory that Anaxagoras and Archelaus seem to have held; some were inspired to say that the first-principles and elements of everything were actually certain partless bodies, and there is one theory which makes bodies come to be from planes, and yet another from numbers)

214. 6 now these thinkers, as I said, not only disagree in saying that blends occur in different ways through the implications of their different first-principles, but there is also an internal discord among the very ones who say that matter is unified and again among those who say that it is divided and discrete.

14. 10 The reason for such a great disagreement among them is the difficulty of the doctrine; for given that it was clear that certain bodies blended with one another almost all philosophers of nature and natural change went in search of its explanation, yet since this proved difficult to discover, and as peculiar difficulties were entailed by each of the explanations offered, divergent theories arose.

14. 16 II So it is no less important that we too should deal with the subject, and uncover what seem to us the more reasonable theories expressed on blending, and explain the reasons why they seem so.

14. 18 Democritus, then, thinks that what is termed blending occurs by juxtaposition of bodies, with the constituents being divided into corpuscles and forming themselves into a mixture by their positioning beside one another; he says that they are not at all blended in reality, but that the apparent blend is a juxtaposition of bodies with one another where each preserves in corpuscular form its own nature, which they had even before the mixture. These bodies seem to be blended since perception is unable to grasp a single one of them because of the minuteness of the juxtaposed bodies. A lover of truth and a philosopher, he did not shrink from stating the consequence for those who say that blends occur in this way. (It is to such a juxtaposition that those who hold that uniform bodies are matter for what comes to be also reduce blends.)

Ἐπίκουρος δὲ φεύγειν βουλόμενος τὸ ὑπὸ Δημοκρίτου ῥηθὲν ἕπεσθαι [βούλεται] τοῖς παραθέσει τῶν κιρναμένων τὴν κρᾶσιν γίνεσθαι λέγουσι παραθέσει μέν τινων σωμάτων καὶ αὐτὸς τὰς κράσεις γίνεσθαι λέγει, ἀλλ' οὐκ αὐτῶν τῶν μιγνυμένων σωμάτων σωζο-
215 μένων ἐν τῇ διαιρέσει, ἀλλ' ἀναλυομένων εἰς τὰ στοιχεῖα καὶ τὰς ἀτόμους, ἐξ ὧν ἕκαστον αὐτῶν συγκείμενόν πως τὸ μὲν οἶνος ἦν, τὸ δὲ ὕδωρ, τὸ δὲ μέλι, τὸ δὲ ἄλλο τι, ἔπειτα ποιᾷ τούτων τῶν σωμάτων, ἐξ ὧν ἦν τὰ κιρνάμενα πρὸς ἄλληλα, συνθέσει γεννώντων τὸ κεκραμένον σῶμα, συμμιγνυμένων <οὐχ> ὕδατός τε καὶ οἴνου, ἀλλὰ τῶν ὑδροποιῶν ἀτόμων, ὡς ἂν εἴποι τις, ταῖς οἰνοποιοῖς, φθορᾷ τε καὶ γενέσει τινῶν τὴν κρᾶσιν γίνεσθαι λέγων· <ἡ> γὰρ εἰς τὰ στοιχεῖα ἀνάλυσις ἑκάστου καὶ ἡ ἐκ τῶν στοιχείων σύνθεσις αὐτῶν <τὸ μὲν γένεσις>, τὸ δὲ φθορά.

τούτων μὲν οὖν οὔτε οἱ τὴν παράθεσιν αἰτιώμενοι τῆς κράσεως οὔθ' οἱ τὴν εἰς τὰ στοιχεῖα ἀνάλυσιν τὴν περὶ τῆς κράσεως σώζουσι πρόληψιν, εἴ γε ἡ μὲν κρᾶσις εἶναι δοκεῖ ἐν ἑνώσει τῶν κεκραμένων, ὡς μηδὲν μόριον τοῦ κράματος ἄμικτον εἶναί τινος τῶν σωμάτων, ἐξ ὧν τὸ κεκραμένον, οὔτε δὲ ἡ παράθεσις ἕνωσις, οὔτε ἡ φθορά τε καὶ ἀναστοιχείωσις μῖξίς ἐστι· ἡ γὰρ μῖξις σωζομένων γενέσθαι δοκεῖ. οὐ γὰρ αὐτὰ μιγνύουσιν οἱ τοῦτον τὸν τρόπον τὴν κρᾶσιν γίνεσθαι λέγοντες, ἀλλ' ἐξ ὧν ἐστι ταῦτα ἐκεῖνα μιγνύουσιν. εἰ δὲ ἐκ τῶν αὐτῶν στοιχείων οἷόν τε καὶ ὕδωρ καὶ οἶνον γενέσθαι τῇ ποιᾷ συνθέσει τε καὶ περιπλοκῇ τῆς τῶν γινομένων ἐξ αὐτῶν διαφορᾶς γεννωμένης, οὐδ' ὅλως ἂν ἔτι κρᾶσις τούτων γίνοιτο τῷ τῇ ἀναστοιχειώσει φθείρεσθαι ταῦτα, καθ' ἃ ἡ κρᾶσις τῶν κιρναμένων γίνεται. γίνεται μὲν γὰρ ἡ κρᾶσις διαφερόντων τινῶν κατὰ ποιότητα, ἀναιρεῖται δὲ ἡ κατὰ τοῦτο διαφορὰ τοῖς ἔχουσιν αὐτὴν ἐν τῇ τῶν αὐτῶν ποιᾷ συνθέσει τῆς συνθέσεως αὐτῶν ἀναιρεθείσης. ὁμοίως δὲ καὶ τοῖς ἐξ ἐπιπέδων γεννῶσι τὰ σώματα ἡ κρᾶσις γίνεται κατὰ ἀναστοιχείωσίν τινα.

---

214. 29 βούλεται secl. Br., om. A[1]: βουλῆς CP: δὲ coni. Br.
215. 3 ἔπειτα ποιᾷ Usener, *Epicurea* p. 207: ἐπὶ τὰ ποιὰ codd.
215. 5 οὐχ suppl. Usener
215. 6 φθορᾷ τε Usener: φθοραῖς codd.
215. 7 ἡ suppl. Usener ἀνάλυσις Usener: ἀναλύσεις codd.
215. 8 τὸ μὲν γένεσις suppl. Usener φθορά Usener: φθεῖρα ACRa: φθεῖραι BP: φθεῖρας S
215. 9 οὔθ' οἱ Ideler: οὗτοι codd.
215. 10 ἀνάλυσιν Ideler: ἀνὰ codd.
215. 12 δὲ Br.: δὴ codd.
215. 22 ὁμοίως coni. Diels, cf. Alex. *de an.* 83. 10: ὁ ἴσως codd., obelo not. Br.: ἴσως Ideler Rex v. infra

14. 28 Epicurus, wanting to evade the consequence described by Democritus for those who say that blending occurs by juxtaposition of the constituents, says that blends occur by juxtaposition of certain bodies, but not of the bodies that are actually mixed and [215] preserved in the division, but of bodies which are dissolved into the elements and atoms from which each of them is somehow composed to be variously wine, water, honey or something else; then by a particular composition of *these* bodies—the source of the related constituents—to form the blended product, it is not water and wine that are mixed together but water-producing atoms, so to speak, with wine-producing ones; thus he describes blends as occuring by the passing-away and coming-to-be of certain bodies, for the dissolution of each constituent into its elements and its composition from the elements is < coming-to-be > and passing-away respectively.

215. 8 Now neither those who offer juxtaposition as an explanation of blending, nor those who explain it by a process of reduction-to-the-elements preserve our preconception about blending—if indeed blending is thought to lie in the unification of the bodies that have been blended so that no part of the blend is unmixed with any of the bodies from which the blended product is formed—when neither is juxtaposition unification, nor passing-away and reduction-to-the-elements mixture; for mixture should occur between bodies that are preserved. Those who say that blending occurs in this way do not mix the bodies themselves but rather mix the bodies from which they are formed. If both water and wine can come from the same elements, with the difference in the things that come to be from them caused by their particular composition and interlinking, then the bodies by which the blending of constituents occurs would really no longer be blended when they are dissolved in their reduction-to-the-elements. Blending occurs among specific bodies differing in quality, and this particular difference is abolished for those who locate it in a specific composition of identical bodies after the composition [of the constituents] has been dissolved. (And similarly for those who create bodies from planes blending occurs by some kind of reduction-to-the-elements.)

εἰ μὲν οὖν ἦν ἔχουσά τι ἥδε ἡ περὶ τῶν ἀρχῶν δόξα εὔλογον, ἴσως ἐπὶ πλέον ἔδει τὴν ἐξέτασιν ποιεῖσθαι τῆς τοῦτον τὸν τρόπον γίνεσθαι λεγομένης κράσεως, ἐλεγχομένης δὲ ἐκείνης περιττὸν τὸ ζητεῖν ἔτι, εἰ κατὰ παράθεσιν μερῶν ἢ μῖξιν ἀτόμων ἢ ἐπιπέδων ἢ ἀμερῶν τινων σωμάτων ἡ κρᾶσις γίνεται. διὸ περὶ μὲν τῆς τῶν τοιαύταις ἀρχαῖς χρωμένων διαφωνίας τε πρὸς ἀλλήλους καὶ δόξης περὶ κράσεως οὐδὲν δεῖ μηκύνειν τὸν λόγον. οὐ γὰρ ἐπίδειξις ἡμῖν ἱστορίας καὶ πολλῆς γνώσεως τὸ προκείμενον, ἀλλ' ἐξέτασίς τε καὶ ζήτησις τοῦ, πῶς ἄν τις λέγων γίνεσθαι τὰς κράσεις τῶν σωμάτων συμφώνως λέγοι ταῖς περὶ αὐτῶν αἰσθήσεσί τε καὶ κοιναῖς προλήψεσι.

216 III Μετέλθωμεν δὲ ἐπὶ τοὺς κοινῶς ἡνῶσθαι τὴν ὕλην λέγοντας καὶ μίαν πᾶσι τοῖς γινομένοις καὶ τὴν αὐτὴν ὑποθέντας, ὧν καὶ αὐτῶν παραιτησάμενοι τὰ ἐπιπολαιότερόν τε καὶ προχειρότερον εἰρημένα τὰς μάλιστα δοκούσας ἔχεσθαί τινος λόγου δόξας ἐξετάσωμεν προχειρισάμενοι. τῶν δὴ ἡνῶσθαι τὴν ὕλην λεγόντων δοκοῦσι μάλιστά τε καὶ περικρατῶς οἱ ἀπὸ τῆς Στοᾶς περὶ κράσεως διαλαμβάνειν. οὔσης δὲ καὶ ἐν τούτοις πολυφωνίας (ἄλλοι γὰρ ἄλλως αὐτῶν τὰς κράσεις γίνεσθαι λέγουσιν), ἡ μάλιστα δοκοῦσα δόξα εὐδοκιμεῖν παρ' αὐτοῖς περὶ κράσεώς ἐστιν ἡ ὑπὸ Χρυσίππου λεγομένη. τῶν γὰρ μετ' αὐτὸν οἱ μὲν Χρυσίππῳ συμφέρονται, οἱ δέ τινες αὐτῶν τῆς Ἀριστοτέλους δόξης ὕστερον ἀκοῦσαι δυνηθέντες πολλὰ τῶν εἰρημένων ὑπ' ἐκείνου περὶ κράσεως καὶ αὐτοὶ λέγουσιν· ὧν εἷς ἐστι καὶ Σωσιγένης, ἑταῖρος Ἀντιπάτρου. οἷς οὐ δυνάμενοι πάντη συμφέρεσθαι διὰ τὴν ἐν τοῖς ἄλλοις διαφωνίαν ἐν πολλοῖς αὐτοῖς λέγοντες εὑρίσκονται μαχόμενα.

ἔστι δὲ ἡ Χρυσίππου δόξα περὶ κράσεως ἥδε· ἡνῶσθαι μὲν ὑποτίθεται τὴν σύμπασαν οὐσίαν, πνεύματός τινος διὰ πάσης αὐτῆς διήκοντος, ὑφ' οὗ συνέχεταί τε καὶ συμμένει καὶ σύμπαθές ἐστιν αὑτῷ τὸ πᾶν, τῶν δὲ μιγνυμένων ἐν αὐτῇ σωμάτων τὰς μὲν παραθέσει μίξεις γίνεσθαι λέγει, δύο τινῶν ἢ καὶ πλειόνων οὐσιῶν εἰς ταὐτὸν συντεθειμένων καὶ παρατιθεμένων ἀλλήλαις, ὥς φησιν, καθ' ἁρμήν, σωζούσης ἑκάστης

215. 32 προλήψεσι Rex: προσλήψεσι codd.
216. 2 ὑποθέντας Ideler: ὑποθέτας ABC?S?Ra: ὑποθέσεως P
216. 3 τὰ ... εἰρημένα scripsi: τὰς ... εἰρημένας codd. Cf. 232. 32-33
216. 4 δόξας coni. Diels: αὐτὰς codd. Br.
216. 5 περικρατῶς Rex: περὶ κράσεως codd. Br.: πρὸ πάντων coni. Br.
216. 16-17 τὸ πᾶν, τῶν Apelt Rodier: τὸ πάντων A¹Ra: τῶν πάντων A²BCPS
216. 17 παραθέσει μίξεις Ideler: παραθέσεις μίξεις codd.
216. 19 ἁρμὴν Br., cf. 219. 3: ὁρμήν ABCPS: ὁρμῆς Ra

215. 23 Now if this theory of principles had anything reasonable to it perhaps we would have to make a more extensive examination of the blending said to occur in this manner. But since the theory has been refuted, it is superfluous to investigate still further whether blending occurs by juxtaposition of parts, or by the mixture of atoms, or planes, or certain partless bodies. There is then no need to lengthen the discussion of the dispute existing among those who employ such principles and of their theory of blending. For our object is not a demonstration of great learning through a long exposition but a critical inquiry as to how an account of the blends of bodies can be given consistent with our perceptions of them and our common preconceptions.

216. 1 III [216] Let us return to those who say that matter is entirely unified and who posit it as one and the same for everything that comes to be, and excusing ourselves from their more superficial and basic statements too, take up the task of examining their opinions that are thought to be particularly significant.

216. 4 Of those who say that matter is unified the Stoics seem particularly and preeminently to discuss blending. They also show a diversity (for they have various views on how blending occurs) but Chrysippus' theory of blending has the best reputation with them. While some of his successors agree with Chrysippus, other who were later able to hear Aristotle's theory actually express many of his views on blending (one such is Sosigenes, a student of Antipater), but being unable to completely conform to them because of their disagreement with him on other points they are found stating views in many respects inconsistent.

216. 14 Chrysippus' theory of blending is as follows: he holds that while the whole of substance is unified because it is totally pervaded by a pneuma through which the whole is held together, is stable, and is sympathetic with itself, yet some of the mixtures of bodies mixed in this substance occur by juxtaposition, through two or more substances being composed into the same mass and juxtaposed with one another "by juncture" as he says, and with each of them

αὐτῶν ἐν τῇ τοιαύτῃ παραθέσει κατὰ τὴν περιγραφὴν τὴν οἰκείαν οὐσίαν τε καὶ ποιότητα, ὡς ἐπὶ κυάμων φέρε εἰπεῖν καὶ πυρῶν ἐν τῇ παρ’ ἀλλήλους θέσει· γίνεσθαι τὰς δέ τινας συγχύσει δι’ ὅλων τῶν τε οὐσιῶν αὐτῶν καὶ τῶν ἐν αὐταῖς ποιοτήτων συμφθειρομένων ἀλλήλαις, ὡς γίνεσθαί φησιν ἐπὶ τῶν ἰατρικῶν φαρμάκων κατὰ σύμφθαρσιν τῶν μιγνυμένων, ἄλλου τινὸς ἐξ αὐτῶν γεννωμένου σώματος· τὰς δέ τινας γίνεσθαι μίξεις λέγει δι’ ὅλων τινῶν οὐσιῶν τε καὶ τῶν τούτων ποιοτήτων ἀντιπαρεκτεινομένων ἀλλήλαις μετὰ τοῦ τὰς ἐξ ἀρχῆς οὐσίας τε καὶ ποιότητας σώζειν ἐν τῇ μίξει τῇ τοιᾷδε, ἥντινα τῶν μίξεων κρᾶσιν ἰδίως εἶναι λέγει. τὴν γὰρ δύο ἢ καὶ πλειόνων τινῶν σωμάτων ὅλων δι’ ὅλων ἀντιπαρέκτασιν ἀλλήλοις οὕτως, ὡς σώζειν ἕκαστον αὐτῶν ἐν τῇ μίξει τῇ τοιαύτῃ τήν τε οἰκείαν οὐσίαν καὶ τὰς ἐν αὐτῇ ποιότητας, λέγει κρᾶσιν εἶναι μόνην τῶν μίξεων. εἶναι γὰρ ἴδιον τῶν κεκραμένων τὸ δύνασθαι χωρίζεσθαι πάλιν ἀπ’ ἀλλή-
217 λων, ὃ μόνως γίνεται τῷ σώζειν ἐν τῇ μίξει τὰ κεκραμένα τὰς αὐτῶν φύσεις.

τὸ δὲ ταύτας τὰς διαφορὰς εἶναι τῆς μίξεως πειρᾶται πιστοῦσθαι διὰ τῶν κοινῶν ἐννοιῶν, μάλιστα δὲ κριτήρια τῆς ἀληθείας φησὶν ἡμᾶς παρὰ τῆς φύσεως λαβεῖν ταύτας. ἄλλην γοῦν φαντασίαν ἔχειν ἡμᾶς τῶν καθ’ ἁρμὴν συγκειμένων, καὶ ἄλλην τῶν συγκεχυμένων τε καὶ συνεφθαρμένων, καὶ ἄλλην τῶν κεκραμένων τε καὶ ἀλλήλοις δι’ ὅλων ἀντιπαρεκτεινομένων οὕτως, ὡς σώζειν ἕκαστον αὐτῶν τὴν οἰκείαν φύσιν. ἣν διαφορὰν φαντασιῶν οὐκ ἂν εἴχομεν, εἰ πάντα τὰ ὁπωσοῦν μιγνύμενα παρέκειτο ἀλλήλοις καθ’ ἁρμήν. τὴν δὲ τοιαύτην ἀντιπαρέκτασιν τῶν κιρναμένων ὑπολαμβάνει γίνεσθαι χωρούντων δι’ ἀλλήλων τῶν κιρναμένων σωμάτων, ὡς μηδὲν μόριον ἐν αὐτοῖς εἶναι μὴ μετέχον πάντων τῶν ἐν τῷ τοιούτῳ κεκραμένῳ μίγματι. οὐκέτι γὰρ ἄν, εἰ μὴ τοῦτο εἴη, κρᾶσιν, ἀλλὰ παράθεσιν τὸ γινόμενον εἶναι.

IV Τοῦ δὲ τοῦτο οἴεσθαι γίνεσθαι πίστεις φέρουσιν οἱ προϊστάμενοι τῆσδε τῆς δόξης τό τε πολλὰ τῶν σωμάτων σώζειν τὰς ἑαυτῶν ποιότητας ἐπί τ’ ἐλαττόνων ἐναργῶν ὄγκων καὶ ἐπὶ μειζόνων ὄντα (ὡς

---

216. 22 θέσει· γίνεσθαι (codd.) τὰς Apelt: θέσει γίνεται, τὰς Br. συγχύσει Ideler: συγχύσεις ABCSRa: συγχύσεις P
217. 4 λαβεῖν ταύτας coni. Br.: λαβόντας codd.
217. 5 ἁρμὴν Br.: ὁρμὴν codd.
217. 6 ἀλλήλοις Apelt: ἀλλήλους codd.
217. 9 ἁρμὴν Br.: ὁρμὴν codd.

preserving the surface of their own substance and quality in such a juxtaposition, as, one will grant, happens with beans and wheat-grains in their juxtaposition; other mixtures occur by total fusion with both the substances and their qualities being destroyed together, as he says happens with medical drugs in the joint-destruction of the constituents and the production of some other body from them; the third type of mixture he says occurs through certain substances and their qualities being mutually coextended in their entirety and preserving their original substance and qualities in such a mixture: this mixture is blending in the strict sense of the term. The mutual coextension of some two or even more bodies in their entirety with one another so that each of them preserves their own substance and its qualities in such a mixture—this, he says, alone of the mixtures is blending; for it is a peculiarity of bodies that have been blended that they can be separated again [217] from one another, and this only occurs through the blended bodies preserving their own natures in the mixture.

217. 2 He tries to establish that these differences in mixture exist by means of the common notions, and says that we take these above all as the standards of truth from nature; at least we have one [sense-] presentation for the bodies that are juxtaposed by juncture, another for those that are fused and destroyed together, and a third for those that are blended and mutually coextended in their entirety so that each of them preserves its own nature. We would not have this difference in presentations if bodies had been juxtaposed by juncture whatever way they were mixed. He assumes that such a mutual coextension of constituents [of blends] occurs as constituent bodies go through one another, so that there is no part of them that does not partake of everything in such a product of mixture through blending; otherwise the result would no longer be blending but juxtaposition.

217. 13 IV The proponents of this theory forward as suasions of their belief that this occurs the fact that many bodies preserve their own qualities in manifestly smaller and larger masses (as can be

ὁρᾶν ἔστιν ἐπὶ τοῦ λιβανωτοῦ, ὃς ἐν τῷ θυμιᾶσθαι λεπτυνόμενος ἐπὶ πλεῖστον τὴν αὑτοῦ φυλάσσει ποιότητα), ἔτι τε τὸ πολλὰ εἶναι, ἃ καθ' ἑαυτὰ μὴ οἷά τε ὄντα ἐπί τι ἐλθεῖν μέγεθος ὑπ' ἄλλων βοηθούμενα ἐπ' αὐτὸ πρόεισι. τὸν γοῦν χρυσὸν ὑπό τινων μιγνυμένων φαρμάκων ἐπὶ πλεῖστον χεῖσθαί τε καὶ λεπτύνεσθαι, ἐφ' ὅσον καθ' αὑτὸν ἐλαυνόμενος οὐκ ἐδύνατο. καὶ ἡμεῖς δέ, ἃ καθ' αὑτοὺς ὄντες οὐκ ἐσμεν οἷοί τε ἐνεργεῖν, σὺν ἄλλοις ἐνεργοῦμεν. τούς τε γὰρ ποταμοὺς διαβαίνομεν ἀλλήλων ἐφαπτόμενοι, οὓς οὐχ οἷοί τέ ἐσμεν διαβαίνειν καθ' αὑτούς. καὶ βάρη τινὰ φέρομεν μετ' ἄλλων, ὧν τὸ ἐπιβάλλον ἡμῖν μέρος μόνοι γενόμενοι φέρειν οὐ δυνάμεθα. καὶ ἄμπελοι δὲ καθ' αὑτὰς ἵστασθαι μὴ δυνάμεναι ἀνίστανται ἀλλήλαις ἐμπλεκόμεναι.

ὧν οὕτως ἐχόντων οὐδέν φασι θαυμαστὸν τὸ καὶ σώματά τινα βοηθούμενα ὑπ' ἀλλήλων οὕτως ἀλλήλοις ἑνοῦσθαι δι' ὅλων, ὡς αὐτὰ σωζόμενα μετὰ τῶν οἰκείων ποιοτήτων ἀντιπαρεκτείνεσθαι ἀλλήλοις δι' ὅλων ὅλα, κἂν ᾖ τινα ἐλάττω τὸν ὄγκον καὶ μὴ δυνάμενα καθ' αὑτὰ ἐπὶ τοσοῦτον χεῖσθαί τε καὶ σώζειν τὰς οἰκείας ποιότητας. οὕτω γὰρ καὶ τὸν κύαθον τοῦ οἴνου κιρνᾶσθαι τῷ ὕδατι τῷ πολλῷ βοηθούμενον ὑπ' αὐτοῦ εἰς τὴν ἐπὶ τοσοῦτον ἔκτασιν.

τοῦ δὲ τοῦθ' οὕτως ἔχειν ὡς ἐναργέσι χρῶνται μαρτυρίοις τῷ τε τὴν ψυχὴν ἰδίαν ὑπόστασιν ἔχουσαν, ὥσπερ καὶ τὸ δεχόμενον αὐτὴν σῶμα, δι' ὅλου τοῦ σώματος διήκειν ἐν τῇ μίξει τῇ πρὸς αὐτὸ σώζουσαν τὴν οἰκείαν οὐσίαν (οὐδὲν γὰρ ψυχῆς ἄμοιρον τοῦ τὴν ψυχὴν ἔχοντος σώματος), ὁμοίως δὲ ἔχειν καὶ τὴν τῶν φυτῶν φύσιν, ἀλλὰ καὶ τὴν ἕξιν ἐν τοῖς συνεχομένοις
218 ὑπὸ τῆς ἕξεως. ἀλλὰ καὶ τὸ πῦρ ὅλον δι' ὅλου χωρεῖν τοῦ σιδήρου λέγουσι, σώζοντος αὐτῶν ἑκατέρου τὴν οἰκείαν οὐσίαν. καὶ τῶν στοιχείων δέ φασι τῶν τεσσάρων τὰ δύο, τό τε πῦρ καὶ τὸν ἀέρα, λεπτομερῆ τε καὶ κοῦφα καὶ εὔτονα ὄντα, διὰ τῶν δύο, γῆς τε καὶ ὕδατος, παχυμερῶν καὶ βαρέων καὶ ἀτόνων ὄντων διαπεφοιτηκέναι ὅλα δι' ὅλων, σώζοντα τὴν οἰκείαν φύσιν καὶ συνέχειαν αὐτά τε καὶ ἐκεῖνα. δηλητήριά τε τὰ φάρμακα καὶ τὰς ὀσμάς, ὅσαι τοιαῦται, ἡγοῦνται κιρνᾶσθαι τοῖς ὑπ' αὐτῶν πάσχουσιν, ὅλα δι' ὅλων παρατιθέμενα. καὶ τὸ φῶς δὲ τῷ ἀέρι ὁ Χρύσιππος κιρνᾶσθαι λέγει. καὶ αὕτη μὲν ἡ περὶ κράσεως δόξα Χρυσίππου τε καὶ τῶν κατ' αὐτὸν φιλοσοφούντων.

---

218. 1 τῆς Ra, cf. Montanari p. 44: om. cett.
218. 6-7 φάρμακα Ideler, cf. Galen *Nat. Fac.* III. 7 (II. 161 K): φθείροντα codd.

seen in the case of incense which, though attenuated when incinerated, preserves its own quality over a very large expanse), and again the fact that many bodies which in themselves are unable to advance to a certain magnitude attain it with the help of other bodies. Certainly gold can, through the admixture of certain chemicals, be melted and spread to an extent that it could not reach when simply hammered. And we perform with others activities which we are unable to do by ourselves; for in contact with one another we cross rivers which we are unable to cross by ourselves, and along with others bear certain weights of which we could not when left alone bear the part that befalls us. And vines which are unable to stand up by themselves stand up when linked to one another.

217. 26 Since this is the case, they say that there is nothing remarkable in the fact that certain bodies when helped by one another are in this way united together in their entirety so that being preserved along with their own qualities they have a complete mutual coextension through one another, even if some of them are rather small in bulk and in themselves unable both to be spread to such an extent and to preserve their own qualities; for in this way also the cup of wine is mixed with a large amount of water and helped by it to such a great extension.

217. 32 They employ as clear evidence that this is the case the fact that the soul which has its own substantiality, just like the body that receives it, pervades the whole of the body while preserving its own substantiality in the mixture with it (for there is nothing in the body possessing the soul that does not partake of the soul); and the same holds for the Nature of plants, as also the State in bodies held together [218] by their State. Also they say that fire passes completely through iron with each of them preserving its own substance. And they say that two of the four elements, Fire and Air, being rare, light, and having tension, completely pervade Earth and Water which are dense, heavy, and lack tension; and that each pair preserves its own nature and continuity. They think that drugs that are deleterious, and all such odours, are blended with the bodies affected by them in a total juxtaposition. Chrysippus also thinks that light is mixed with air. This is Chrysippus'
218. 10 theory of blending and that of the philosophers who follow him.

V

Θαυμάσαι δ' ἄν τις αὐτῶν, πῶς ταῖς κοιναῖς ἐννοίαις δεῖν χρῆσθαι λέγοντες πρὸς τὰς τῶν τιθεμένων ἀποδείξεις, ὡς οὔσαις φυσικοῖς τῆς ἀληθείας κριτηρίοις, πάσαις μᾶλλον ἢ ταύταις χρῶνται πρὸς τὰς θέσεις τῶν οἰκείων δογμάτων. τὰ γοῦν περὶ κράσεως ὑπ' αὐτῶν λεγόμενα οὐ μόνον οὐ προσχρῆται ταῖς φυσικαῖς ἐννοίαις, ἀλλὰ καὶ πλεῖστον ὅσον ἀποδεῖ. τό τε γὰρ σῶμα διὰ σώματος χωρεῖν ὅλον ὅλῳ παρεκτεινόμενον οὐ μόνον ‹οὐ› προσπίπτει κατὰ τὰς κοινὰς ἐννοίας, ἀλλὰ καὶ ἀδύνατον εἶναι προείληπται. φυσικὴ γὰρ ἔννοια τὸ τὸ πλῆρες μηκέτ' ἐν αὐτῷ δύνασθαι δέχεσθαί τι. μὴ γὰρ ἔτι πλῆρες εἶναι δύνασθαι τὸ χώραν ἔχον ἐν αὑτῷ δεκτικὴν ὁμοίου τινός σαφήνειαν ἔχει, δι' ἣν φυσικήν τε καὶ κοινὴν πρόληψιν ἤδη δοκεῖ τισιν εὔλογον, εἶναί τι τῶν σωμάτων δεκτικόν, ὃ καλοῦμεν τόπον. πῶς γὰρ ἄν τις μὴ φωνὰς κενὰς βουλόμενος λέγειν ἐπινοήσαι σῶμά τι πλῆρες αὑτοῦ καὶ μηδὲν ἔχον ἐν αὑτῷ κενὸν διάστημα, δεχόμενον ἐν αὑτῷ ἄλλο σῶμα ὁμοίως πλῆρες αὑτοῦ;

εἰ μὲν γὰρ διὰ πόρων τινῶν χωρεῖν αὐτὰ δι' ἀλλήλων λέγει, ἢ κενοὺς ἐρεῖ τοὺς πόρους τοὺς δεχομένους τὸ σῶμα ‹ἢ› ἄλλα τὰ ἐν τῷ σώματι κενά. εἰ δὲ πλήρεις ἄλλου τινὸς σώματος, εἰ μὲν ὑπεξιόντος τούτου καὶ παραχωροῦντος τῆς χώρας τῷ ἐπεισκρινομένῳ σώματι, οὐδ' οὕτως ἔσται τὸ σῶμα, καθὸ σῶμα, δεχόμενόν τι σῶμα ἐν αὑτῷ, εἴ γε ὑποχωρεῖ θάτερον θατέρῳ, εἰ δὲ ὑπομένοντος, πῶς οἷόν τε τοὺς πεπληρωμένους πόρους σώματος ἐν αὐτοῖς ἄλλο τι δέχεσθαι σῶμα; καὶ γὰρ διὰ τῆς αἰσθήσεως γνώριμον, ὅτι μηδὲ τὰ λεπτότατα τοῦ σώματος ἀγγεῖα πεπληρωμένα οἷά τέ ἐστι δέξασθαι σῶμα ἐν αὐτοῖς τῷ κεκωλῦσθαι καὶ ἐπιπεφράχθαι τὰς τῆς μεταστάσεως αὐτῶν ὁδούς. ἔτι δ' εἰ διὰ πεπληρωμένων τῶν πόρων ἡ πάροδος τοῖς σώμασι, περιττόν ἐστι τὸ πόρους εἶ-
219 ναι λέγειν. δέχοιντο γὰρ ἂν ἄλλα σώματα καὶ χωρὶς τῶν πόρων. καθόλου γὰρ ἡ διὰ τῶν πόρων πάροδος οὐ κρᾶσίς ἐστιν, οἵαν ἀξιοῦσιν, ἀλλὰ παράθεσις, ὡς αὐτοὶ λέγουσι καθ' ἁρμήν, εἰ μή τις πᾶν τὸ σῶμα πόρους ποιοίη. οὕτως γὰρ μόνως ἔσται τὸ διὰ τῶν πόρων διερχόμενον διὰ παντὸς ἐρχόμενον τοῦ σώματος, εἰ πᾶν εἴη πόρος· ἀλλὰ μὴν ἄτοπον τοῦτο. τινὸς

218. 12-13 πάσαις μᾶλλον ἢ ταύταις scripsi: οὐ πάσαις μ.ἢ.τ. Br.: οὔσαις μᾶλλον ἢ [ἢ Bruns] ταύταις ABCPS v. infra
218. 15 ἀποδεῖ Apelt, cf. Alex. *de an.* 9. 1, 5: ἅπαν δεῖ codd.
218. 16 οὐ suppl. Apelt
218. 19-20 σαφήνειαν ἔχει, δι' ἣν Diels: ὤφθησαν ἔχειν δεῖ ἣν codd., obelo not. Br. v. infra
218. 21 τι Apelt: τὸ codd. ὃ Br.: ὂν codd.
218. 25 ἢ codd. Rex: οὐ Br. ‹ἢ› Rex ἄλλα τὰ codd. Rex: ἀλλ' οὕτω Br.
218. 28 αὑτῷ scripsi, cf. 218. 18, 23: αὐτῷ codd.
219. 3 ἁρμὴν ABCS: ἀρμὴν PR: κατὰ ἁρμήν (Br.) correxi

218. 10 V One might wonder how they could say that the common notions, as the natural standards of truth, should be used in demonstrating theories, and yet use every notion but these in establishing their own doctrines. Certainly their statements on blending not only fail to be based on natural notions but are in fact as far removed from them as it is possible to be. For bodily interpenetration with full coextension not only fails to strike us as in accord with the common notions but is also preconceived to be impossible. It is, indeed, a natural notion that what is full can no longer receive anything in itself; for it is obvious that that which has space in it capable of receiving another body cannot yet be full, and on account of this natural and common preconception some think it reasonable that there be something receptive of bodies which we call place. For how could someone who did not wish to talk nonsense think that any body, full of itself and with no empty interval in it, received in itself another similarly full body?

218. 24 For if he says that they go through one another by means of pores he will either say that the pores which receive the body are empty, or that there are some other empty spaces in the body. Now if [the pores] are full of some other body, but if this body departs and provides space for the body that is blended in, then the body *qua* body will not even in this way be receiving body in itself, if one of the bodies withdraws before the other; but if the body remains, how could the pores that are filled with body receive some other body in themselves? For it is known through perception that not even the narrowest vessels of the body can receive a body in them when they are filled because their paths for displacement are impeded and blocked. But again if the entrance into bodies is through filled pores it is superfluous to say that pores [219] exist; for bodies could then receive other bodies even without the pores. For, in general, entrance through pores is not blending of the kind they uphold but juxtaposition "by juncture," to use their precise term—unless one made the whole body into pores. For only if all of it were a pore would the body passing right through the pores go through the whole body. But that is ridiculous, for the

γὰρ ὁ πόρος (καὶ ἄλλο τὸ ἔχον καὶ τὸ ἐχόμενον), εἴ γε ὁ πόρος διάστημά τί ἐστι κενὸν τοῦ ἔχοντος τοὺς πόρους σώματος. ὥστ' εἰ πᾶν εἴη <πόρος>, οὕτως δ' ἂν οὐκ ὂν εἴη τι· μὴ ὂν δ' οὔτ' ἂν κιρνῷτο ἔτι οὔτ' ἂν πόρους <ἔχοι>.

VI Εἰ δ' οὕτω μὲν οὔ φασι τὰ σώματα ἄλληλα δέχεσθαι, καθὸ δὲ μεστά ἐστι αὐτά, χωρεῖν δι' ἀλλήλων αὐτά φασιν, πρῶτον μὲν ἐπιζητήσαι τις ἄν, τί δήποτε οὐ τὸ τυχὸν σῶμα [τὸ τυχὸν] πάντῃ συναυξητικὸν τοῦ ὁμοίου. τὰ γὰρ ποσὰ [τὰ] κατὰ τὴν πρὸς ἄλληλα σύνθεσιν τὴν τοιαύτην τὸ ἐξ αὐτῶν μεῖζον ἑκατέρου ποιεῖ τῶν συγκειμένων. γραμμαί τε γὰρ κατὰ σημεῖον ἀλλήλαις συντεθεῖσαι τὸ μῆκος αὔξουσιν (οὕτως γὰρ συντιθέμεναι ἐξ ἀνάγκης ἕξουσί τι παρὰ τὴν ἀφήν), ἐπίπεδά τε τὸ αὐτὸ ποιεῖ, ἂν κατὰ τὴν γραμμὴν συντεθῇ, τό <τε> σῶμα, τῷ τριχῇ τε καὶ πάντῃ διεστάναι, καθ' ὃ ἐὰν ἄλλῳ ὁμοίως αὐτῷ διεστῶτι συντεθῇ, ἐξ ἀνάγκης συναύξει τοῦτο. εἰ δὴ τοῦτο μὲν οἰκεῖον τοῖς σώμασιν καὶ ἴδιον αὐτῶν, οἱ δὲ λέγοντες σῶμά τι διὰ σώματος χωρεῖν καὶ ἔλαττόν ποτε καὶ ἴσον τὸ ἐξ ἀμφοῖν ποιεῖν ἀναιροῦσι τοῦτο, ἀναιροῖεν ἂν τὴν τοῦ σώματος φύσιν. συναναιρεῖται γὰρ ἀναιρουμένῳ τῷ ἰδίῳ τινὸς τὸ πρᾶγμα, οὗ ἦν ἴδιον τὸ ἀναιρούμενον.

συμβήσεται δὲ κατ' αὐτοὺς καὶ κενὸν γίνεσθαι τὸν κατεχόμενον τέως τόπον ὑπὸ τοῦ οὕτως ἔν τινι γινομένου σώματος, ὡς μηδένα προσεπιλαμβάνειν ἄλλον τόπον τὸ σῶμα, ἐν ᾧ τοῦτο γίνεται, ἀλλ' ἀρκεῖσθαι τῷ καὶ πρὸ τῆς τούτου μίξεως ὑπ' ἐκείνου πεπληρωμένῳ. ὁ γὰρ ὑπ' αὐτοῦ κατεχόμενος τόπος ἔσται κενὸς τῶν δύο σωμάτων εἰς τὸν θατέρου χωρησάντων τόπον. τί γὰρ ἔσται τὸ ἐξ ἀνάγκης γινόμενον ἐν τῷ τοῦ εἰς τὸ σῶμα μεταστάντος τόπῳ; ὅσα δ' αὖ πάλιν ἐν τῇ πρὸς ἄλληλα μίξει τῶν κιρνᾶσθαι λεγομένων αὔξεται, αὐτόθεν ἤδη προσπίπτει τὸ μηδὲ

219. 7 lacunam post εἴη Br.: πόρος supplevi: <πόρος, διάστημα πᾶν ἂν εἴη> coni. Br.
219. 8 pr. ὄν coni. Br.: ἂν codd. alt. οὔτ' Schwartz: δ' οὐδ codd.
219. 9 ἔχοι suppl. Schwartz
219. 10-11 ἐπιζητήσαι τις ἄν scripsi, cf. 225. 18, Alex. *de an.* 60. 22, *Quaest.* 72. 17, *Metaph.* 59. 28: ἐπιζητήσαι τις Apelt Br.: ἐπιζητῆσαν codd.
219. 11 [τὸ τυχὸν] coni. Ideler Rex: τοῦ τυχόντος coni. Br.
219. 12 τὰ secl. Schwartz
219. 16 <τε> Rex τῷ coni. Br.: τὸ σῶμα, τὸ codd.
219. 29 μηδὲ coni. Br.: μὴ τὸ codd.

pore belongs to something (both the container and the contained are distinct), if, that is, the pore is an empty interval of the body that has the pores. So if it were completely ‹ a pore › then it would not be anything, and as nonexistent it would no longer be blended or ‹ have › pores.

219. 9 VI If they deny that bodies receive one another in this way, but say that insofar as they are full they go through one another, one might first inquire why any given body does not contribute to an increase in the size of a similar body in all dimensions; for by such a mutual composition *quanta* make their compound greater than each of the components. Lines, for example, increase their length when compounded with one another at their points (for in such a compound they will necessarily have something beyond contact), and planes will do the same if they are compounded line by line, as will body by its three complete dimensions necessarily contribute to increasing the size of another body of similar dimensions with which it may be compounded. If this is peculiar to, and a *proprium* of, bodies, and if those who say that a body goes through body and makes the compound both less somehow and equal [to the constituents] abolish this, they would abolish the nature of body; for the subject of which the abolished property was a *proprium* is abolished along with its abolished *proprium*.

19. 22 It will follow according to them* that the place [p2] until then occupied by the body [B] which comes to be in some body [A] in this way also becomes void, so that the body [A] in which it [B] comes to be occupies no additional place but is sufficed by the place [p1] filled by it [A] even before its mixture with this body [B]. For the place [p2] occupied by it [B] will be void when the two bodies have moved into the place of one. For what is there that will necessarily come to be in the place [p2] of the body [B] which is transferred into the body [A]?

19. 28 When again an increase occurs in the intermixture of the bodies said to be blended, it thereupon strikes us that the body [B]

* In the next three paragraphs assume that body A in one place (p 1) is to receive body B initially occupying another place (p 2).

σῶμα εἶναι τὸ σῶμα δεδεγμένον ἐν ταὐτῷ, εἴ γε ἄλλην τινὰ χώραν προσείληφε καὶ οὐκ ἠρκέσθη τῇ χώρᾳ τῇ τοῦ λεγομένου αὐτὸ ἐν αὑτῷ δέχεσθαι σώματος. ὅτι δὲ μὴ σῶμά ἐστι τὸ δεύτερον ἐν αὑτῷ τὸ πρῶτον δεχόμενον [τὸ] σῶμα, δηλοῖ καὶ ἡ γενομένη διαίρεσις ὑπ' ἀλλήλων τῶν
220 κιρναμένων σωμάτων. αὐτὰ γὰρ διαιροῦντα διΐστανται· διαιρούμενα δὲ χώραν ἑαυτοῖς παρασκευάζει, ὡς μὴ δυνάμενα διὰ συνεχοῦς καὶ ἀδιαιρέτου τοῦ ὑποκειμένου σώματος εἴς τι προσελθεῖν.

καθόλου δὲ [εἰ] ἔδει μὲν τὸ δεχόμενον σῶμα ἐν αὑτῷ ἄλλο σῶμα, μηδέν τι μεῖζον γίνεσθαι· τοῦτο γάρ ἐστι τὸ δέξασθαι ἐν αὑτῷ τι. ἐπὶ μηδενός τινος σωμάτος ἡ μῖξις τῶν σωμάτων ἴσον τηρεῖ τὸν ὄγκον ἑνὶ τῶν μιγνυμένων. ἐφ' ὧν γὰρ ἴσον δοκεῖ μένειν, ἐπ' ἐκείνων οὐ σώματων μῖξίς ἐστι, ἀλλ' ἢ εἶδός ἐστι καὶ ὕλη τὰ λαμβανόμενα, ὡς ἡ ψυχὴ καὶ τὸ σῶμα, ἢ σῶμα καὶ πάθος, ὡς ὁ σίδηρος καὶ ἡ θερμότης, ἢ μεταβολή τινος γίνεται οἷον ἔς τι ἄλλο, ὡς ἐπὶ τῆς τέφρας. οὔτε δὲ τὰ πάθη οὔτε τὰ εἴδη σώματα, οὐδ' ἂν ἄλλο σῶμα ἐν αὑτοῖς δέχοιτο.

καὶ γὰρ εἰ οἷά τε εἴη τὰ σώματα ἄλληλα δέχεσθαι, οὐδ' ἂν τῆς ἀντιπεριστάσεως ἔτι δέοι πρὸς τὴν κίνησιν αὐτοῖς· οὐ γὰρ δὴ χωρὶς κινήσεως σῶμα δίεισι διὰ σώματος. πρὸς δὲ τούτοις, πῶς οὐκ ἄλογον καὶ τὸ λέγειν μὲν βραχύτατον σῶμα παρισοῦσθαί τε καὶ παρεκτείνεσθαι τῷ μεγίστῳ, ὡς τὸν κύαθον τοῦ οἴνου πολλοῖς ὕδατος μέτροις ἴσον γινόμενον κατὰ μέγεθος αὐτῷ; πῶς δ' οὐκ ἄτοπον τὸ πειρᾶσθαι κατασκευάζειν διὰ τοῦ τὸν λιβανωτὸν καὶ τὰ ἄλλα τὰ θυμιώμενα ἐν τῇ θυμιάσει ἐπὶ πλεῖον ὁρᾶσθαι χεόμενα; ὁ μὲν γὰρ λιβανωτὸς καὶ τὰ ὁμοίως τούτῳ θυμιώμενα καὶ εἰς ἄλλο τι σῶμα μεταβάλλοντα λεπτότερον,

---

219. 30 ταὐτῷ scripsi, cf. Aristot. *Phys.* 209a 6-7 (ἐν ταὐτῷ [sc. τόπῳ]): αὐτῷ codd.
219. 31 αὑτῷ scripsi: αὐτῷ codd.
219. 32 αὑτῷ scripsi: αὐτῷ codd.
219. 33 [τὸ] coni. Br.
220. 1 αὐτὰ coni. Br., cf. ἑαυτοῖς (220. 2): τὰ codd.
220. 3 τι scripsi: τὸ codd. προσελθεῖν] obelo not. Br. qui πρόσω ἐλθεῖν coni. [εἰ] coni. Ideler Rex
220. 4 σῶμα, μηδὲν obelo not. Br.; [εἰ] ἔδει, <εἰ> ἕν τι δεχόμενον σῶμα ἐν αὑτῷ ἄλλο σῶμα <μηδὲν μεῖζον γίνεται> coni. Br.
220. 9 ἔς τι Apelt: ἐστὶν codd.
220. 10 οὐδ' ἂν obelo not. Br.
220. 11 αὑτοῖς scripsi, cf. 218. 32: αὐτῷ codd.
220. 14 μὲν codd. Rex: ἓν Diels Br.
220. 16 πῶς δ' οὐκ Diels: οὕτως δ' ὃν codd.
220. 18 χεόμενα Diels: καιόμενα codd.

received in the same place [as A, p1] is not even body, if indeed it [B] has occupied some other place and is not sufficed by the place [p1] of the body [A] said to receive it [B] in itself.

219. 32 And that the second body [B] which receives the first body [A] in itself is not even body is clear from the mutual division of constituents; [220] for in dividing one another they create a cleavage, and in being divided provide space for themselves because they cannot make any advance through the underlying body that is continuous and undivided.

220. 3 In general a body that receives another body in itself must become no greater; for that is what "to receive something in itself" means. There is no body for which the mixture of bodies preserves a mass equal to one of the constituents; for where it seems to remain equal there is no mixture of bodies but the cases involve either form and matter (as soul and body) or body and quality (as iron and heat) or a given change occurs into some other thing, as happens with ash. But neither qualities nor forms are bodies and certainly would not receive another body in themselves.

220. 11 Again if bodies were able to receive one another there would no longer be any need of mutual-replacement for their motion; for certainly body will not pervade body without motion.

220. 13 Furthermore, how irrational also to say that a minute body would be made equal to and be extended with a very large one, as when the pitcher of wine becomes equal to many measures of water in its quantity. How absurd to attempt to establish this through the fact that incense and bodies that are similarly incinerated are seen to be melted over a considerable expanse in the incineration; for while incense, and bodies that are similarly incinerated and change into some other more attenuated body, in

οὕτως τὴν ἐπὶ πλέον δέχεται χύσιν, τὰ δὲ κιρνάμενά τε καὶ μιγνύμενα καὶ κατ' αὐτοὺς σώζοντα τὴν οἰκείαν φύσιν κιρνᾶται, ὡς καὶ χωρισθῆναι δύνασθαι πάλιν. ὥστε οὐδὲν αὐτοῖς εἰς παραμυθίαν τῶν ἑπομένων ἀτόπων τοῖς κενῶς λεγομένοις ἡ τούτων συντελεῖ παράθεσις.

VII Τό τε οὖν σῶμα διὰ σώματος χωρεῖν, ᾧ χρῶνται πρὸς τὴν τῆς κράσεως ἀπόδοσιν, ψεῦδός τε καὶ παρὰ τὰς κοινὰς προλήψεις καὶ τὰ φυσικὰ κατ' αὐτοὺς τῆς ἀληθείας κριτήρια, καὶ πολὺ μᾶλλον ἔτι τούτου τοιοῦτον τὸ σώματά τινα λέγειν, δύο ἢ καὶ πλείω, δύνασθαι πάντῃ παρεκτεινόμενα αὑτοῖς καὶ μιγνύμενα δι' ὅλων αὐτά τε σώζεσθαι τὰ ἐξ ἀρχῆς μένοντα, καὶ σώζειν τὰς οἰκείας ποιότητας. τούτῳ γὰρ ἡ κρᾶσις αὐτοῖς τῆς συγχύσεως διαφέρει, ὅτι ἐπὶ μὲν τῆς συγχύσεως ἕν τι τὸ ἐκ τῶν συγχεομένων γίνεται, μηδενὸς τῶν ἐν τῇ συγχύσει μήτε κατὰ τὴν οὐσίαν σωζομένου, μήτε κατὰ τὰς ποιότητας, ἐπὶ δὲ τῆς κράσεως ἑκάτερον τῶν ἐν τῷ κεκραμένῳ σωμάτων ἔτι σώζεται καὶ κατὰ τὸ ὑποκείμενον καὶ κατὰ τὰς ποιότητας, καίτοι δι' ὅλων ἀλλήλοις κεκραμένων τῶν σωμάτων. ὃ λέγουσι μὲν βουλόμενοι σώζειν τὸ δύνασθαι καὶ τὰ κεκραμένα χωρίζεσθαι πάλιν ἀλλήλων. ἀδυνάτου δ' ὄντος τοῦ λεγομένου, ἀδύνατον ἂν εἴη κατ' αὐτοὺς ἢ τὸ τὴν κρᾶσιν διὰ πάντων εἶναι ἢ τὸ δύνασθαι χωρίζεσθαι τὰ κεκραμένα.

εἰ μὲν γὰρ ὅλα δι' ὅλων
221 τὰ κεκραμένα μέμικται καὶ μὴ ἕτερον αὐτῶν ἐν τῷ μίγματι ἄμικτον θατέρου μόριον ἔχει, ἀδύνατον αὐτῶν ἑκάτερον ὑπὸ ἰδίας ἐπιφανείας περιέχεσθαι· πᾶν γὰρ μόριον αὐτῶν, τὸ ὑπὸ οἰκείας ἐπιφανείας περιεχόμενον, ἄμικτον ἔσται θατέρου. οὐ γὰρ οἷόν τε τὴν τοῦ οἴνου ἐπιφάνειαν ὕδατος εἶναι, ἢ τὴν τοῦ ὕδατος οἴνου, ὥστε οὕτως οὐκ ἔσται δι' ὅλων μῖξις ἡ κρᾶσις, ἀλλ' εἶεν ἂν παράθεσιν τὴν κρᾶσιν μορίων μορίοις λέγοντες· ὃ φυλασσόμενοι ἄλλο φασὶ μῖξιν καὶ ἄλλο κρᾶσιν εἶναι.

εἰ δὲ μηδὲν μόριον κατ' οἰκείαν περιγραφήν τε καὶ ἐπιφάνειαν εἴη τῶν μεμιγμένων, ἀλλ' εἴη πᾶν ὁμοιομερὲς γεγονὸς τὸ σῶμα, οὐκέτι μὲν ἂν εἴη παράθεσις, ἀλλὰ δι'

---

220. 27 αὑτοῖς scripsi: αὐτοῖς codd. v. infra
220. 29 τούτῳ Rodier, coni. Br., cf. 221. 12: τοῦτο codd.
221. 5 ὥστε Apelt: ὡς ὁ codd.
221. 6 ὃ Apelt: οἱ codd.

this way undergo melting that leads to expansion, the constituents of blends and mixtures are also, according to them, blended while preserving their own character so that they can also be separated again. Hence the association of these examples in no way contributes to diverting us from the ridiculous consequences of their meaningless statements.

220. 23 VII So the theory of body going through body which they employ in establishing blending is both an error and against common preconceptions and what according to them are the natural standards of truth; yet much more so than this is the view that certain bodies, two or even more, while being mutually coextensive in all dimensions and totally mixed can themselves be preserved in their original character and retain their own qualities.

220. 29 For they say that blending and fusion differ in that with fusion a unity emerges from the bodies that are fused, while none of the bodies in the fusion is preserved either in substance or in qualities, while with blending each of the bodies in the mixture is preserved both in substance and qualities, though the bodies have been blended with one another in their entirety. They say this because they also want to preserve the capacity of the blended bodies to be separated again from one another. Were this account impossible, it would be impossible, according to them, for blending to be complete, or for the bodies that have been blended to be capable of separation.

220. 37 Now if the blended bodies are totally mixed [221] and neither of them has a part unmixed with the other, it is impossible for either to be contained by its own surface; for each of their parts, insofar as it is surrounded by its own surface, will be unmixed with the other; for it is impossible that the surface of wine should be that of water, or the surface of water that of wine, so that blending will not be total mixture but they would turn out to be describing as juxtaposition the blending of parts with parts. They avoid this and say that mixture and blending are distinct.

221. 7 But if no part of the bodies that had been mixed existed with its own shape and surface but the compound had become totally uniform there would no longer be juxtaposition but total blending,

ὅλων κρᾶσις· οὐ μὴν ἔτι σῴζοιτο ἂν τὰ ἐξ ἀρχῆς σώματα τῶν μεμιγμένων, ἀλλ' εἴη ἂν συγκεχυμένα τε καὶ συνεφθαρμένα. εἰ δὴ ἔδει τὰ χωρισθησόμενα σῴζεσθαι καὶ μὴ συγκεχύσθαι (τούτῳ γὰρ ἡ κρᾶσις κατ' αὐτοὺς τῆς συγχύσεως διαφέρει), ἀναγκαῖον δὲ τοῖς δι' ὅλων μεμιγμένοις συγκεχύσθαι, ἀδύνατον τὰ δι' ὅλων μεμιγμένα κατ' αὐτοὺς χωρίζεσθαι δύνασθαι.

ἔτι δέ, εἰ ἀνάγκη μὲν τὰ κεκραμένα δι' ὅλων μεμῖχθαι, τὰ δὲ δι' ὅλων μεμιγμένα ἀδύνατον μὴ συγκεχύσθαι, τὰ δὲ συγκεχυμένα τε καὶ συνεφθαρμένα οὐχ οἷόν τε αὐτὰ σῴζεσθαι, οὐδ' ἂν ἕξεις σῴζοιντο αὐτῶν, εἴ γε ἓν μέν τι τὸ γεγονὸς ἐκ τῶν συγκεχυμένων τε καὶ συνεφθαρμένων. ἀνάγκη δὲ τὸ ἓν σῶμα ὑπὸ μιᾶς, ὥς φασιν, ἕξεως συνέχεσθαι, ὥστε καὶ κατὰ τοῦτο ἂν ἀχώριστα ἀλλήλων εἴη τὰ κεκραμένα κατ' αὐτούς.

εἰ δὲ κατὰ μὲν τὰ λεγόμενα ὑπ' αὐτῶν ἀχώριστα ἀλλήλων ἀναγκαῖον εἶναι τὰ κεκραμένα (οὐ γὰρ δὴ οἷόν τε τὴν δι' ὅλων κρᾶσιν γενέσθαι χωρὶς συμφθάρσεως, ἀχώριστα δέ φασιν εἶναι τὰ συνεφθαρμένα), ὁρῶμεν δὲ ἐπ' ἐνίων χωριζόμενα, δῆλον ὡς οὐκ ἂν ἡ κρᾶσις γένοιτο κατὰ τὸν ὑπ' αὐτῶν εἰρημένον τρόπον.

VIII Πρὸς δὲ τούτοις, εἰ διαιρούντων ἄλληλα τῶν κιρναμένων ἡ κρᾶσις γίνοιτο (διὰ τοῦτο γὰρ ἐν τοῖς ὑγροῖς ἡ κρᾶσις μάλιστα, ὅτι ἐστὶν εὐδιαίρετα ταῦτα καὶ ῥᾳδίως θάτερον διὰ θατέρου δίεισι διαιροῦν αὐτό, ὡς ὁρᾶται καὶ ὁ οἶνος ὁ εἰς τὸ ὕδωρ ἐπιχεόμενος, κιρνάμενος αὐτῷ), εἰ μὲν οὖν διαιροῦντα ἄλληλα ἀδιαίρετα ἀλλήλων καταλείπει τινὰ μέρη, οὐκέτ' ἂν εἴη ἐκεῖνα τὰ μέρη κεκραμένα (ἀμιγῆ γὰρ καὶ ἄμικτα ἀναγκαῖον εἶναι τὰ μὴ διῃρημένα μέρη, εἴ γε ἡ κρᾶσις καὶ ἡ κατ' αὐτὴν μῖξις κατὰ τὴν διαίρεσιν γίνεται καὶ ταύτην ὅρον ἔχει· οὕτω τε πάλιν γίγνοιτο ἂν ἡ κρᾶσις [γινομένη] παραθέσει, ἀλλ' οὐ δι' ὅλων ἀλλήλων τῶν κεκραμένων διηκόντων), εἰ δὲ μηδὲν ἀδιαίρετον ἀλλήλων ἐν τῇ κράσει καταλελοίπασιν, εἴη ἂν πάντῃ διῃρημένα καὶ οὐκ εἰς μέρη, ἀλλ' εἰς διαιρέσεις ἡ τομὴ γε-

221. 19 συνέχεσθαι Apelt: συνελέσθαι (συνελέχθαι B) codd.
221. 22 οὐ γὰρ δὴ Br.: εἰ γὰρ μὴ codd.
221. 25 εἰ Ideler: ἢ codd.
221. 29 ἀδιαίρετα Ideler: διαιρετὰ codd.
221. 33 γινομένη seclusi δι' ὅλων ἀλλήλων scripsi: ὅλων ἀλλήλων δι' ὅλου codd. v. infra

as the original bodies that had been mixed would no longer be preserved but would be fused and destroyed together.

21. 11 Now if the bodies that were to be separated had to be preserved and not fused (this is how they distinguish between blending and fusion), while it is necessary for the bodies that are totally mixed to be fused, then it is impossible, by their theory, for bodies that are totally mixed to be separated.

21. 15 And further if it is necessary that the blended bodies be totally mixed, while it remains impossible for bodies that are totally mixed not to be fused, and for bodies that are fused and destroyed together to be themselves preserved, then not even their States would be preserved, if, that is, a single thing is created from the bodies that are fused and destroyed together. A unified body must be held together by one State (to use their term) so that in this respect also the bodies that have been blended would be inseparable from one another.

21. 20 Now if it necessarily follows from their statements that bodies that have been blended be inseparable from one another (for it is impossible that the total blend should occur without their joint destruction, and they say that the bodies destroyed together are inseparable), but we do see that in some cases they are separated, clearly blending would not occur in the manner described by them.

21. 25 VIII Furthermore, if blending occurred through the constituents dividing one another (for blending occurs this way particularly in liquids because they are easily divisible and one [constituent] will easily pervade the other and divide it, as we see in the case of
21. 28 wine poured into water and blended with it), then if the bodies dividing one another leave behind some mutually undivided parts, those parts would not be blended (for the undivided parts must be unmixable and unmixed—if, that is, blending and mixture *qua* blending occur by division and have this as their defining characteristic; thus blending would again occur by juxtaposition and certainly not through the bodies that have been blended pervading
21. 34 one another totally). On the other hand, if the constituents leave no undivided remainder in the blend they would be divided through and through, and not into parts but into divisions, if no particle

γονυῖα, εἴ γε μηδὲν αὐτῶν μόριον παρὰ τὴν διαίρεσιν καταλέλειπται. καὶ
222 ἔτι εἰ ἕκαστον, εἰς ἃ διῄρηται ἕκαστον, ἐκ τούτων συντιθεμένων πάλιν γίνεται, εἴη ἂν τὰ οὕτως διῃρημένα ἐκ διαιρέσεων, οὐκ ἐκ μερῶν οὐδὲ ἐκ σωμάτων συγκείμενα. ἡ γὰρ διαίρεσις οὐ σῶμα, ἀλλὰ πάθος σώματος.

ἕπεται δὲ τοῖς λέγουσιν ἐπ' ἄπειρον τὴν τομὴν καὶ τὸ [ἢ] ἀδύνατον λέγειν πάντῃ διῃρῆσθαί τι σῶμα ἐνεργείᾳ, οὕτως δὲ καὶ μίγνυσθαι δι' ὅλων τῷ εἰς ἄπειρα ἐνεργείᾳ διῃρῆσθαι τὰ σώματα.

εἰ μὲν γὰρ λέγουσιν ἐπ' ἄπειρον εἶναι διαιρετὰ τὰ σώματα τῷ μηδέποτε ἐπιλείπειν τὴν τομήν, ἀλλ' ἀεὶ ἐκ τῶν τεμνομένων περιλείπεσθαί τι τέμνεσθαι δυνάμενον, οὐχ οἷόν τε ἔσται σῶμά τι πάντῃ διῃρῆσθαι ὡς μηκέτι ὑπολείπεσθαί τι ἐξ αὐτοῦ τομὴν ἀναδέξασθαι δυνάμενον. εἰ δὲ τοῦτο, οὐδέποτε ἂν εἴη τὰ κιρνάμενα δι' ὅλων κεκραμένα, εἰ γὲ κίρναται μὲν ἀλλήλοις, καὶ παρεκτείνεται διαιροῦντα ἄλληλα <ἀδύνατον δὲ οὕτως ἄλληλα> διῃρηκέναι, ὡς μὴ ὑπολείπεσθαί τινα αὐτῶν μέρη μὴ διῃρημένα. κατὰ γὰρ τὰ μὴ διῃρημένα οὐδέπω ἂν ἀλλήλοις εἴη κεκραμένα.

εἰ δὲ λέγοιεν ἐπ' ἄπειρον εἶναι τὰ σώματα διαιρετά, τῷ εἰς ἄπειρον δύνασθαι διῃρῆσθαι [τὸ] πᾶν διῃρημένον σῶμα, κατ' αὐτοὺς εἴη ἂν εἰς ἄπειρα ἐνεργείᾳ διῃρημένα τὰ κεκραμένα ἀλλήλοις. εἰ γὰρ πάντῃ κέκραται, πάντῃ διῄρηται. πάντῃ δὲ διῃρημένα <τὰ μιγνύμενα εἰς ἄπειρα διῄρηται> καὶ εἰ μὲν τὰ εἰς ἃ διῄρηται μεγέθη καὶ αὐτά, εἴη ἂν ἕκαστον τῶν οὕτως μιγνυμένων ἄπειρον (τὸ γὰρ ἐξ ἀπείρων μέγεθός τι καὶ διάστασιν ἐχόντων συγκείμενον ἄπειρον· γίνεται γὰρ ἑκάτερον τῶν κεκραμένων ἐξ ἀπείρων μέγεθός τι ἐχόντων, εἴ γε εἰς ἃ τέμνεταί τι καὶ συγκεῖσθαι αὐτὸ ἐκ τούτων ἀναγκαῖον, συμβαίνοιτο <δ'> ἂν οὕτω καὶ πλείω ἂν εἶναι σώματα ἄλλα ἄπειρα), εἰ δὲ μὴ μεγέθη τὰ ὑπολειπόμενα εἰς ἃ ἡ τομὴ τῶν πάντῃ διῃρημένων (οὐ γὰρ δὴ ἐλάχιστά

---

222. 1 διῄρηται scripsi: διαιρεῖται codd.
222. 4 [ἢ] coni. Ideler,
222. 5-6 διῃρῆσθαι bis scripsi: διαιρεῖσθαι codd. v. infra τῷ Apelt: τὸ codd. ἐνεργείᾳ (οὕτως ... ὅλων) <ἢ> τὸ Br.
222. 9 σῶμά τι scripsi, cf. 222. 5: σώματι codd. πάντῃ coni. Br., cf. 222. 5, 17, Aristot. *de gen. et. corr.* 316a 15, 23, 25: πάντι codd. διῃρῆσθαι scripsi: διαιρεῖσθαι codd.
222. 11 γε Rex: δὲ codd. Br. κίρναται B, κίρνανται cett.: εἰ γε κίρναται μόνα, <ἃ> ἀλλήλοις καὶ παρεκτ. coni. Br. qui loc. obelo not.
222. 12 <ἀδύνατον δὲ οὕτως ἄλληλα> coni. Br. qui lacunam indicavit
222. 15 διῃρῆσθαι scripsi: διαιρεῖσθαι codd. v. infra τὸ secl. Schwartz
222. 17 τὰ μιγνύμενα εἰς ἄπειρα διῄρηται supplevi: lacunam ind. Ideler Br.
222. 22 <δ'> Br.

of them remains beyond the division. [222] Then if each body is reconstituted through the composition of the bodies into which it has been divided, the bodies that have been divided in this way would be composed neither of parts nor bodies, but divisions; for a division is not a body but an attribute of a body.

222. 4 The consequence for those who describe division as continuing to infinity is that it is impossible to speak of a body being in actuality divided* through and through and of bodies being in this way also totally mixed by being in actuality divided into infinites.[1]

222. 6 For if they say that bodies are divisible to infinity through the division never letting up, but there always being a residue from what is undergoing division capable of being divided, then it will be impossible for a body to be divided* through and through so that it leaves no remainder able to undergo division. But if this is so, the constituents would never be totally blended, if they *are* blended with one another and extended in dividing one another, ‹ yet incapable of so dividing one another › that no undivided parts are left as a remainder; for they would remain unmixed with one another with respect to the undivided parts.

222. 14 But if they claimed that bodies were divisible to infinity in that every body that has been divided can be actually divided* to infinity, then according to them the bodies that have been blended with one another would be actually divided* into infinites; for if there is blending through and through there must be division through and through.

222. 17 But ‹ constituents › that are divided* through and through ‹ are divided* into infinites ›, and if the things into which they are divided are also themselves magnitudes, each of the things being mixed in this way would be infinite (for the sum of infinites with size and extension is infinite; for each of the bodies that have been blended is formed of infinite bodies that have some size, if something must be composed of just the things into which it is divided, and this would result in there being still further additional infinites).

222. 22 But if the remainder produced by the division of the bodies divided* through and through are not magnitudes (for they cer-

[1] The Greekless reader should realise that at 222. 1, 5, 6 here, and below at lines 9, and 15 I have emended the text to read the perfect passive tense of "divide" where the manuscripts read the present passive. Those apparent English present passives marked with an asterisk are in fact present perfects; they refer to a completed state of division.

τινα ἐροῦσιν εἶναι καὶ ἀδιαίρετα σώματα), εἴη ἂν αὐτοῖς τὸ μέγεθος οὐκ ἐκ μεγεθῶν συγκείμενον, ὃ ἕπεται τῷ καὶ τῆς γραμμῆς μέρη λέγειν τὰ σημεῖα.

IX Πῶς δὲ σωζόντων ἐστὶ τὴν περὶ κράσεως κοινὴν πρόληψιν τὸ λέγειν καὶ τὴν ἕξιν τοῖς ἔχουσιν αὐτὴν μεμῖχθαι, καὶ τὴν φύσιν τοῖς φυτοῖς καὶ τὸ φῶς τῷ ἀέρι καὶ τὴν ψυχὴν τῷ σώματι, εἴ γε κιρνᾶσθαι μὲν προείληπται <τὰ> κατ' ἰδίαν πρὸ τῆς κράσεως ὑφεστάναι δυνάμενα; διὰ τοῦτο γοῦν καὶ αὐτοί φασι τὰ κεκραμένα χωρίζεσθαι πάλιν δύνασθαι, καὶ ταύτῃ διαφέρειν κρᾶσιν συγχύσεώς τε καὶ φθάρσεως. οὔτε ἕξις τις χωριστὴ τοῦ ἔχοντος αὐτήν, ὡς καθ' αὑτὴν εἶναι δύνασθαι, οὔτε ἡ τῶν φυτῶν φύσις χωρὶς φυτῶν ὑποσταίη ποτ' ἄν. τό τε φῶς πῶς οἷόν τε ἐπινοῆσαι κεχωρισμένον τῶν διαφανῶν σωμάτων; ἀλλ' οὐδ' ἡ ψυχὴ τοιοῦτον, ὡς οἴονται, εἴ γε εἶδος ἔνυλον οὐχ οἷόν τε εἶναι χωρὶς ὕλης τε καὶ σώματος.

ἀλλ' οὐδὲ τὸ πῦρ τῷ σιδήρῳ, καθά φασι, μίγνυται, ὥσπερ οὐδὲ τοῖς χυμοῖς
223 οὐδὲ τοῖς ξύλοις. ὅλως γὰρ ἄτοπον τὴν ὕλην τῷ εἴδει μίγνυσθαι λέγειν. ὕλη δὲ πυρὸς τὰ καιόμενά τε καὶ πεπυρωμένα πάντα, ἀλλ' ἡ μὲν ἄφθαρτος, ἡ δ' οὔ. διὸ καὶ μέχρι πολλοῦ σβεννύμενά τινα ταὐτὸ εἶδος τῷ ἐξ ἀρχῆς δύναται φυλάττειν, οὐ μὴν ἀμείωτα πάντῃ· καὶ γὰρ τούτων ὑπὸ τοῦ πυρὸς ἀναλίσκεταί τι καὶ φθείρεται. διὸ καὶ ταῦτα χρονίζοντα ἐν αὐτῷ πλέον ἀπόλλυταί τε καὶ τοῦ οἰκείου εἴδους ἐξίσταται.

X Πῶς δ' οὐκ ἄτοπον καὶ τὸ λέγειν ἡνῶσθαι τὴν σύμπασαν οὐσίαν πνεύματός τινος διὰ πάσης αὐτῆς διήκοντος, ὑφ' οὗ συνέχεταί τε καὶ συμμένει τὸ πᾶν καὶ συμπαθές ἐστιν αὑτῷ; τὴν γὰρ κυριωτάτην αἰτίαν τῆς τοῦ παντὸς ἑνώσεως οὐκ εἰδότες (αὕτη δ' ἐστὶν ἡ τοῦ θείου τε καὶ κυκλοφορητικοῦ καὶ αἰθερίου σώματος φύσις, ἥτις περιέχουσα πᾶσαν τὴν ἔνυλόν τε καὶ παθητὴν καὶ μεταβλητὴν οὐσίαν τῇ συνεχεῖ τε καὶ διηνεκεῖ κινήσει καὶ ἄλλοτε ἀλλοίᾳ σχέσει πρὸς αὐτὰ τὰς εἰς ἄλληλα τῶν ἐν γενέσει σωμάτων μετα-

222. 25 ὃ . . . τῷ scripsi: ᾧ . . . τὸ codd. Br. v. infra
222. 29 τὰ suppl. Schwartz
222. 34 οἴονται coni. Br.: οἷόν τε codd.
223. 4 ἀμείωτα Ideler: ἀμείω τὰ codd.
223. 10 αὕτη Ideler: αὐτὴ codd.

tainly will not say that there are some minimal and indivisible bodies) then their magnitude would not be composed of magnitudes—also a consequence of the theory that points are parts of the line.

222. 26 IX How can they maintain the common preconception about blending and say that even the State itself is mixed with the things that have it, and that their Nature is mixed with plants, light with air, and the soul with the body, if, that is, it is preconceived that bodies capable of existing individually before the blend are blended? Certainly it is for this reason that they themselves say that bodies that have been blended can be separated again, and that they can thereby distinguish blending from fusion and destruction. But no State is separable from what has it so as to be capable of independent existence, nor could the nature of plants exist apart from them. How could one conceive of light as separable from transparent bodies? Neither could soul be like this, as they think, if enmattered form cannot exist apart from matter and body.

222. 35 Nor is fire mixed with iron, as they say, as not even with fluids [223] or wood. For it is in general absurd to say that matter is mixed with form; everything that is burnt and heated by fire is the matter of fire, but while the former kind [of matter] is indestructible, this kind is not. So things that have quite considerably deteriorated can preserve the same form that they originally had, though they are in every respect diminished; for they are somewhat expended and destroyed by the fire. Thus only after remaining in [the fire] for a long time are these bodies too finally destroyed and expelled from their own form.

223. 6 X Surely it is absurd also to claim that the whole of substance is unified by a pneuma which pervades it completely, and through which the whole is held together, is stable and is sympathetic with itself? For they do not know the fundamental explanation for the unification of the whole (this is the nature of the divine body, in circular motion and composed of aether, which holds together and preserves the whole by surrounding the whole enmattered, affectable and alterable substance with continuous and everlasting motion, and by causing in a fixed order the interchange of bodies that come to be by its different states towards them at different

βολὰς ἐν ὡρισμένῃ τάξει ποιουμένη συνέχει καὶ σῴζει τὸ πᾶν), ταύτην μήτ' ἰδόντες ἑαυτοῖς, μήτε τοῖς ἰδοῦσιν ἀκολουθῆσαι δυνηθέντες, διὰ τὸ ὑπὸ πολλῶν τινων δοξῶν προειλῆφθαι δεσμοῖς τισι καὶ ὑλικαῖς αἰτίαις καί τινι πνεύματι διὰ πάσης τῆς οὐσίας διήκοντι ἀνατιθέασιν αὐτοῦ τὴν ἕνωσιν.

ἥτις δόξα, πρὸς τῷ κεχρῆσθαι τῷ ψευδεῖ σῶμά τι διὰ σώματος διήκειν, καὶ διὰ τῶν ἄλλων, ὅτι ἐστὶ ψευδής, ἤδη ἐλέγχεται· φανερῶς γὰρ ἔνια τῶν σωμάτων οὐχ οἷά τε ἐνεργείᾳ τι ὂν ἔχειν πνεῦμα ἐν αὑτοῖς· τὸ γοῦν ὕδωρ τοσοῦτον ἀποδεῖ τοῦ ἔχειν διὰ παντὸς αὑτοῦ μεμιγμένον πνεῦμά τι, ὡς μηδ' ἂν τὸ τυχὸν ἐν αὐτῷ γένηταί ποτε, ἢ ἀποπνιγέντος ἐν αὐτῷ ζῴου τινὸς τῶν ἀναπνευστικῶν ἢ καὶ δι' ἄλλην αἰτίαν τινὰ γενόμενον, μηδ' ἐπ' ὀλίγον ἐν αὐτῷ δύνασθαι μένειν, ἀλλὰ παραχρῆμα μετὰ βίας ἀναφέρεσθαί τε καὶ ἐκκρίνεσθαι, κἂν ἐν βυθῷ γενόμενον τύχῃ. τούτου δ' οὕτως ἔχοντος, πῶς ἂν ἔτι ἀληθὲς εἴη τὸ πᾶν ἡνῶσθαί τε καὶ συνέχεσθαι, πνεύματός τινος διὰ παντὸς διήκοντος αὐτοῦ;

ἔπειτα δ' εὔλογον μὲν ἦν, ὁποίαν τὴν ἀπὸ τοῦ πνεύματος συνοχὴν γινομένην ἐν πᾶσιν εἶναι τοῖς σώμασιν· οὐχ οὕτως δ' ἔχει. τῶν γὰρ σωμάτων τὰ μέν ἐστι συνεχῆ, τὰ δὲ διωρισμένα. διὸ εὐλογώτερον, ἕκαστον αὐτῶν ὑπὸ τοῦ οἰκείου εἴδους συνέχεσθαί τε καὶ ἡνῶσθαι λέγειν πρὸς ἑαυτό, καθό ἐστιν αὐτῶν ἑκάστῳ τὸ εἶναι, τὴν <δὲ> συμπάθειαν αὐτῶν σῴζεσθαι τὴν πρὸς ἄλληλα διά τε τὴν τῆς ὕλης κοινωνίαν καὶ τὴν τοῦ περικειμένου θείου σώματος αὐτῷ φύσιν, ἢ τῷ διὰ τοῦ πνεύματος δεσμῷ· τίς γὰρ καὶ ὁ τόνος τοῦ πνεύματος, ὑφ' οὗ συνδούμενα τήν τε συνέχειαν ἔχει τὴν πρὸς τὰ οἰκεῖα μέρη καὶ συνῆπται τοῖς παρακειμένοις. βιαζόμενον μὲν γὰρ ὑπό τινος τὸ πνεῦμα, διὰ
224 τὴν πρὸς τοῦτο εὐφυΐαν. τῷ μηδεμίαν ἀντίβασιν ἔχειν τῷ κινοῦντι δι' εὐπάθειαν δύνασθαι, ὑπὸ τῆς ἀθρόας κινήσεως ἰσχύν τινα λαμβάνει, εὐπαθὲς δὲ ὂν κατὰ τὴν οἰκείαν φύσιν. ὑγρὸν δ' ἐστὶ καὶ εὐδιαίρετον, ὡς καὶ τῶν ἄλλων, ἐν οἷς ἂν ᾖ τοῦτο μεμιγμένον, κατὰ τοῦτο μάλιστά τε καὶ ἀρίστην γίνεσθαι τὴν διαίρεσιν. διὰ τοῦτο γοῦν οἱ μὲν κενόν τι αὐτὸ ᾠήθησαν εἶναι καὶ φύσιν τινὰ ἀναφῆ, οἱ δὲ πολλὰ ἔχειν ἐν αὐτῷ κενά.

---

223. 18 τῷ (bis?) Br.: τὸ codd.
223. 22 ποτε Ideler: τότε codd.
223. 23 γενόμενον scripsi, cf. 223. 25: γενομένου codd.
223. 32 δὲ suppl. Br. alt. τὴν Apelt: τῇ codd.
223. 34 τῷ ... δεσμῷ; τίς Von Arnim: τὸ ... πνεύματος δεσμώτης codd. Br.
224. 1 δι' Br.: δ' codd.
224. 2 εὐπαθὲς coni. Schwartz, cf. εὐφυΐα (224. 1) et v. 233. 4: ἀπαθὲς codd. Br.
224. 5 ἀρίστην coni. Diels Rodier: ἀρέστην codd. Br.

223. 14 times)—not being aware of this for themselves nor, because of their many dogmatic preconceptions about it, capable of following those who are aware of it, they instead attribute its unity to certain bonds and material causes, and a pneuma which pervades the whole of substance.

223. 18 In addition to relying on the false notion that body goes through body this theory stands refuted for other reasons, since it is false.

223. 19 Clearly some bodies are incapable of having "pneuma"* actually present in them. Water is so far from having some "pneuma" completely mixed with it that no "pneuma" at all would ever arise in it, either when some animal capable of respiration was choked within it, or when it occurred for some other reason; nor could any "pneuma" remain in it to even a slight degree without being immediately removed and separated by force, even if it should chance to occur in its depths. Since this is true, how could it still be true that the whole is unified and held together through a "pneuma" pervading the whole of it?

223. 27 Further, it would be reasonable that whatever sort of continuity results from pneuma should exist in all bodies. But this is not the
223. 30 case; for some bodies are continuous, others discrete. It is more reasonable, then, to describe each as unified with itself and held together by its own form, insofar as each has being, while their mutual sympathy is preserved because of their participation in matter, and because of the nature of the surrounding divine body, rather than through the bond of pneuma.

223. 34 For what is the tension of pneuma through which things are bound together and have continuity with their related parts, and are connected with juxtaposed bodies? Pneuma is forced by something [224] because of its good disposition towards it to assume some power through its motion in a mass, in that it has not capacity to resist its mover because of its malleability, being malleable in its own nature. It *is* moist and easily-divided, so that with other things with which it has been mixed division is therefore especially viable. Hence some have thought that it was something void and of an intangible nature, while others that there were many empty spaces in it.

* I.e. pneuma in its literal sense of "breath"; see Comm. on 223. 19-27.

καὶ γὰρ εἰ τοῦ μὴ διαπίπτειν, ἀλλὰ συμμένειν τὰ σώματα, αἴτιον τὸ συνέχον αὐτὰ πνεῦμα, τὰ διαπίπτοντα τῶν σωμάτων δῆλον ὡς οὐκ ἂν ἔχοι τὸ πνεῦμα τὸ συνδέον. πῶς δ' ἂν τὴν ἀρχὴν ἡ διαίρεσις σῴζοιτο τῶν σωμάτων, εἴ γε ἡ μὲν διαίρεσις χωρισμός ἐστι τῶν ἡνωμένων, μένει δὲ κατ' αὐτούς τὰ ἡνωμένα [ὁμοίως ἀλλήλοις] πάντα, κἂν διαιρεθῇ; πῶς δ' οὐκ ἂν πάντα φαίνοιτο τὰ παρακείμενα ἀλλήλοις καὶ ῥᾳδίως ἀλλήλων χωρίζεσθαι δυνάμενα <συνέχεσθαι> ὑπ' αὐτοῦ καὶ ὁμοίως ἀλλήλοις ἡνῶσθαι [λέγειν] τοῖς συνεχέσι τε οὖσι καὶ χωρὶς διαιρέσεως μὴ δυναμένοις τινὸς ἀλλήλων χωρισθῆναί ποτε;

πρὸς δὲ τούτοις, εἰ τὸ πνεῦμα γεγονὸς ἐκ πυρός τε καὶ ἀέρος διὰ πάντων πεφοίτηκε τῶν σωμάτων <τῷ> πᾶσιν αὐτοῖς κεκρᾶσθαι καὶ ἑκάστῳ αὐτῶν ἐκ τούτου ἠρτῆσθαι τὸ εἶναι, πῶς ἂν ἔτι ἁπλοῦν τι εἴη σῶμα; πῶς δ' ἄν, εἰ ὕστερον τὸ ἔκ τινων συγκείμενον τῶν ἁπλῶν, τὸ πῦρ ἂν καὶ ὁ ἀὴρ εἴη, ἐξ ὧν μιγνυμένων τὸ πνεῦμα γίνεται, οὗ χωρὶς ἀδύνατον εἶναί τι σῶμα; εἰ γὰρ ἐξ ἐκείνων μὲν ἡ γένεσις τῷ πνεύματι, ἀδύνατον δὲ ἐκείνων εἶναί τι χωρὶς πνεύματος, οὔτ' ἂν ἐκείνων τι εἴη πρὸ τῆς τοῦ πνεύματος γενέσεως, οὔτ' ἂν τὸ πνεῦμα γίνοιτο, οὐκ ὄντων, ἐξ ὧν ἡ γένεσις αὐτῷ. πῶς δ' ἄν τις ἐν τῷ ψυχρῷ ἐνεργείᾳ τι θερμὸν εἶναι λέγοι; τίς δὲ καὶ ἡ εἰς τὸ ἐναντίον ἅμα κίνησις αὐτοῦ, καθ' ἣν συνέχει τὰ ἐν οἷς ἂν ᾖ, ὃν ὥς φασι, πνεῦμα κινούμενον ἅμα ἐξ αὑτοῦ τε καὶ εἰς αὑτό; καὶ κατὰ τί εἶδος κινήσεως γίνεται; κατ' οὐδὲν γὰρ οἷόν τ' ἐστὶ νοῆσαί τι ἅμα εἰς τὰ ἐναντία κινούμενον καθ' αὑτό.

XI

Ἀλλὰ γὰρ ἐοίκασιν, ὡς εἶπον, οἱ οὕτω λέγοντες τῷ μὴ δυνηθῆναι διαλαβεῖν τῷ λόγῳ τῆς ὕλης τὸ εἶδος, καθὸ εἶδος ἕκαστόν τ' ἐστὶ τῶν ὄντων ὅ ἐστι, καὶ πρὸς τἆλλα διαφορὰν ἔχει, κατὰ τὴν ὕλην οὐδὲν ἀλλήλων διαφέροντα, καθὸ εἶδος ἔχει, καὶ τῷ σῴζε-

224. 11 ὁμοίως ἀλλήλοις seclusi   τὰ ἡνωμένα πάντα scripsi: τῶν ἡνωμένων ὁμ. ἀλλ. πάντα codd.: πάντων ἡνωμένων, ὁμ. ἀλλ., πάντα, κ.τ.λ. Br.   v. infra

224. 13 συνέχεσθαι supplevi, v. infra   [λέγειν] coni. Ideler

224. 15 τῷ suppl. Ideler

224. 24 αὐτοῦ Apelt: αὐτοῖς; οὐ codd.   συνέχει Ideler: συνέχειν codd.

224. 25 κινούμενον Apelt: κινουμένου codd. αὑτοῦ... αὑτό scripsi: αὐτοῦ... αὐτό codd.   v. infra

224. 6 Now if the pneuma that binds them together explains why bodies do not fall apart but remain together, clearly those bodies which do fall apart will not have pneuma as their binding agency. And how would the division of bodies be at all preserved if division is a separation of bodies that are unified, yet according to them all of the bodies that are unified are stable even if they have been divided? And would not all the bodies that are juxtaposed with one another and easily separable from one another appear < to be held together > by it and unified with one another in the same way as continuous bodies that are incapable of ever being separated from one another without some division?

24. 14a Another point: if pneuma is a product of fire and air and pervades all bodies by being mixed with all of them, and by each being dependent on it for its existence, how could it still be an uncompounded body? How, if the product of uncompounded bodies is secondary, could it be Fire and Air from the mixture of which evolves the pneuma without which it is impossible for any body to exist? For if pneuma comes to be from these bodies, but none of them can exist without it, no part of them would exist before pneuma comes to be, nor would pneuma come to be without the existence of the bodies from which it takes its coming-to-be.

24. 22 And how could anyone say that there is something actually hot in what is cold?

24. 23 Also, what is its [pneuma's] motion in opposite directions at the same time, by which it holds together the bodies in which it is present, being, as they say, pneuma simultaneously moved into and out of itself? Now by what kind of motion does this occur, when it is impossible to conceive of any by which something is in itself in motion in opposite directions at the same time?

24. 27 XI Now, as I said, with this doctrine they seem, by their inability to distinguish in theory between form and matter—in that everything that exists has its identity in its form and differs with respect to other things which in themselves are indistinguishable in matter, insofar as it has form, and in that these things are pre-

σθαί τε καὶ μένειν ταῦτα, ἅ ἐστι, πνεύματι ὡς διὰ πάντων διήκοντι ἀνάπτειν τό τε εἶναι ἑκάστου καὶ τὸ σώζεσθαί τε καὶ συμμένειν.

αἰτιάσαιτο δ' ἄν τις εὐλόγως αὐτῶν ἐνταῦθα τοῦ λόγου γενόμενος καὶ τὸ δύο ἀρχὰς τῶν πάντων λέγοντας εἶναι ὕλην τε καὶ θεόν, ὧν τὸν μὲν ποιοῦντα εἶναι
225 τὴν δὲ πάσχουσαν, μεμῖχθαι τῇ ὕλῃ λέγειν τὸν θεόν, διὰ πάσης αὐτῆς διήκοντα καὶ σχηματίζοντα αὐτήν, καὶ μορφοῦντα καὶ κοσμοποιοῦντα τούτῳ τῷ τρόπῳ. εἰ γὰρ θεὸς κατ' αὐτοὺς σῶμα, πνεῦμα ὢν νοερόν τε καὶ ἀΐδιον, καὶ ἡ ὕλη δὲ σῶμα, πρῶτον μὲν ἔσται πάλιν διῆκον σῶμα διὰ σώματος, ἔπειτα τὸ πνεῦμα τοῦτο ἤτοι τι τῶν τεσσάρων τῶν ἁπλῶν ἔσται σωμάτων, ἃ καὶ στοιχεῖά φασιν, ἢ ἐκ τούτων σύγκριμα (ὥς που καὶ αὐτοὶ λέγουσιν· καὶ γὰρ ἀέρος καὶ πυρὸς ὑφίστανται τὴν οὐσίαν ἔχειν τὸ πνεῦμα), ἤ, <εἰ> ἄλλο τι εἴη, ἔσται τὸ θεῖον αὐτοῖς σῶμα πέμπτη τις οὐσία χωρὶς ἀποδείξεώς τινος καὶ παραμυθίας λεγομένη τοῖς πρὸς τὸν μετὰ τῶν οἰκείων τιθέμενον τοῦτο ἀντιλέγουσιν ὡς λέγοντα παράδοξα. εἰ δὲ ἢ τῶν τεσσάρων τι εἴη ἤ τι ἐξ ἐκείνων σύγκριμα, ἔσται τὸ ἐκ τῆς ὕλης γεννώμενον σῶμα πρὸ τοῦ γενέσθαι πεφοιτηκὸς δι' αὐτῆς καὶ τεκνοῦν ἐξ ἐκείνης ὁμοίως τοῖς ἄλλοις καὶ ἑαυτό. ἔτι τε ὕστερον ἂν ὁ θεὸς τῆς ὕλης εἴη, εἴ γε πᾶν μὲν τὸ ἔνυλον σῶμα τῆς ὕλης ὕστερον. τὸ γὰρ ἐκ τῆς ἀρχῆς ὕστερον, ὁ δὲ θεὸς τοιοῦτον σῶμα· οὐ γὰρ δὴ τῇ ὕλῃ ὁ αὐτός. τοιοῦτος δὲ ὢν εἴη ἂν μέχρι φωνῆς ἀΐδιος αὐτοῖς μόνης· εἰ γὰρ γέγονε (γέγονε δὲ ἐκ τῆς ὕλης), εἴτε τι τῶν ἁπλῶν ἐστι σωμάτων, εἴτε ἐκ τούτων σύγκριμα.

πρὸς δὲ τούτοις ἐπιζητήσαι τις ἄν, εἰ τῶν ἐκ τῆς ὕλης γενομένων οἷόν τε δημιουργὸν λέγειν τὸν διαπεφοιτηκότα τῆς ὕλης καὶ ὄντα ἐν αὐτῇ θεόν. φέρουσι μὲν γὰρ εἰς τὴν κατασκευὴν τοῦδε τὸ μὴ ὁμοίως

224. 34 ποιοῦντα Br.: ποιοῦν codd.
225. 1 λέγειν Apelt: λέγει codd.
225. 8 εἰ suppl. Apelt
225. 9 λεγομένη Apelt: λεγομένης codd.
225. 10 λέγοντα Ideler: λέγοντι codd.
225. 12 τεκνοῦν Apelt: τέκνων codd.
225. 13 ἑαυτό Apelt: ἑαυτῷ codd.
225. 16 μέχρι φωνῆς ἀΐδιος αὐτοῖς Schwartz: μέχρι φωνῆς αἰδοῖος αὐτοῖς A: μέχρι φωνῆσαι δύο αὐτοῖς Ra: μέχρι φωνῆσαι δ' αὐτοῖς BCLPS v. Montanari pp. 45-46

served and stable as they are—to attribute the preservation, being, and stability of each [of the latter] to a pneuma which allegedly pervades everything.

24. 32 Entering the argument at this point one might reasonably challenge them with also claiming the existence of two universal principles, matter and God, of which the latter is active, [225] the former passive; and with saying that God is mixed with matter and pervades the whole of it, in this way shaping and forming it and creating the universe. For if God is on their view body—an intelligent and eternal pneuma—and matter is body, first there
225. 5 will again be body going through body; then this pneuma will certainly be either one of the four uncompounded bodies which they say are also elements, or a compound of them (as of course they themselves say; for they certainly suppose that pneuma has the substance of air and fire), or, if it is something else, the divine body will be a fifth substance for them here presented without proof or support for opponents who claim that the philosopher who established this theory *with* appropriate support offered an incredible doctrine. But if it were one of the four bodies or a compound of them, then the body that is produced from matter will have pervaded it before it comes to be and will generate itself too from it just like other things. Again, God would be secondary to matter if all enmattered body is secondary to matter; for what is derived from a principle is secondary to it, and God is such a body since he is not identical with matter. Were he so he would be eternal for them in name alone; for if he has come to be (and he has come to be from matter) he is either one of the uncompounded bodies or a compound of them.

25. 18 Again one might enquire if it is possible to describe the God that has gone through matter and exists in it as a craftsman of what comes to be from matter. For they offer in support of this the fact

τοῖς φύσει γινομένοις γίνεσθαι τὰ κατὰ τὰς τέχνας. τὰ μὲν γὰρ ἀπὸ τῆς φύσεως ἀποτελέσματα οὐκ ἐπιπολῆς, ἀλλὰ δι' ὅλων εἰδοποιεῖταί τε καὶ διαπλάττεται, καὶ τὰ ἔνδον αὐτῶν γλαφυρώτατα πεφιλοτέχνηται, τὰ δὲ τῶν τεχνῶν διαμεμόρφωται, ὡς ἐπὶ τῶν ἀνδριάντων ἔχει· τὰ γὰρ ἔνδον τούτων ἀδιάπλαστα. διὰ δὴ τοῦτό ἔφησαν τῶν μὲν γινομένων κατὰ τέχνην ἔξωθεν εἶναι καὶ κεχωρισμένον τὸ ποιοῦν, ἐπὶ δὲ τῶν γινομένων φύσει ἐν τῇ ὕλῃ εἶναι τὴν δύναμιν τὴν μορφοῦσάν τε καί γεννῶσαν αὐτά.

ταῦτα δὲ οὐχ ὁρῶμεν τοῖς γινομένοις συνᾴδοντα. ἐπὶ γὰρ πάντων τῶν γινομένων φύσει ἔξωθέν τί ἐστι τὸ τὴν ἀρχὴν τῆς γενέσεως αὐτῶν ἔχον καὶ τῆς πρώτης ἐν τῇ ὕλῃ μεταβολῆς αἴτιον. τά τε γὰρ ἁπλᾶ σώματα τῆς τε εἰς ἄλληλα μεταβολῆς ἔξωθεν ἔχει τὴν αἰτίαν. ψῦξις γὰρ καὶ θερμότης ἡ γινομένη διὰ τὴν τῶν οὐρανίων σωμάτων ἄλλοτε ἀλλοίαν πρὸς αὐτὰ σχέσιν τῆς γενέσεως αὐτοῖς καὶ τῆς φθορᾶς καὶ τῆς εἰς ἄλληλα μεταβολῆς αἴτια. ἀλλὰ καὶ τῶν φυτῶν καὶ τῶν ζῴων ἕκαστον ὑπό τινος ἔξωθεν
226 ὄντος κεχωρισμένου τῆς ὕλης, ἐξ ἧς ἕκαστον αὐτῶν γίνεται, τὴν ἀρχήν τε τῆς γενέσεως λαμβάνει καὶ τὰ κατὰ γῆς πηγνύμενα πάντα καὶ τὰ ἐν τοῖς μετεώροις συνισταμένα [κεχωρισμένα] κεχωρισμένον ἔχει τὸ ποιοῦν. οὔτε γὰρ τὸ θερμαινόμενον καὶ διὰ τοῦτο μεταβάλλον αὐτὸ θερμαίνει· ψυχρὸν δὲ καθὸ θερμαίνεται. ἀλλὰ καὶ οἷς ἡ γέννησις ἀπὸ σπέρματος, ἔξωθεν τὸ γεννῶν διὰ τῆς προέσεως τοῦ σπέρματος. εἰ δ' οὕτως τὰ ἐκ τῆς ὕλης γινόμενα γίνεται, πῶς ἂν ἔτι ὁ θεὸς αἴτιος εἴη διὰ τοῦ τῇ ὕλῃ μεμῖχθαί τε καὶ κεκρᾶσθαι; τοῦ γὰρ οὕτως ἐν τοῖς γινομένοις φύσει ποιοῦντος ἄλλο τὶ πρῶτόν ἐστιν αἴτιον ἔξωθεν ὄν. οὐδὲν δὲ ποιητικὸν πρῶτον τοῦ θεοῦ.

ἐοίκασι δὲ δι' ὧν λέγουσιν εἶδος τῆς ὕλης λέγειν τὸν θεόν. εἰ γὰρ οὕτως ὁ θεὸς μέμικται τῇ ὕλῃ κατ' αὐτούς, ὡς ἐν τοῖς ζῴοις ἡ ψυχὴ τῷ σώματι, καὶ ἡ δύναμις τῆς ὕλης ἐστὶ ὁ θεὸς (φασὶ γὰρ τὴν ὕλην ποιὰν τῇ ἐν αὐτῇ δυνάμει), εἶδός πως ἂν λέγοιεν αὐτῆς τὸν θεόν, ὡς τὴν ψυχὴν τοῦ σώματος καὶ τὴν δύναμιν τοῦ δυνάμει. ἀλλ' εἰ τοῦτο,

---

225. 25 ἔφησαν coni. Apelt Rodier: φησάντων codd.: φασιν Br.
226. 2 κατὰ γῆς coni. Apelt: κατ' αὐτῆς codd. post λαμβάνει lacunam Br.
226. 3 κεχωρισμένα seclusi; susp. Br.
226. 6 προέσεως Apelt: προθέσεως codd.
226. 7 αἴτιος Rodier: αὐτῶν codd. αἴτιος post ἂν suppl. Diels Br.
226. 13 ποιὰν scripsi, v. *Hermes* 101 (1973) 278-282: ποιεῖ codd.: ποιεῖν Ideler Br. τῇ Apelt: τὴν codd. πῶς Apelt: ὡς codd.

that artistic products do not come to be in the same way as what comes to be by nature; for the products of nature are not superficial but are completely formed and moulded and their interior parts are precisely crafted with care, while those of the arts are formed as in the case of statues where the inner parts are not moulded. Therefore they claimed that what creates artistic products was external and separate, while with things that come to be by nature the power that forms and generates them was present in the matter.

25. 27 But we see that this does not accord with what happens; for with everything that comes to be through nature there is something external containing the principle of its coming to be and the cause of the first change in matter. The uncompounded bodies have the cause of their change into one another outside them, for cooling and heating which occur on account of the varying state of the heavenly bodies towards them are the cause of their coming to be, destruction, and change into one another. Each plant and animal derives [226] the source of its coming to be from something external and separate from the matter from which each of them comes to be, and bodies that are frozen on the earth and that are compounded in the upper air have their creator separate. Neither does something that is heated, and thereby changed, itself heat; it is cold insofar as it is heated. Also with what is brought to birth from semen the begetter is external because of his emission of semen. If things that come to be from matter come to be in this way, how could God still be the cause through being mixed and blended with matter? For among things that come to be by nature there is another external cause prior to what creates in this way; yet there is nothing creative that is prior to God.

26. 10 The implication of their statements seems to be that God is the form of matter; for if, according to them, God is mixed with matter as the soul is with the body among animals, and God is the power of matter (for they say that matter is characterised by the power in it), they would in a sense mean that God is its form, as the soul is of the body and the power of what is in potentiality.

Sem. Griekse Letterkunde
Blandijnberg 2
Gent

πῶς ἂν ἔτι ἡ ὕλη ἀνείδεος εἴη κατὰ τὸν αὐτῆς λόγον, εἴ γε τὸ συμμένειν αὐτῇ καὶ εἶναι ὕλῃ παρὰ τῆς οὔσης ἐν αὐτῇ δυνάμεως;

μάλιστα δ' ἐν τῇ ἐκπυρώσει φαίνεται κατ' αὐτοὺς ὁ θεὸς τῆς ὕλης εἶδος ὤν, εἴ γε ἐν τῷ πυρί, ὃ μόνον ἐστὶ κατ' αὐτοὺς τότε, ἡ ὕλη καὶ ὁ θεὸς [τῆς ὕλης] σώζονται μόνοι. εἴη γὰρ ἂν ὁ θεὸς τότε εἶδος τὸ ἐπὶ τῇ ὕλῃ τοῦ πυρός. εἰ δὲ τοῦτο, μεταβάλλει δὲ τὸ πῦρ εἰς ἄλλα τινὰ σώματα, τὸ εἶδος ἀλλάσσον εἴη ἂν ὁ θεὸς φθειρόμενος τότε, εἴ γε κατὰ φθορὰν τοῦ προϋπάρχοντος εἴδους ἡ μεταβολὴ εἰς ἄλλο σῶμα τῇ ὕλῃ γίνεται. καὶ εἰ τῆς τοιαύτης μεταβολῆς ὁ θεὸς αἴτιος, εἴη ἂν ὁ θεὸς κατ' αὐτοὺς φθείρων ἑαυτόν, οὗ τί ἂν ἀτοπώτερον ῥηθείη ποτ' ἄν;

πῶς δ' οὐκ ἀνάξια τῆς θείας προλήψεως τό τε τὸν θεὸν διὰ πάσης τῆς ὑποκειμένης πᾶσιν ὕλης κεχωρηκέναι λέγειν καὶ μένειν ἐν αὐτῇ, ὁποία ποτ' ἂν ᾖ, καὶ τὸ προηγούμενον ἔχειν ἔργον, τὸ ἀεί τι γεννᾶν τε καὶ διαπλάσσειν τῶν ἐξ αὐτῆς γενέσθαι δυναμένων, καὶ ποιεῖν τὸν θεὸν δημιουργὸν σκωλήκων τε καὶ ἐμπίδων, ἀτέχνως ὥσπερ κορόπλαθόν τινα τῷ πηλῷ σχολάζοντα καὶ πᾶν τὸ δυνάμενον ἐξ αὐτοῦ γενέσθαι τοῦτο ποιοῦντα;

πρὸς δὲ τούτοις εἰ τὰ κιρνάμενα ἀλλήλοις σώματα ἀντιπάσχειν ὑπ' ἀλλήλων ἀνάγκη (διὰ τοῦτο γὰρ οὐδέτερον αὐτῶν φθείρεται, ὅτι ἑκάτερον αὐτῶν πάσχον ὑφ' ἑκατέρου ἐν τῷ πάσχειν ἀντιποιεῖ), τὰ δὲ δι' ἀλλήλων χωροῦντα σώματα κιρνᾶται ἀλλήλοις, εἴη τ' ἂν [ἄλληλα] καὶ ὁ θεὸς κιρνάμενος τῇ ὕλῃ, εἰ δὲ τοῦτο, καὶ ἀντιπάσχων ὑπ' αὐτῆς· οἷς ἕπεται τό τε τὸν θεὸν πάσχειν καὶ τὸ τὴν ὕλην ποιεῖν. [ἀλλὰ ταῦτα]

XII 'Αλλὰ ταῦτα μὲν εἰπεῖν προήχθην διὰ τοὺς ἀντιλέγοντας μὲν 'Αριστοτέλει περὶ τοῦ πέμπτου σώματος καὶ τοῖς μόνοις κατ' ἀξίαν τῶν
227 θείων εἰρημένοις ἐνίστασθαι πειρωμένους διὰ φιλοτιμίαν, τῆς δὲ ἀτοπίας τῶν ὑφ' αὐτῶν λεγομένων μηδὲ τὴν ἀρχὴν συνιέντας, οἷς καὶ τὰ κυριώτατα καὶ μέγιστα τῶν κατὰ φιλοσοφίαν δογμάτων ἤρτηται καὶ τὴν κατασκευὴν ἀπὸ τοῦ θαυμαστοῦ δόγματος ἔχει τοῦ 'σῶμα χωρεῖν διὰ σώματος'· ὅ τε γὰρ περὶ κράσεως αὐτοῖς λόγος οὐκ ἐν ἄλλῳ τινί, ἀλλὰ καὶ τὰ περὶ ψυχῆς ὑπ' αὐτῶν λεγόμενα ἐντεῦθεν ἤρτηται ἥ τε πολυθρύλλητος αὐτοῖς

226. 18 τῆς ὕλης del. A: om. a
226. 33a-34 καὶ ὁ θεὸς . . . ἀλλὰ ταῦτα = 227. 23-24; locos coniunxit Brinkmann *RM* 57 (1902) 448-491; lacunas post ἄλληλα et ἀλλὰ ταῦτα ind. Br. v. infra
226. 34: 'Αλλὰ ταῦτα Brinkmann: ταῦτα codd.
226. 35 μόνοις Schwartz: νόμοις codd.

But if this is so, how could matter still be formless in its own essence, if its being and stability is derived from the power present in it?

26. 16 Particularly in the conflagration does God appear, according to them, to be the form of matter, if matter and God are the only things preserved in the fire which at that time is, on their view, the only thing existing; for God would then be the form belonging to the matter of fire. If this is so, and if Fire changes into some other bodies, God would be the alternating form through being destroyed at that time, if, that is, change into another body occurs for matter by the destruction of the preexistent form. If God is the cause of such change he would, according to them, be self-destructive—and what view could be more absurd than this?

26. 24 Surely it demeans our preconception of the deity to say that God pervades the whole of the matter underlying everything and remains in it, whatever it may be like, and has as his premeditated task the perpetual generation and moulding of anything that can come to be from it; and for them to make God a craftsman of grubs and gnats, simply devoting himself like a modeler to clay, and making everything that can be created from it?

26. 30 Again, if the bodies that are being blended with one another must be reciprocally acted on by one another (this is why neither is destroyed, since the one acted on by the other reacts in the process of being acted on) and if the bodies that go through one one another are blended together, then God too would be blended with matter, and thereby also acted on by it—from which it follows that God is acted on while matter acts.

26. 34 XII I was provoked into this argument by denials of Aristotle's theory of the fifth body, and ambitious attempts to resist the only theories worthy [227] of divine things, made by opponents unaware of even the source of the stupidity of their statements, when their central and major philosophical beliefs depend on and take their support from the remarkable belief that body goes through body. For their theory of blending does not rely on something else, but their views on the soul depend on *it*, and their notorious Fate and

εἱμαρμένη καὶ ἡ τῶν πάντων πρόνοια τῇδε πίστιν λαμβάνουσιν, εἴ γε ὁ περὶ ἀρχῶν τε καὶ θεοῦ <λόγος> καὶ ἡ τοῦ παντὸς ἕνωσίς τε καὶ συμπάθεια πρὸς αὐτό. πάντα γὰρ αὐτοῖς ταῦτ᾽ ἐστὶν ὁ διὰ τῆς ὕλης διήκων θεός.

τούτου δὲ τοῦ σῶμα χωρεῖν διὰ σώματος, ἐξ οὗ σχεδὸν πάσης τῆς φυσιολογίας αὐτοῖς ἀνήρτηται τὰ πείσματα, παρά τε τὰς κοινὰς προλήψεις λεγομένου καὶ παρὰ τὰς ἁπάντων δόξας τῶν φιλοσόφων, τὴν πίστιν κατ᾽ αὐτοὺς λαμβάνει ὡς ἂν ἀπὸ ἐναργοῦς τοῦ τὸν σίδηρον, ὅταν ᾖ πεπυρωμένος [μὲν] μὴ αὐτὸν ἐξάπτεσθαί τε καὶ πυροῦσθαι λέγειν ὁμοίως τοῖς οἷς ὕλη τῷ πύρι, ἀλλὰ διὰ παντὸς αὐτοῦ χωρεῖν τὸ πῦρ ὑπολαμβάνειν μετὰ τῆς ὕλης ἐκείνης, ἐφ᾽ ἧς ὂν γειτνιᾶσαν τῷ σιδήρῳ ἐθέρμαινέ τε καὶ ἐξῆψεν αὐτόν.

καίτοι πῶς οὐ πρὸς τοῖς ἄλλοις καὶ τοῦτο παντάπασιν ἄτοπον τὸ τὴν τῆς φλογὸς ὕλην, μὴ κατέχουσαν τὴν ἔξωθεν χορηγίαν, μὴ καταναλίσκεσθαι, σώζεσθαι δὲ ἐπὶ τοσοῦτον λέγειν ἐν τῷ πυρὶ τῷ μεμιγμένῳ τῷ σιδήρῳ, ἕως ἂν ὁ σίδηρος πεπυρωμένος μένῃ, ἐναργοῦς ὄντος τοῦ τὴν φλόγα συνεχοῦς δεῖσθαι τῆς κατὰ τὴν ὕλην χορηγίας, ὡς ἔχουσαν ἐν τῷ γίνεσθαι τὸ εἶναι διὰ τὴν τῆς ὑποβεβλημένης ὕλης αὐτῇ ταχεῖαν φθοράν. [om.] εἰ γὰρ μὴ ὁ σίδηρος ὕλη ποτὲ τῷ πυρί, τίνα χρὴ τὴν καιομένην ὕλην λέγειν εἶναι; ὅλως δὲ ἐπεὶ ὁ σίδηρος ὑπό τινος ἐξάπτεται πυρὸς ὄντος ἐφ᾽ ὕλης τινός, εἰ διὰ τοῦ σιδήρου δίεισι τὸ πῦρ, δεῖ φυλάσσον αὐτὸ τὴν ὕλην, ἐφ᾽ ἧς ὂν ἐπλησίασε τῷ σιδήρῳ, χωρεῖν δι᾽ αὐτοῦ, ἀλλ᾽ οὔτε τὰ ξύλα οὔτε οἱ ἄνθρακες οὔτε τις ἄλλη τῶν ὑποκειμένων ὕλη τῷ πυρὶ τῷ πυροῦντι τὸν σίδηρον, ἐν τῷ σιδήρῳ γίνεται. λείπεται δή, χωριζόμενον αὐτὸ τῆς ὕλης γίνεσθαι ἐν αὐτῷ. ἀλλ᾽ εἰ χωρίζοιτο, ἀναγκαῖον αὐτὸ ἐν τῇ μεταβάσει ὕλην τινὰ μεταλαμβάνειν, καὶ ἔστιν οὐκ ἄλλη τις ἢ ὁ σίδηρος αὐτός. καθόσον γὰρ ἡ ὑγρότης ἐστὶν ἐν αὐτῷ, κατὰ τοσοῦτον ὕλη τῷ
228 πυρὶ γίνεται, σκληρότερος γοῦν ὁ πυρούμενος σίδηρος μετὰ τὴν σβέσιν τοῦ

227. 7 τῇδε Rex: δὲ τὴν codd.: τήνδε coni. Br.
227. 8 εἴ γε coni. Apelt: εἴτε Ra: ἔτι τε cett. Br. ὁ codd.: τὸ Br. ὁ ... <λόγος> Von Arnim: ὁ <λόγος> Apelt
227. 10 τοῦ Br.: τὸ codd.
227. 13 ἂν Apelt: αἱ codd.
227. 14 μὲν secl. Br.
227. 15 τῷ πύρι scripsi, cf. 227. 25, 29, 227. 33-228. 1: τὸ πῦρ codd.
227. 23-24 v. 226. 33-34
227. 27 εἰ Apelt: ἢ codd.
227. 28-29 ἧς ὂν Schwartz: ὅσον codd. οὔτε τὰ Br.: οὐ τά τε codd.
227. 29 pr. οὔτε Br.: οὐδὲ codd.
227. 30 δή, χωριζόμενον Br.: διαχωριζόμενον codd.
227. 33 αὐτὸς Ideler: αὐτή codd.

their universal Providence gain conviction in this way if indeed their < theory > of principles and God, as well as the unification and sympathy of everything depend on it; for the God that pervades matter is all of these things for them.

7. 10 Now while this notion of body going through body, on which the credibility of their whole theory of nature depends, is offered contrary to common preconceptions and the views of all philosophers, it gains conviction, according to them, from it being allegedly clear that when iron is heated by fire it is not itself kindled and burnt, as in the cases where there is matter for the fire, but the fire is understood to pervade the whole of it, along with that matter to which it belonged when, adjacent to the iron, it burnt and kindled it.

7. 17 Now, other reasons aside, surely it is also completely absurd that the matter of the flame which does not possess outside support is not used up, but is said to be preserved in the fire that is mixed with the iron, as long as the iron is stable in its heated state, although it is clear that the flame needs continuous support from its matter, because its being rests in its coming-to-be through the rapid destruction of its supporting matter. [227. 23-25] For unless at some stage the iron were matter for the fire what should one describe as the burnt matter?

7. 26 In general, since the iron is kindled by a particular fire which depends on specific matter, then, if the fire goes through the iron, it must go through it whilst protecting the matter on which it depended when it was adjacent to the iron; but neither pieces of wood, nor coal, nor any other matter supporting the fire which burns the iron come to be within the iron. So we are left with the fact that fire must come to be in the iron when separated from its matter. But if it is separated, it must acquire new matter in the change, and there is none except the iron itself. For to the extent that there is moisture in it, [228] it becomes matter for the fire; certainly the iron that is heated becomes harder after its ex-

πρώτου γίνεται, ὡς ἀναλισκομένης ὑπὸ τοῦ πυρὸς τῆς ὑγρότητος τῆς ἐν αὐτῷ, καὶ ἐξάπτεταί τι μέχρις ἂν ὑγρότης ᾖ τις ἐν αὐτῷ, ὥσπερ οὖν καὶ ξύλα. ἀλλὰ τούτων μὲν ἅλις.

XIII Ἐπανίωμεν δὲ ἐπὶ τὸν ἐξ ἀρχῆς λόγον. εἰ γὰρ τὰ εἰρημένα οὕτως περὶ κράσεως παντελῶς ἄτοπα, ἀναγκαῖον ἄλλως πως τὰς κράσεις γίνεσθαι λέγειν. λείπεται δὲ παρὰ ταύτας, ὧν ἄξιον ποιεῖσθαι λόγον τινά, <ἡ> ὑπ' Ἀριστοτέλους εἰρημένη δόξα. εἴπωμεν δὴ καὶ περὶ ταύτης, καὶ δείξωμεν τίς ποτ' ἐστίν, ἐπεὶ μηδὲ γνώριμός ἐστι τοῖς πολλοῖς τῶν φιλοσοφούντων διὰ συντομίαν τῶν περὶ αὐτῆς εἰρημένων ὑπὸ τοῦ φιλοσόφου.

πρῶτον μὲν ὁρίσωμεν, τίνα ἐστὶν ἀλλήλοις κιρνᾶσθαι λεγόμενά τε καὶ δυνάμενα. ὁρισθὲν γὰρ ἡμῖν τοῦτο ἀποσκευάσεται πολλὰ τῶν ἐνοχλούντων τὸν περὶ κράσεως λόγον. ἔστι δὴ ἡ μῖξίς τε καὶ ἡ κρᾶσις ἐν τοῖς καθ' αὑτὰ ὑφεστάναι φύσιν ἔχουσι. διὸ καὶ δοκεῖ δύνασθαι χωρίζεσθαι πάλιν ἀλλήλων τὰ μεμιγμένα ὅτι ἐκ τούτων συνῆλθεν. εἰ δὲ μηδὲν ἄλλο χωριστὸν παρὰ τὴν οὐσίαν, οὐσιῶν ἂν ἡ μῖξίς τε καὶ ἡ κρᾶσις εἴη. εἰ δὲ τοῦτο, οὐκ ἂν Ἀναξαγόρας εἴη καλῶς λέγων, πάντα ἐν πᾶσιν μεμῖχθαι. οὔτε γὰρ τὰ πάθη χωριστά, ὡς καθ' αὑτὰ εἶναι δύνασθαι (διὸ οὐδὲ μῖξίς τε καὶ κρᾶσις ἂν εἴη παθῶν πρὸς ἄλληλα, ἢ παθῶν πρὸς τὰ ὧν ἐστι πάθη), ἀλλ' οὐδὲ τὰ εἴδη τῶν οὐσιῶν. καὶ γὰρ εἰ οὐσίαι καὶ ταῦτ', ἀλλ' οὐ καθ' αὑτὰ συνυφίστανται. οὐ γὰρ οἷόν τε χωρὶς ὕλης εἶναί τι αὐτῶν. ἀλλ' οὐδὲ ἡ ὕλη μικτή. οὐδὲ γὰρ αὕτη χωρὶς εἴδους ἐνεργείᾳ ποτ' ἐστίν. οὔτ' οὖν ἐν ἄλλοις τισίν, ἢ ἐν οὐσίαις ἡ μῖξίς τε καὶ κρᾶσις, οὔτε ἄλλαι τινὲς τῶν οὐσιῶν εἰσὶ μικταὶ παρὰ τὰς χωριστάς τε καὶ κατ' ἰδίαν ὑφεστάναι δυναμένας καὶ κατεχούσας τόπον. τοιαῦτα δὲ τὰ σώματα.

τούτου τοίνυν διωρισμένου μετὰ ταῦτα ἄξιον ἐπιστῆσαι, πότερον ταὐτόν ἐστι μῖξίς τε καὶ κρᾶσις, ἢ διαφορὰν ἔχει τινά. ἔοικε δὴ διαφέρειν, ᾗ τὸ μὲν κοινότερόν ἐστιν ἡ μῖξις, ἡ δὲ κρᾶσις ἰδικώτερον. ποιὰ γὰρ μῖξις ἡ κρᾶσις. τῶν γὰρ μίξεων ἡ μέν τις κατὰ παράθεσιν τῶν οὐσιῶν καὶ ἁφὴν γίνεται, ἣν

228. 3 τι scripsi, cf. 223. 5: τε codd.
228. 7 ἡ suppl. Ideler
228. 12 ἀποσκευάσεται Diels: ἀποσκευάσαι τε codd.
228. 15 χωριστὸν Apelt, Rodier: χωρὶς τῶν codd.
228. 23 οὔτε Br.: οὐδὲ codd.

tinction than before, since the moisture in it is expended by the fire, and it is kindled as long as there is some moisture in it—as

228. 4 also with pieces of wood. But enough of this.

228. 5 XIII Let us return to our original statement; for if the accounts of blending so far described are completely absurd, some other explanation of how blends occur must be given. Apart from the preceding theories the one worth expounding is that held by Aristotle. Let us also describe it and explain its character, since it is not known to the majority of philosophers on account of the brevity of its treatment by the philosopher himself.

228. 10 First, let us define what things are termed capable of blending together, for with such a definition we will disperse many of the issues that confuse the theory of blending. Now mixture and blending occur among things that are naturally independent substances; and bodies that have been blended are considered capable of being separated from one another because they are combined from this source. If there is nothing else separable except substance, mixture and blending will be of substances. So Anaxagoras was wrong to say that everything is mixed with everything, for neither are qualities separable so that they can be independent (hence there is not even mixture and blending of qualities with one another, nor of qualities with the things of which they are the qualities) nor are forms [separable] from substances. For even if they are also substances they do not exist with equal independence, as it is not possible for any of them to exist without matter. Yet neither is matter mixed, for it does not exist in actuality apart from form. So neither among things other than substances is there mixture and blending, nor are any other substances mixed beyond those that are separable and capable of existing individually and that occupy place—namely physical bodies.

228. 25 Given this definition, it is next worth establishing whether mixture and blending are the same or whether they are different. They certainly seem to be different in that mixture is the more generic, blending the more specific; for blending is a kind of mixture. Among mixtures one type occurs by the juxtaposition and

λέγομεν [τῆς μίξεως] γίνεσθαι κατὰ σύνθεσιν (οὐ πᾶσα μὲν γὰρ σύνθεσις μῖξις· σύνθεσις μὲν γὰρ καὶ τῶν ὁμοίων τε καὶ ὁμοειδῶν γίνεται, ἡ δὲ μῖξις ἐκ διαφερόντων τε καὶ ἐν διαφέρουσι· διὸ ὁ μὲν τῶν πυρῶν σωρὸς κατὰ μόνην σύνθεσιν, ὁ δὲ τῶν πυρῶν τε καὶ κυάμων ἤδη τῇ συνθέσει καὶ τὴν μῖξιν προσείληφεν), ἡ δὲ ὡς κρᾶσις μῖξις γίνεται, οὐ σωζομένων ἔτι τῶν μιγνυμένων καὶ οὕτως ἀλλήλοις παρακειμένων, ἀλλ' ἑνουμένων κατὰ τὸ ὑποκείμενον. διὸ ἐν τοῖς εὐορίστοις τε καὶ ὑγροῖς ἡ ὡς κρᾶσις μῖξίς ἐστιν. ὥσπερ δὲ οὐχ ἡ τῶν τυχόντων σύνθεσις μῖξις ἦν, οὕτως οὐδὲ
229 ἡ τῶν τυχόντων ὑγρῶν ἕνωσις κρᾶσίς τε καὶ μῖξις. οὐ γὰρ ὕδωρ ὕδατι κιρνᾶται, καίτοι ἑνούμενον αὐτῷ, οὐδὲ ἔλαιον ἐλαίῳ, ἀλλ' οὐδὲ ἔλαιον ὕδατι· ἀλλὰ τοῦτο μὲν διὰ γλισχρότητα.

ὅταν δὲ ᾖ τοιαῦτα τὰ ἑνούμενα, ὡς ἔχειν ποιότητας, καθ' ἃς τὰ μιγνύμενα ποιεῖν τε καὶ πάσχειν ὑπ' ἀλλήλων ἐστὶν οἷά τε (ἡ γὰρ κρᾶσις ἐν τοῖς ποιεῖν τε καὶ πάσχειν ὑπ' ἀλλήλων δυναμένοις), τότε καὶ τούτων κρᾶσις γίνεται. διὰ τοῦτο γὰρ ἄμικτον πᾶν τὸ θεῖον σῶμα τὸ ποιητικὸν ὂν μηκέτ' ἀντιπάσχειν ὑπὸ τοῦ πάσχοντος ὑπ' αὐτοῦ δύναται. μόνα γὰρ τὰ ἔνυλα πάσχει. διὸ ταῦτα μόνα τῶν ποιούντων ἀντιπάσχει ὅσα τῆς αὐτῆς κεκοινώνηκεν ὕλης. ταῦτα δὲ τῶν τῆς αὐτῆς ὕλης κεκοινωνηκότων ὑπ' ἀλλήλων ἀντιπάσχειν οἷά τε, ὅσα ἐναντίωσιν ἔχει τινὰ πρὸς ἄλληλα. οὐ γὰρ τὸ τυχὸν ὑπὸ τοῦ τυχόντος οἷόν τε πάσχειν· οὐδὲ γὰρ τὸ ὅμοιον ὑπὸ τοῦ ὁμοίου. ἀπαθῆ γὰρ ὑπ' ἀλλήλων τὰ ὅμοια. ἐν διαφέρουσι γὰρ καὶ ὑπὸ διαφερόντων τὸ ποιεῖν τε καὶ πάσχειν γίνεται. ἀλλ' οὐδὲ τὰ ἕτερα πάντῃ πάσχειν τι ὑπ' ἀλλήλων. οὐδὲ γὰρ παθεῖν οἷόν τε φωνὴν ὑπὸ γραμμῆς, τῷ μηδὲν ὑπ' ἀλλήλων ἀντιπάσχειν, <εἰ μὴ διὰ τὸ> κοινὴν ὑποκεῖσθαί τινα φύσιν τῶν ἑκατέρων παθῶν ἐπιδεκτικήν. διό ἐστιν ἐν τούτοις τὸ ποιεῖν τε καὶ πάσχειν, ὅσα τὴν αὐτὴν ὕλην ὑποκειμένην ἔχοντα ἐναντίωσίν τινα ἔχει πρὸς ἄλληλα. οὔτε γὰρ ὑπὸ χρώματος τὸ γλυκύ, καθὸ γλυκύ, πάθοι ποτ' ἄν, οὔτε ὑπὸ θερμοῦ τὸ ξηρόν, ἀλλ' οὐδὲ ὑπὸ ψυχροῦ τὸ ὑγρὸν καθὸ ὑγρόν, ὅτι

---

228. 30 τῆς μίξεως seclusi
229. 8 δύναται Ideler: δύνασθαι codd.
229. 12 οὐδὲ Br.: οὔτε codd.
229. 16 <εἰ μὴ διὰ τὸ> coni. Br. qui post ἀντιπάσχειν lacunam ind.: ἀντιπάσχειν <μηδὲ> Ideler
229. 19 οὔτε Br.: οὐδὲ codd.

contact of substances, and we say that this occurs by composition (not all composition is mixture, for composition occurs among bodies that are similar and have similar forms while mixture occurs from and among different bodies—so the pile of grains is a mere composition, while the pile of grains and beans has also added mixture to what is already composition), while a second type of mixture occurs as blending when the constituents are no longer preserved and juxtaposed together in this way but unified in substrate. Hence mixture *qua* blending occurs among very malleable and moist bodies, and just as the composition of random bodies was not mixture [229] so the unification of random moist bodies is not blending and mixture; for water is not blended with water, although unified with it, nor oil with oil, but not even oil with water, in this latter case because of its viscosity.

229. 3 Whenever the bodies that are unified are of the kind that have qualities by which constituents can reciprocally interact (for blending occurs among bodies that can reciprocally interact) then their blending also occurs. For this reason the whole divine body, which is active, is unmixed and unable to be reciprocally acted on by the body acted on by it; for only enmattered bodies can be acted on. Hence only where active bodies share in the same matter are they reciprocally acted on. Of bodies that do share in the same matter it is only those that have some mutual contrariety that can interact; for one random body cannot be acted on by another, nor can similar bodies be acted on by one another because similar bodies are mutually unaffected. Interaction in fact occurs among different bodies and through the agency of different bodies but not even in all cases do these interact; sound, for example, cannot be acted on by line, since they cannot interact ‹ except through › having a common underlying nature that can receive the qualities of each of them. Interaction, then, occurs among bodies with the same underlying matter and a mutual contrariety; for Sweet would never be acted on by Colour, insofar as it is sweet, nor Dry by Hot, nor Moist, insofar as it is moist, by Cold, because these are not

μὴ ἔστι ταῦτα ἀλλήλοις ἐναντία, ἀλλ' ἔστι τὸ ποιεῖν τε καὶ πάσχειν τοῖς ἐναντίοις τε καὶ τοῖς μεταξύ. οὐ γὰρ μόνον τὸ θερμὸν ὑπὸ τοῦ ψυχροῦ καὶ τὸ ψυχρὸν ὑπὸ τοῦ θερμοῦ μεταβάλλεταί τε καὶ πάσχει, οὐδὲ τὸ ὑγρὸν ὑπὸ τοῦ ξηροῦ καὶ τὸ ξηρὸν ὑπὸ τοῦ ὑγροῦ, ἀλλὰ καὶ τὰ μεταξὺ ἑκατέρου τῶν ἄκρων, ὅτι κατὰ μῖξίν ἐστι τῷ πρὸς ἑκάτερον τῶν ἐναντίων [τὸ] μεταξὺ τὸν θατέρου τῶν ἐναντίων σώζειν λόγον. τοῖς γὰρ προειρημένοις συμβέβηκε τὴν αὐτὴν ὕλην ἔχουσιν καὶ ἐναντίωσιν κατὰ τὸ εἶδος ἔχειν πρὸς ἄλληλα. τῆς γὰρ ὕλης ἴδιον τὸ τὴν αὐτὴν εἶναι τῶν ἐναντίων ἐπιδεκτικήν. πάντα γὰρ τὰ ἐν γενέσει τε καὶ φθορᾷ σώματα ἐπὶ τῆς αὐτῆς ἀλλήλοις ἐστὶν ὕλης, ἡ δὲ διαφορὰ αὐτοῖς κατά τε εἴδη καὶ συμβεβηκότα πάθη.

τὰ γοῦν τέσσαρα σώματα, ἃ στοιχεῖα λέγεται, ἔστι ταῦτα· γῆ καὶ ὕδωρ, ἀήρ τε καὶ πῦρ, ἃ ὕλην τὴν αὐτὴν ἀλλήλοις ἔχοντα κατὰ τὰ εἴδη τὴν πρὸς ἄλληλα σώζει διαφοράν. διαφέρουσι γὰρ ἀλλήλων τῷ τὸ μὲν αὐτῶν θερμόν τ' εἶναι καὶ ξηρόν (ἐν γὰρ τούτοις τὸ εἶναι τῷ πυρὶ πυρί· ἐστὶ μὲν γὰρ καὶ ἐν τῇ ὕλῃ αὐτῷ τὸ εἶναι τῇ ταῦτα
230 δεδεγμένῃ, ἀλλ' ἐν τούτοις ὡς πῦρ ἐστὶ καὶ τῶν ἄλλων διαφέρον τι· κατὰ γὰρ τὴν ὕλην τὰ αὐτὰ ἀλλήλοις τὰ τέσσαρα), τὸ δέ τι αὐτῶν ἐστιν ὑγρόν τε καὶ θερμόν (ἐν τούτοις γὰρ πάλιν τὸ εἶναι τῷ ἀέρι), τὸ δὲ ψυχρόν τε καὶ ὑγρόν (τῷ γὰρ ὕδατι ἐν τούτοις τὸ εἶναι), τὸ δὲ ξηρόν τε καὶ ψυχρόν (τοιοῦτον γὰρ ἡ γῆ). ἃς ἔχοντα διαφοράς τε καὶ ἐναντιώσεις πρὸς ἄλληλα, ποιοῦντά τε καὶ πάσχοντα εἰς ἄλληλα ποιεῖ τὰς μεταβολάς, τῆς ὕλης μεταβαλλούσης εἰς τὸ τῶν ποιούντων κρατοῦν. ὅταν γάρ τι αὐτῶν πλεονάσαν τινὸς τῶν ἐναντίωσιν ἐχόντων πρὸς αὐτὸ κατισχύῃ ταῖς ἐναντιώσεσι ταῖς ἐν αὐτῷ ἐπὶ τοσοῦτον, ὡς ἀποβάλλουσαν τὴν ὑποκειμένην ὕλην, ἣν πρότερον εἶχε ποιότητα μεταλαβεῖν τὸ τοῦ κρατήσαντος εἶδος, τότε γίνεται τοῦ μὲν οὕτω μεταβάλλοντος φθορά, γένεσις δὲ <τοῦ> εἰς ὃ γένονεν ἡ μεταβολή. αἱ μὲν οὖν γενέσεις καὶ φθοραὶ τῶν σωμάτων αἱ γινόμεναι κατὰ μεταβολὴν τὴν κατὰ τὸ ποιεῖν τε καὶ πάσχειν τοῦτον γίνονται τὸν τρόπον.

---

229. 21a ante ἐναντίοις lacuna 15 litterarum A: 20 litterarum BC ἔστι τὸ ποιεῖν τε καὶ πάσχειν τοῖς F, Montanari p. 38: <ἔστι τὸ ἀντιπάσχειν μόνον ἐν τοῖς> coni. Br.
229. 25 τὸ seclusi
229. 29 αὐτοῖς F, Montanari p. 38: αὐτῆς codd. Br.
230. 1 ὡς πῦρ ἐστι . . . διαφέρον τι coni. Apelt: ὡς πυρί τε . . . διαφέροντι Br.: ὥσπερ εἴτε . . . διαφέροντι codd.
230. 6 ποιεῖ τὰς Apelt: ποιεῖται codd.
230. 8 πρὸς αὐτὸ (Br.) κατισχύῃ (Apelt): πρὸς αὐτὸν καὶ ἰσχύει codd.: αὐτὸ καὶ ἰσχύον F, Montanari p. 37 v. infra
230. 11 <τοῦ> coni. Br.

29. 21 mutually opposite, but reciprocal interaction occurs among opposites and also their intermediaries. That is, not only is Hot changed and acted on by Cold, and Cold by Hot, and Wet by Dry and vice versa, but also the points between each of the extremes, because in mixture the intermediary with respect to one of the opposites can preserve the character of the other opposite. It follows then that the bodies described here have the same matter and a mutual contrariety in form, as it is a *proprium* of matter that it can remain identical in receiving both of a pair of opposites, since all bodies that come to be and pass away belong to the same matter as one another, while they are differentiated by forms and accidental qualities.

29. 30 Certainly the four bodies known as elements are of this sort; Earth and Water, Air and Fire, have the same matter and are differentiated by their forms. They differ in that one is hot and dry (these are what Fire is, for while its being rests in the matter which has received them [230] yet in these qualities lies its difference as Fire from the other elements too; for in matter the four elements are identical), another is moist and hot (these, again, are what Air is), another cold and moist (these are what Water is) and a fourth is dry and cold (such is Earth). Having these mutual differences and contrarieties they produce changes by reciprocal interaction, with the matter changing into the dominant agent. When one of them overpowers one of the bodies that has a property opposite to its own, and prevails by its inherent contrarieties to the point where the underlying matter loses its former quality in exchange for the form of the dominant body, then the body that changes in this fashion passes away, while the body into which the change has occurred comes to be. In this way bodies that come
30. 13 to be by change involving interaction come to be and pass away.

XIV Ἐπεὶ δὲ ἄλλο γένεσις καὶ φθορά, καὶ ἄλλο κρᾶσις (ἡ γὰρ κρᾶσις σωζομένων τινῶν φθορὰ εἶναι δοκεῖ· οὔτε γὰρ ὑπὸ τοῦ ἑτέρου φθαρέντος καὶ μεταβάλλοντος εἰς τὸ κρατῆσαν κρᾶσις, ἀλλὰ <καὶ> τοῦ μὲν αὔξησις, τοῦ δὲ φθορά — διὸ οὐδὲ ἡ τροφὴ λέγεται κιρνᾶσθαι καὶ τῷ τρεφομένῳ — οὔτ', εἰ ἀμφότερα παντάπασιν φθαρείη τὰ μιγνύμενα, τῶν μεμιγμένων κρᾶσις ἔτι· σωζομένων γάρ τινων φθορὰ ἡ κρᾶσις), ἀκόλουθον ἂν εἴη τὴν τῆς κράσεως πρὸς τὴν γένεσίν τε καὶ φθορὰν εἰπεῖν παραβολήν.

κοινὸν μὲν γὰρ αὐταῖς ἥ τε κατὰ τὴν ὕλην κοινωνία, καθ' ἣν ὥσπερ φθείρεταί τι, οὕτως καὶ κιρνᾶται, καὶ ἡ διὰ ἐναντίωσιν διὰ τοῦ ποιεῖν τε καὶ πάσχειν μεταβολή. ἴδιον δὲ κράσεως πρὸς τὴν γένεσίν τε καὶ φθορὰν τὸ ἐπὶ μὲν τούτων τέλεον θάτερον εἰς θάτερον μεθίστασθαι καὶ μεταλαμβάνειν τὰς ἐναντιώσεις, καθ' ἃς παθὸν ἀπέβαλε τέλεον, ἃς εἶχε πρὸ τούτου, ἐπὶ δὲ τῶν κιρναμένων οὐ τοῦτον ἔτι γίνεται τὸν τρόπον. ἀλλ' ὅταν πλείω σώματα τοῦ ποιεῖν καὶ πάσχειν ὑπ' ἀλλήλων ἔχοντα δύναμιν ἀλλήλοις συνέλθῃ, ἔχῃ δὲ οὕτως, ὡς μὴ ὑπερέχειν κατὰ τὰς ἐναντιώσεις ἄλλο ἄλλου δύνασθαι ὡς φθεῖραν αὐτὸ εἰς τὴν ἑαυτοῦ μεταστῆσαι φύσιν, τότε ταῦτα διὰ τὴν τῶν δυναμένων ἰσότητα, καθ' ἃς ποιεῖ καὶ πάσχει, ὁμοίως ὑπ' ἀλλήλων ἀντιπάσχοντα μέχρι τούτου πρόεισιν, ἕως ἀποβαλόντα κατὰ τὰς ἐναντιώσεις τὰς ὑπεροχάς, δι' ἃς ἑτερά τε ἦν ἀλλήλων καὶ ἐναντία, μίαν ἐξ ἁπασῶν τῶν δυνάμεων γεννήσει ποιότητα, ἑνωθείσης καὶ μιᾶς γενομένης τῆς ἀμφοτέροις ἢ πᾶσιν αὐτοῖς ὕλης ὑποκειμένης.

γίνεται δὲ ἡ τοιαύτη μῖξις, ἣν κρᾶσιν καλοῦμεν, ἐν τοῖς ὑγροῖς τῶν σωμάτων. καὶ γὰρ ἡ ἐν τοῖς ξηροῖς γίνεσθαι δοκοῦσα

230. 15 φθορὰ Montanari p. 39: μῖξις codd. Br. v. infra
230. 16 καὶ supplevi, cf. 213. 8-9, 218. 15-17, 234.1-2.
230. 17 τῷ τρεφομένῳ codd. Rex: τὸ τρεφόμενον Br.
230. 18 τῶν μεμιγμένων scripsi: τὰ μεμιγμένα Apelt Br.: μὴ μιγνύμενα codd.
230. 19 φθορὰ Montanari, v. 230. 15: μῖξις codd. Br.
230. 20 εἰπεῖν παραβολὴν F, Montanari p. 40: λέγειν διαφορὰν post φθορὰν suppl. Br.
230. 27 ἔχῃ Br.: ἔχει codd.
230. 31 κατὰ τὰς ἐναντιώσεις τὰς ὑπεροχάς F, Montanari p. 37: τὰς κατὰ ἐναντιώσεις ὑπεροχάς Apelt Br.: κατὰ τὰς ἐναντιώσεις ὑπεροχάς codd.

230. 14 XIV Since coming-to-be and passing-away differ from blending (for blending is held to be the passing away of bodies that are preserved; for neither is there blending through the destruction of one of the bodies and its change into the dominant constituent but in fact the one increases while the other passes away—hence nutriment is not even said to be blended with the thing that is nourished—nor are the bodies that have been mixed blended if both constituents are completely destroyed, since blending is the passing away of bodies that are preserved) we should next make a comparison between blending and both coming-to-be and passing-away.

230. 20 Now common to both processes is their participation in matter, in respect to which both passing-away and blending similarly occur, as well as alteration caused by interaction through contrariety. Peculiar to blending in relation to coming-to-be and passing-away is that with the latter one thing completely changes into another and takes in exchange the contrarieties by which it was acted on in completely losing its former properties, while with the constituents of blends there is a different process. When several bodies with the potentiality for reciprocal interaction are compounded with one another and are in the state where one cannot exceed the other by its contrariety so as to destroy it and change it into its own nature, then by the equality of the powers by which they interact these bodies are in turn acted on by one another in the same way, and advance to the point where they lose that superiority in their contrarieties by which they differed and were opposite, and create one quality from all the powers, while the matter underlying both or all of them is unified and becomes one.

230. 34 The type of mixture that we term blending involves moist bodies. Certainly the apparent blending that occurs among dry bodies

κρᾶσις οὐ γίνεται χωρὶς ὑγρότητος. ἑνοῦται γὰρ τὰ ὑγρὰ συντιθέμενα καὶ
231 οὐ φυλάσσει τὰς οἰκείας ἐπιφανείας, ἐὰν μὴ ᾖ γλισχρόν. ὅταν δὴ τοῖς κατὰ τὴν ὕλην διὰ τὴν ὑγρότητα κατὰ τὴν παράθεσιν ἑνουμένοις σώμασιν προσῇ καὶ τὸ ἐναντίωσιν ἔχειν ἰσοκρατῆ πρὸς ἄλληλα, γίνεται αὐτοῖς κατὰ τὰς ποιότητάς τε καὶ δυνάμεις ἕνωσις, ὅπερ ἐστὶν ἡ κρᾶσις. ἑκάτερον γὰρ τῶν οὕτως ἐχόντων διὰ μὲν τὴν ἐνυπάρχουσαν ἐν αὐτοῖς ἐναντίωσιν καὶ τὴν τῆς ὕλης κοινωνίαν μεταβάλλει τε καὶ πάσχει, διὰ δὲ τὴν ἰσότητα τῶν ἐναντιώσεων οὐκ ἐπὶ τοσοῦτον ποιεῖ τὴν μεταβολήν, ὡς ταὐτὸν γενέσθαι θατέρῳ θάτερον. ἀντιπάσχον γὰρ τὸ ποιοῦν ὑπὸ τοῦ πάσχοντος πρὸς αὐτοῦ καὶ μὴ μένον ὅμοιον αὐτῷ, παύεται τοῦ ποιεῖν ἔτι δύνασθαι πρὸ τοῦ κρατῆσαι τέλεον αὐτοῦ.

καὶ τοῦτ᾽ ἐστὶν ἡ κρᾶσις· ἡ γὰρ διὰ τοῦ ποιεῖν καὶ πάσχειν τῶν παρακειμένων ἀλλήλοις σωμάτων διὰ μεταβολῆς χωρὶς φθορᾶς αὐτῶν τινος ἕνωσις.

XV

Συνεργεῖ δὲ τοῖς ὑγροῖς πρὸς τὴν θᾶττω μεταβολήν τε καὶ κρᾶσιν καὶ τὸ εὐδιαίρετον. διαιροῦντα γὰρ ἄλληλα πρὸ τῆς ἑνώσεως καὶ κατὰ μικρὰ παρατιθέμενα ἀλλήλοις, ῥᾷον καὶ θᾶττον ἀντιπάσχοντα ὑπ᾽ ἀλλήλων, ταχέως ἕν τι γίνεται σῶμα καὶ κατὰ τὸ ὑποκείμενον καὶ κατὰ τὴν ποιότητα, ἐνεργείᾳ μὲν οὐδὲν ὂν τῶν μεμιγμένων, δυνάμει δὲ πάντα, τοσοῦτον ἀπολειπόμενον τοῦ καὶ ἐνεργείᾳ σώζειν τὰ ἐν αὐτῷ μεμιγμένα, ὅσον διὰ τοῦ ποιεῖν τε καὶ πάσχειν ἀφήρηται διὰ τῆς δυνάμεως αὐτῶν ἑκάστου. διὸ καὶ ὀλίγης δεῖται βοηθείας [τῆς] πρὸς αὖ τὴν εἰς τελειότητα μεταβολὴν ὅσα δι᾽ αὐτῶν τοιαῦτα χωρεῖ μὴ πάντη ἀχώριστα τὰ κεκραμένα. προσθήκης γὰρ δεῖταί τινος εἰς τελειότητα, ἀλλ᾽ οὐ παντελοῦς γενέσεώς τε καὶ μεταβολῆς.

διὸ καὶ προφερομένων αὐτοῖς τινῶν, ἃ πρὸς τὴν τοιαύτην τελειότητα συνεργεῖν ἐστιν οἷά τε, ῥᾳδίως ἐπὶ

---

231. 3 προσῇ καὶ F, Montanari p. 38: προσήκῃ Br.: προσῆκε codd.

231. 17 πάντα Rodier: πᾶν τι Ideler Br.: παντὶ codd. τοῦ Apelt: τὸ codd.

231. 19 τῆς seclusi

231. 20-21 v. infra

231. 20 αὖ τὴν Apelt: αὐτὴν codd., obelo not. Br. τελειότητα scripsi, cf. 231. 21, 232. 29: ταύτην codd. Br. ὅσα δι᾽ αὐτῶν scripsi, cf. ὅσαι τοιαῦται 218. 7: ὅς ἦν γε αὐτῶν codd. Br.

231. 21 πάντῃ ἀχώριστα Apelt: πάντα χωριστὰ codd. Br. τὰ κεκραμένα coni. Br.: τὰ πεπραγμένα codd.

does not occur without moistness, as moist bodies are unified and compounded [231] and do not retain their own surfaces unless they are viscous. Whenever bodies that are materially unified in juxtaposition because of their moistness have in addition a mutual contrariety of equal power the unification of their qualities and powers occurs, and this is blending. Each of the bodies in such a state changes and is acted on because of its inherent contrariety and its participation in matter, and because of the equality of the opposites they do not cause such an extensive change that one becomes identical with the other. The agent, when acted on in turn by the body affected by it, and not remaining like itself, loses its capacity to act before it can completely overpower it.

231. 10 Blending, then, can be defined as the unification through interaction of bodies juxtaposed with one another by means of an alteration that excludes their passing-away.

231. 12 XV Contributing to the rapid alteration and blending of moist bodies is their easy divisibility; for they divide one another before being unified, and are juxtaposed together as corpuscles, thus interacting more easily and more quickly, and they rapidly become one body both in substrate and quality—a body not in actuality any of the bodies that have been mixed, but in potentiality all of them, and deprived from preserving in actuality the bodies mixed in it only to the extent that it is removed from them by the interaction caused by their individual powers. For this reason just slight assistance is required for a change into ‹ actuality › where bodies that have been blended are of the kind that go through one another without being completely inseparable; for they do require some addition to be actualised but not absolute coming-to-be and change.

231. 22 Thus when certain things are added to them which can contribute to such an actualisation they easily recede to their own nature,

Sem. Griekse Letterkunde
Blandijnberg 2
Gent

τὴν οἰκείαν ἀνατρέχει φύσιν, οὐ τοῦ πρότερον ὄντος ὕδατος ἢ οἴνου [τοῦ] ἐκκρινομένου πάλιν (οὐδὲ γὰρ τὴν ἀρχὴν ἐσώζετο ἐν τῷ μίγματι, παράθεσις γὰρ ἄν, οὐ κρᾶσις ἦν· ἐν γὰρ τοῖς κεκραμένοις ἕν τε καὶ ὁμοιομερὲς πᾶν τὸ γεγονὸς ἐκ τῆς κράσεως), ἀλλὰ δυναμένου [ἀπὸ] τοῦ τοιούτου κράματος εἰς ὕδωρ <ἢ οἶνον> [μὲν] ῥᾳδίως μεταβάλλειν <καὶ> μεταβάλλοντος εἰς τὸ τὴν ἀρχὴν μὴ μιχθὲν ἐκ τοιούτων ἐπὶ τοιαύτης ποιότητος.

ὡς γὰρ ἐπὶ τοῦ γάλακτος ὄντος ὁμοιομεροῦς σώματος, ἔχοντος δὲ ἐν αὐτῷ [τι] δυνάμει καὶ ὑγρόν τι καὶ στερεόν, ὁ πεπυρωμένος ἐμβληθεὶς λίθος ἑκάτερον αὐτῶν ἐξ αὐτοῦ διέκρινε γεννήσας τρόπον τινά, καὶ τὸ μὲν αὐτοῦ τυρὸν ἐποίησε, τὸ δὲ ὀρρόν, οὐ μέρος τι ἐνυπάρχον αὐτῷ 232 ἐνεργείᾳ χωρίσας, ἀλλ' ἀπὸ παντὸς μορίου γεννήσας αὐτῶν ἑκάτερον, οὕτως ὑποληπτέον ποιεῖν καὶ τὴν καθιεμένην σπογγίαν εἰς τὸ ἀγγεῖον τὸ ἔχον ἐν αὐτῷ κεκραμένον οἶνον ὕδατι. τῇ γὰρ οἰκείᾳ ποιότητι ἀπὸ παντὸς τοῦ μίγματος γεννῶσα τὸ δυνάμενον ῥᾳδίως ὕδωρ καθαρὸν ὑπ' αὐτοῦ γενέσθαι ὑπὸ τοσαύτης τε καὶ τοιαύτης δυνάμεως χωριζόμενον ἀναφέρει. ὡς γὰρ ἐπὶ τοῦ γλεύκους, ὁμοιομεροῦς ὄντος σώματος, ἡ ὀλίγη πέψις προσελθοῦσα ἐξ ἅπαντος αὐτοῦ γεννᾷ τε καὶ διακρίνει πνεῦμά τε καὶ οἶνον (ὅτι γὰρ μὴ πρότερον ἦν ἐν αὐτῷ ταῦτα ἐνεργείᾳ, δῆλον ἐκ τοῦ μὴ δύνασθαι μὲν οὕτως ὑγρῷ κατέχεσθαι ἐνεργείᾳ πνεῦμα, 9a ὡς εἴρηται καὶ πρὸ ὀλίγου, ἀλλὰ τὸ γενόμενον ἐκκρίνεσθαι παραχρῆμα πᾶν ἐπὶ τῆς πέψεως), οὕτως ὑποληπτέον καὶ τὰ <ἀπο>χωρίζοντα τὰ κιρνάμενα, ἐξ ὧν κέκραταί τινα, οὐκ ἐνυπάρχοντα ἐνεργείᾳ χωρίζειν, ἀλλὰ μεταβάλλοντα δυνάμει τινὶ καὶ τελειοῦντα τὰ δυνάμει διὰ τὴν κρᾶσιν ἐνυπάρχοντα αὐτοῖς. ὡς γὰρ ἐπ' ἐκείνων ἐπιπόλαιός τινος γένεσις γίνεται (οὐ γὰρ τοῦ γάλακτος εἰς ἐναντίας τινάς, αἷς ἔχει, μεταβαλόντος ποιότητας τὸ μὲν ὀρρὸς τὸ δὲ τυρὸς γίνεται, ἀλλὰ τὰ ὑπάρχοντα αὐτῷ δυνάμει οὕτως, ὡς ὀλίγης τινὸς πρὸς τὸ ἐνεργείᾳ εἶναι δεῖσθαι βοηθείας, εἰς τὸ ἐνεργείᾳ εἶναι γίνεται), οὕτως μεταβολὴν ἡγητέον γίνεσθαι καὶ ἐπὶ τῶν κεκραμένων.

---

231. 24 τοῦ secl. Br.
231. 27 ἀπὸ secl. Br.
231. 28-29 ἢ οἶνον supplevi, cf. 231. 24   μὲν seclusi   καὶ supplevi   μεταβάλλοντος [sc. τοῦ κράματος] εἰς τὸ τὴν ἀρχὴν (cf. 232. 19) μὴ μιχθὲν Diels: μεταβαλόντος εἰς τοῦτο ἀρχὴν codd., obelo not. Br.
231. 31 τι secl. Br., om. F, Montanari p. 36
232. 5 χωριζόμενον coni. Apelt: χωρίζον codd. Br.
232. 7 γεννᾷ τε F, Montanari p. 36, Apelt Br.: γεννᾶται codd.
232. 10 τὰ <ἀπο>χωρίζοντα τὰ κιρνάμενα scripsi: τὰ χωρίζοντα ἀπὸ (ἐπὶ coni. Br.) τῶν κεκραμένων codd. Br.
232. 15 τὰ ὑπάρχοντα Schramm: τῶν ὑπαρχόντων codd. Br.

not because the original water and wine is separated out again (they were not, that is, preserved in their original state in the mixture, since that would just be juxtaposition and not blending; for where bodies have been blended the whole product of the blend is one and uniform) but through such a blend being easily able to change into water < or wine >, < and > changing into what was not originally mixed from such [constituents] with such a quality.

31. 30 For just as a heated stone cast into milk, a uniform body containing in potentiality something both moist and solid, separates each of them from it, and in some way creates them, making the one into cheese, the other into whey, not through separating a part actually inherent in the milk but by creating each of them [232] from every part, so must the action of the sponge dipped into the pitcher holding wine blended with water be understood; for by its own quality it creates from the whole mixture the water that can be easily purified by it, and reconstitutes it as separate through the extent and nature of its power.

232. 6 For as the onset of a slight fermentation in must, which is a uniform body, creates and separates from the whole of it both air and wine (clearly these bodies were not in it in actuality beforehand, since it is impossible that air should actually be contained by water, as we said just above, but the product is separated out as a whole just when the fermentation takes place); so must it be understood that agents which separate constituents from which blends have been formed also do not separate what is actually inherent in [blends] but cause an alteration by a specific force, and actualise bodies that as a result of blending are present in them in potentiality.

32. 13 Just as in the preceding cases there is a superficial coming-to-be of something (for it is not through the milk changing into qualities opposite to its own that it becomes one part whey one part cheese, but rather the bodies present in potentiality in it in such a way that they require slight assistance to reach actual being, come into actual being), so must the change that also occurs in the case of bodies that have been blended be understood.

ἕπεται δὲ τῇ ἐπὶ τῶν κεκραμένων τοιαύτῃ μεταβολῇ [ἣ] τὸ διακρίνεσθαι δύνασθαι δοκεῖν αὐτά, ὅτι καὶ τὴν ἀρχὴν ἐκ μίξεως τοιούτων ἡ γένεσις τῷ ταῦτα πάσχοντι σώματι. καὶ ἔστιν ὁ χωρισμὸς λεγόμενος τῶν κεκραμένων οὔτε ὅμοιος τῷ ἐπὶ τῶν ἀλλήλοις παρακειμένων, οὔτ' αὖ τῷ πάλιν τῶν κατὰ φθορὰν καὶ γένεσιν καὶ τὴν εἰς τοὐναντίον μεταβολὴν ἀποκρινομένων, ὡς ἐξ ὕδατος ὁρῶμεν εἰς ἀέρα γινομένην ἀπόκρισιν, ἀλλ' ἔστι τὸ γινόμενον μεταξὺ τούτων. οὔτε γὰρ ἐνεργείᾳ ἐνυπάρχοντα διακρίνεται οὔτε εἰς τοὐναντίον τῷ ὑποκειμένῳ μεταβάλλοντα ἀποκρίνεται. ἐπὶ μὲν γὰρ τούτων ταὐτὸν κατὰ τὸ εἶδος μένει μετὰ τὴν ἀπόκρισιν τὸ ὑπομένον τῷ πρὸ τοῦ, μόνον κατὰ ποσὸν ἐλαττούμενον, ἐπὶ δὲ τῶν κεκραμένων οὐχ ὁμοίως, τῷ ἕκαστον τῶν ὄντων δυνάμει ἐν τῷ ἐκ τοῦ κράματος γεγονότι σώματι ἐκκρίνασθαι, μεταβάλλον εἰς τὴν τελειότητα, ἧς ἀφῃρέθη διὰ τοῦ ἐπ' ἴσης αὐτὰ ἀντιπαθεῖν ὑπ' ἀλλήλων, δι' ὃ πάθος καὶ ἐπὶ τούτων τὸ χωρίζεσθαι κατηγορεῖται.

εἰ οὖν τά τε διὰ τῆς τοιαύτης μεταβολῆς τελειούμενα οὐ γενέσθαι, ἀλλὰ χωρίζεσθαι λέγειν εὔλογον, τά τε ἄλλα περὶ τῆς γενέσεως τῆς κράσεως εἰρημένα ἀκόλουθά τε ταῖς ὑποκειμέναις ἀρχαῖς καὶ συνῳδὰ ταῖς μεταβολαῖς τε καὶ γενέσεσι τῶν σωμά-
233 των καὶ σώζοντα τὰς περὶ τῆς κράσεως κοινὰς προλήψεις, μόνος Ἀριστοτέλης εἴη ἂν τὸν ἀληθῆ περὶ τῆς κράσεως ἀποδεδωκὼς λόγον.

ὅτι δὲ <τὰ> κιρνάμενα διαιροῦντά τε πρῶτον ἄλληλα καὶ τῇ κατὰ μικρὰ παραθέσει εὐπαθέστερα γινόμενα ἑνοῦταί τε καὶ ὁμοιοῦται κατὰ τὸ εἶδος καὶ τὰς ὁμοιότητας, καὶ κατὰ τὴν αἴσθησίν ἐστι γνώριμον. ἕως γὰρ ἔτι μένει τῶν κιρναμένων ἑκάτερον ἐπὶ τῆς οἰκείας οὐσίας, ἡ ὑπ' ἀλλήλων γινομένη διάκρισις αὐτῶν καὶ ἀπὸ τῆς τῶν χρωμάτων διαφορᾶς γίνεται γνώριμος· ὑγρὰ γὰρ ὄντα καὶ εὐδιαίρετα καὶ ἀνόμοια ὑπὸ τῆς πληγῆς τῆς κατὰ τὴν ἐπίχυσιν θατέρῳ θάτερον ἀρχὴν λαβόντα τοῦ διαιρεῖν ἄλληλα, ὄντα καὶ ἀνισόρροπα, εἰ οὕτως τύχοι, τὸ μὲν βαρύτερον αὐτῶν διαιροῦν φέρεται κάτω, τοῦ δὲ κουφοτέρου τὰ μόρια ὑποχωροῦντα τοῖς βαρυτέροις ἐπιπολῆς αὐτῶν σπεύδει γενέσθαι. ἣν τοπικὴν μετάβασιν αὐτῶν ἵστησι φθάνουσαν ἡ κατὰ πάθη μεταβολή, ἑνοῦσα αὐτὰ τῇ κατὰ πάντα ὁμοιότητι, ὃ καὶ αὐτὸ τῇ ὄψει γίνεται γνώριμον.

---

232. 18 ἣ secl. Apelt Br., om. F, Montanari p. 36
233. 3 τὰ suppl. Br. διαιροῦντα Schwartz: διαιρεῖταί codd.
233. 4 εὐπαθέστερα F, Montanari p. 37, Apelt Br.: ἀπαθέστερα codd., cf. 224. 2
233. 8 γὰρ F, Montanari p. 37, coni. Br.

32. 18 Such an alteration in these bodies entails that they be held capable of dissociation, because in fact the body that was affected by them originally came to be from their mixture. And what is termed the separation of blended bodies neither resembles that involving bodies juxtaposed together, nor again that involving bodies dissociated in passing-away and coming-to-be and the change into an opposite (as we see when Air is separated from Water), but the process lies between these. For neither are bodies present in actuality dissociated, nor by changing to the opposite of the substrate are they separated. In these cases the residue after the dissociation stays the same in form as before and is only decreased in quantity, but with the bodies that have been blended the difference is that each of the things in potentiality in the body produced from the blend is separated out, changing into the actuality of which it was deprived through the process of balanced reciprocal interaction; and because of this characteristic "being separated" is also predicated of these bodies.

32. 31 So if it is reasonable to describe the bodies actualised by such a change not as coming to be but as being separated, and if the other statements made about the process of blending follow our basic principles, are consonant with how bodies change and come to be, [233] and preserve the common preconceptions about blending, only Aristotle will have propounded the true theory of blending.

233. 2 Also it is known by perception that the constituents first divide one another, and by their juxtapositon as corpuscles become malleable and are unified and assimilated in form and likeness. For as long as each constituent still remains with its own substance, their subsequent dissociation can also be recognised by their difference in colour; for as moist, easily-divided, and dissimilar bodies their division of one another starts from the pressure caused by one overflowing on the other, and if they should also happen to be unequally balanced the heavier moves downward in dividing, while the parts of the lighter body yield to the heavier one and readily cover their surface. The change in qualities which unifies them in total similarity stops the preceding change of place—a fact also known by sight.

XVI Ἐπεὶ δὲ πρὸς κατασκευὴν τοῦ σῶμα διὰ σώματος χωρεῖν χρῶνται καὶ ταῖς τῶν ζῴων διὰ τῆς τροφῆς αὐξήσεσιν (ἡ γὰρ αὔξησις τῇ προσθήκῃ τῆς τροφῆς <γίνεται ὥστε εἰ> πάντῃ καὶ αὔξεται τὰ αὐξανόμενα, πάντῃ ἂν ἡ τροφὴ προσκρίνοιτο τῷ προϋπάρχοντι σώματι· οὐχ οἷόν τε δὲ πάντῃ προσκριθῆναι τὴν μὴ διὰ παντός τε καὶ πρὸς πᾶν ἐνεχθεῖσαν· ἀλλ' εἰ διὰ παντὸς ἡ τροφὴ τοῦ σώματος φέρεται, σῶμα οὖσα, ἀναγκαῖον εἶναι δόξει, σῶμα διὰ σώματος χωρεῖν, εἴ γε μόνως οὕτως οἷόν τε τὴν ἐκ τῆς τροφῆς προσθήκην παντὶ τῷ τρεφομένῳ γίνεσθαι σώματι), ὅτι οὖν μὴ οὕτως τὸ τρέφεσθαί τε καὶ αὔξεσθαι γίνεται (οὐδὲν γὰρ τῶν ἀδυνάτων γίνεται) φέρ' εἴπω ἐκθέμενος τὴν Ἀριστοτέλους δόξαν τὴν περὶ αὐξήσεώς τε καὶ τροφῆς.

ποιήσομαι δὲ τὸν λόγον ἐπὶ τῶν κυρίως αὔξεσθαι λεγομένων, ἐπὶ τούτων γὰρ καὶ ἡ ἀπορία πρόεισιν· ἔστι δὲ ταῦτα τὰ διὰ τοῦ τρέφεσθαι τὴν προσθήκην λαμβάνοντα. τρέφεται δὲ ὅσα τὴν θρεπτικὴν ἔχει δύναμιν ἐν αὑτοῖς, τοιαῦτα δὲ τὰ φυτὰ καὶ τὰ ζῷα. ὅσα δὴ τοῖς οὕτως αὐξανομένοις ὑπάρχει καὶ ὅσα δεῖ φυλάττειν τοὺς περὶ αὐξήσεως λέγοντας, ἀνάγκη ταῦτα πρῶτον ἐκθέσθαι· ταῦτα γὰρ ἡμῖν ὄντα γνώριμα συντελέσει πρὸς τὴν τῶν ῥηθησομένων κρίσιν.

τὰ δὴ αὐξανόμενα πάντα προσιόντος γέ τινος αὐτοῖς ἔξωθεν αὔξεται, εἴ γε διὰ τοῦ τρέφεσθαι τὸ αὔξεσθαι γίνεται, ἔτι τε <οὐ> κατὰ μόριον τὴν ἐπίδοσιν λαμβάνει. οὐ γὰρ <καθ'> ἕν τι ἡ αὔξησις τοῖς αὐξανομένοις κατὰ φύσιν, ὥσπερ οἷς ἔξωθέν τι προσ-
234 πλάσσεται (οὐ γὰρ τῶν αὐξομένων τινὸς ὁ πούς αὔξεται μόνος), ἀλλὰ καὶ πάντων τῶν μορίων ἐπίδοσις κατ' ἀναλογίαν γίνεται. καὶ γὰρ ταὐτὸν ὑπομένον τε καὶ σῷζον τὴν οἰκείαν φύσιν τὸ αὐξόμενον τὴν προσθήκην λαμβάνει. οὔτε γὰρ μὴ ὑπομένοντος τοῦ αὔξεσθαι λεγομένου αὔξησις τὸ

233. 15 ἡ codd. Von Arnim: εἰ Br.
233. 16 αὔξησις secl. Br. <γίνεται ὥστε εἰ> Von Arnim
233. 17 προϋπάρχοντι A[2]Ra: ὑπάρχοντι A[1]BPSC Br. v. infra
233. 18 μὴ Br.: μίαν codd.
233. 23 φερ' εἴπω Schramm: φέρε καὶ ὧν codd.: καὶ ὧν obelo not. Br. qui φερ' εἴπωμεν coni.
233. 28 ὑπάρχει Br.: ὑπάρχειν codd. δεῖ Ideler: δὴ codd.
233. 32 οὐ suppl. Br.; fortasse <οὐ> κατὰ μόριον <τι> <καθ'> coni. Br., cf. κατὰ πᾶν 234. 26
234. 2 ταὐτὸν Apelt: τρίτον codd.

33. 14 XVI As they support the notion of body going through body by also invoking the growth of animals through nutriment (for growth does ‹ occur › by the addition of nutriment in every part, ‹ so that if › things that grow do grow in all parts, nutriment would be assimilated by the preexisting body in all its parts; but it is impossible for the implanted [nutriment] to be totally assimilated except throughout the whole body and in every part; yet if the nutriment pervades the whole body as a body, it will be held necessary that body go through body if only in this way can growth through nutriment occur throughout the whole of the body that is nourished), yet because nourishment and growth do not occur in this way (for nothing impossible happens), let me then state and explain Aristotle's doctrine of growth and nutriment.

33. 24a I shall devote the discussion to those things that are strictly speaking said to grow, for with them we shall certainly make progress with our problem. Now these are bodies that grow through being nourished, and only bodies with their own nutritive faculty are nourished, namely plants and animals. First, then, we must describe the character of things that grow in this way, and what must be retained by a theory of growth; for since these [facts] are known to us they will contribute to our assessment of the theory to be forwarded.

33. 30 Everything that grows grows only when something enters it from outside, if indeed it grows through being nourished, and furthermore is ‹ not › added to in part. For growth is not in an individual [feature] for bodies that grow by nature, as it is with things to which something is affixed externally [234] (when something grows its foot will not grow by itself!), but there is in fact proportional addition to all parts. Now the body that grows remains the same and preserves its own nature when added to; for if what is said to grow is not stable the result will not be growth ‹ even

γινόμενον <εἰ καὶ> εἰς ὄγκον μεταβάλλει μείζονα (διὰ τοῦτο γὰρ οὔτε τῶν κιρναμένων τι αὔξεσθαι λέγεται <οὔτε τῶν> μεταβαλλόντων εἰς ἄλλο τι σῶμα ὂν ἐπ' ὄγκου μείζονος, ὡς ὁρῶμεν γινόμενον ἐπὶ τῆς ἐξ ὕδατος εἰς ἀέρα μεταβολῆς, ἀλλ' ἐνταῦθα μὲν γένεσις καὶ φθορά, ὕδατος μὲν φθορά, γένεσις δὲ ἀέρος, ἐκεῖ δὲ μῖξίς τε καὶ κρᾶσις), οὔτε τὸ τυχὸν τῶν εἰς τὴν αὔξησιν συντελούντων αὔξεται, ἀλλὰ τὸ σωζόμενόν τε αὐτῶν καὶ ὑπομένον, καὶ ἔτι τὴν αὐξητικήν τε καὶ θρεπτικὴν ἔχον δύναμιν ἐν αὐτῷ. διὰ ταῦτα γὰρ οὐχ ἡ τροφὴ αὐξανόμενον, καίτοι καὶ αὐτὴ σῶμα καὶ γινομένη τῆς αὐξήσεως αἰτία, ἀλλ' ᾧ ἡ τροφὴ προσκρίνεται. ἡ μὲν γὰρ μεταβάλλει, τὸ δὲ σώζεται· καὶ ἡ θρεπτικὴ δύναμις, ἡ μεταβλητικὴ δὴ καὶ προσκριτικὴ τῆς τροφῆς, ἐστὶν ἐν τούτῳ.

<τούτων> τοίνυν τοῖς αὐξανομένοις ὑπαρχόντων ζητεῖται, ἐπεὶ ἡ αὔξησις προστιθεμένῳ γίνεται, τί ἐστι <τὸ> προστιθέμενον; ἢ γὰρ σῶμα ἢ ἀσώματον εἶναι δεῖ τὸ προστιθέμενον. οὔτε δὲ ἀσωμάτῳ τὴν αὔξησιν εἶναι λέγειν εὔλογον, οὔτε σώματι εἶναι δοκεῖ. ἀσωμάτῳ μὲν <οὐ>, ὅτι οὐκ ἀσώματος ἡ τροφή, ἀκολουθήσει τε τοῖς κατὰ ἀσωμάτου τινὸς προσθήκην τὴν αὔξησιν εἶναι λέγουσι κενόν τι ποιεῖν ἀφωρισμένον, εἴ γε τὸ μὲν αὐξηθὲν μείζονα κατέχει τόπον· οὗ κατεῖχε δὲ τόπον τὸ αὐξῆσαν αὐτό· ὃν γὰρ ἐπιλαμβάνει τὸ αὐξανόμενον, οὗτός ἐστι πρὸ τῆς αὐξήσεως αὐτοῦ κενός, εἴ γε μόνων σωμάτων τὸ κατέχειν τινὰ τόπον. εἰ δὲ σώματι λέγοιτο ἡ αὔξησις γίνεσθαι προστιθεμένῳ, ἔσται σῶμά τε διὰ σώματος χωροῦν καὶ γινόμενον <ἐν> σώματι, εἴ γε καθὸ μὲν αὔξεταί τι, κατὰ τοῦτο τὴν προσθήκην λαμβάνει, κατὰ πάντα δὲ τὸ αὐξόμενον αὔξεται, ὥστε κατὰ πᾶν καὶ προσθήκην λήψεται. ἀλλ' εἰ σῶμα ὂν κατὰ πᾶν ἑαυτὸ τὴν προσθήκην λαμβάνει, δεήσει διὰ παντὸς αὐτὸ χωροῦν τὸ προστιθέμενον αὐτῷ (τοῦτο δ' ἐστὶν ἡ τροφή) τὴν αὔξησιν αὐτοῦ ποιεῖσθαι. εἰ γὰρ λέγει τις διά τινων κενῶν γενέσθαι τὴν δίο-

---

234. 5 <εἰ καὶ> coni. Br. qui lacunam posuit μεταβάλλει coni. Br.: μεταβάλλειν codd.
234. 6 <οὔτε τῶν> coni. Apelt Rodier
234. 7 ἐπὶ coni. Br.: ἐκ codd.
234. 10 ὑπομένον Br., cf. 236. 1, 5: ὑποκείμενον codd.
234. 14 δὴ Br.: δὲ codd.
234. 15 τούτων suppl. Br.
234. 16 προστιθεμένῳ Br.: προστιθεμένου codd. τὸ suppl. Diels
234. 19 <οὐ> coni. Apelt
234. 20 λέγουσι κενόν τι ποιεῖν ἀφωρισμένον (-ων Montanari) εἴ γε τὸ μὲν αὐξηθὲν μείζονα κατέχει τόπον F, Montanari p. 39, cf. Aristot. *de gen. et corr.* 321a 6: λέγουσι κατέχειν τόπον codd., obelo not. Br.: λέγουσι <τὸ αὐξῆσαν αὐτὸ μὴ λέγειν> κατέχειν τόπον· [οὗ] coni. Br.
234. 24 δὲ Apelt: τε codd. ἐν supplevi, cf. 219. 23, 24, 27: γινόμενον σώματι codd., obelo not. Br.: δεχόμενον σῶμά τι coni. Br.: προσγινόμενον σώματι Schramm

if it › changes into a larger volume (hence none of the constituents of a blend is said to grow, nor anything that changes into a body of greater volume, as we see in the case of the change from Water to Air; in the latter instance there is coming-to-be and passing-away of Air and Water respectively, in the former mixture and blending), nor does *any* of the [bodies] contributing to growth grow, but only that which is preserved and stable and also has its own faculty for growth and nourishment. Hence nutriment is not increased, though it is itself body and acts as the cause of growth, but rather that by which the nutriment is assimilated; for this is preserved while the nutriment changes, and it has the nutritive faculty which is able to transform and assimilate nutriment.

34. 15 Since this is the character of bodies that grow we must determine, since growth occurs through an addition being made to a body, what it is that is added. The addition must be corporeal or incorporeal, but it does not seem reasonable to say that there is growth either by the incorporeal or by body.

34. 19 It is not by the incorporeal because nutriment is not incorporeal, and because to say that growth is an addition of something incorporeal will entail claiming that it creates a separate void, if indeed what has grown occupies a larger place; but the cause of its growth does not occupy place, for the place occupied by the body that grows is empty prior to its growth, if the occupancy of place is a unique property of bodies.

34. 23 But if growth were said to occur by the addition of body, then there will be body going through body and coming to be ‹ in › body, if something grows insofar as it is added to, and if the body that grows does so in all its parts so that it is also added to everywhere. But if, being itself a body, it is added to in its entirety, then the thing added to it (the nutriment, that is) goes through the whole of it and causes its increase. For to someone who says that the nutriment passes through certain empty spaces we must

δον τῆς τροφῆς, ἀνάγκη τούτῳ λέγειν, πᾶν τὸ τρεφόμενον σῶμα εἶναι κενόν, εἰ καθ' ὃ μὲν ἡ πάροδος τῆς τροφῆς, κενὸν κατὰ τοῦτο, κατὰ πάντα δ' αὐτοῦ ἡ πάροδος τῆς τροφῆς, εἰ δὴ κατὰ πάντα αὔξεται.

τῷ δὴ μέλλοντι φυλάττειν τε τὰ τοῖς αὐξανομένοις ὑπάρχοντα καὶ λύσειν τὰ ἀπορούμενα ἀναγκαῖον πρῶτον, τί ποτέ ἐστι τὸ αὐξόμενον λαβεῖν, καὶ κατὰ 235 ποῖα μόρια τῶν αὐξομένων τῆς προσθήκης γινομένης τὰ ὅλα αὔξεται. ἐπεὶ τοίνυν ἐν τοῖς αὐξομένοις ἐστὶ τῶν μερῶν τὰ μὲν ἀνομοιομερῆ, τὰ δὲ ὁμοιομερῆ (καὶ ἀνομοιομερῆ μὲν τὰ ἐκ διαφερόντων μερῶν συνεστῶτα, ὡς πρόσωπον καὶ χείρ, ὁμοιομερῆ δὲ σάρξ τε καὶ ὀστᾶ, μῦς καὶ αἷμα καὶ φλέψ, καὶ ὅλως ὧν τὰ μόρια τοῖς ὅλοις ἐστὶ συνώνυμα), καὶ σύγκειται ἐκ τῶν ὁμοιομερῶν τὰ ἀνομοιομερῆ, ἀλλ' οὐκ ἀνάπαλιν (οὐ γὰρ ἐκ προσώπου ἡ σάρξ, ἀλλ' ἐκ σαρκῶν τε καὶ ὀστῶν τὸ πρόσωπον), ἡ αὔξησις κατὰ τὴν τοῖς ὁμοιομερέσι προσθήκην γίνεται. τούτοις γὰρ αὐξομένοις ἕπεται καὶ ἡ τῶν ἀνομοιομερῶν ἐπίδοσις. οὐ γὰρ ἡ προσκρινομένη τροφή, πρόσωπον ἢ χεὶρ γινομένη, προσκρίνεται, ἀλλ' εἰς σάρκα καὶ ὀστοῦν καὶ τῶν ἄλλων ὁμοιομερῶν ἕκαστον μεταβάλλουσά τε καὶ προσκρινομένη τούτοις τῆς τοῦ παντὸς αὐξήσεως σώματος αἰτία γίνεται. εἰ τοίνυν ἡ μὲν αὔξησις τοῦ αὐξανομένου σωζομένου τε καὶ μένοντος, αὔξεται δὲ τὰ ὁμοιομερῆ, δῆλον ὅτι ταῦτα σώζεσθαί τε καὶ μένειν δεῖ.

ἀκόλουθον τοίνυν τούτοις ἰδεῖν, πῶς τις λέγων ὑπομένειν ταῦτα ἀληθεύσει. καθὸ γὰρ ὑπομένειν οἷόν τε ταῦτα, κατὰ τοῦτ' αὐτῶν καὶ τὴν αὔξησιν ἀναγκαῖον εἶναι λέγειν. πᾶν δὴ τὸ ἐξ ὕλης τε καὶ εἴδους συγκείμενον καθ' ἑκάτερον τῶν ἐξ ὧν ἐστιν εἶναι λέγεται· καὶ γὰρ κατὰ τὴν ὕλην καὶ κατὰ τὸ εἶδος. ἐκ τούτων δὲ καὶ τὰ ὁμοιομερῆ, καθ' ἃ λέγομεν τὴν τῶν ζῴων αὔξησιν γίνεσθαι. διὸ καὶ ταῦτ' ἔχει τὸ διττόν. καὶ λέγεται ἕκαστον αὐτῶν ποτὲ μὲν κατὰ τὴν ὕλην, ποτὲ δὲ κατὰ τὸ εἶδος. ὅταν μὲν γὰρ λέγωμεν τὴν σάρκα ῥεῖν καὶ εἶναι ἐν συνεχεῖ ἀποκρίσει τε καὶ προσκρίσει, κατὰ τὴν ὕλην τὴν σάρκα ταῦτα πάσχειν λέγομεν, ὅταν δ' αὖ πάλιν λέγωμεν τὴν σάρκα τὴν αὐτὴν μένειν, αὐτὸ τὸ εἶδος καὶ τὴν κατὰ τὸ εἶδος σάρκα λαβόντες ταῦτα αὐτῇ μαρτυροῦμεν. πάντα γὰρ τὰ ἐν ὕλῃ τὸ εἶναι ἔχοντα κατὰ μὲν τὴν ὕλην ἄλλοτε ἄλλα γίνεται, διὰ τὸ μὴ μένειν ταύτην κατὰ τὸν ἀριθμὸν μίαν, συνεχῶς κατὰ τὰ μόρια μεταβάλλουσάν τε καὶ

---

234. 30 τούτῳ Br.: τοῦτο codd.
235. 6-7 ἐκ προσώπου Rex: εἰς πρόσωπον Br.: ἐκ προσώπων codd.: ἔστι πρόσωπον sive ἐκ προσώπου Apelt
235. 14 δεῖ Diels: διὰ codd.
235. 24 αὐτὸ τὸ εἶδος Ra Rodier: ἀπὸ τοῦ εἴδους cett. Br.

respond that the whole body that is nourished is void, if the passage of the nutriment occurs only where it is void and if this passage is over all [parts], given that it grows in all [parts].

34. 32 If we would retain the character of bodies that grow and also solve these difficulties, we must first consider the nature of the body that grows, [235] and in what parts of growing bodies whole units
35. 1 grow when an addition is made. Since some of the parts of bodies that grow are non-uniform, like face and hand, while flesh and bones are uniform, and marrow, blood, and vein, and in general things whose parts bear the same name as the whole), and the non-uniform parts are composed from the uniform, while the converse is not true (for flesh is not composed of the face, but the face of flesh and bones) then growth occurs by an addition to the uniform parts. For when these grow, an addition to the non-uniform parts also follows, since the nutriment that is assimilated is not assimilated by becoming face or hand but by changing into flesh and bone and each of the uniform parts, and through being assimilated by them causes the growth of the whole body. If, then, growth is a characteristic of a body that is preserved and stable as it grows, and if it is the uniform parts that grow, clearly they must be preserved and stable.

35. 14 We must next see how it could be true to describe them as stable when it must be granted that they can only grow insofar as they can be stable. Now the whole compound of form and matter is said to be with respect to each of the things of which it is composed—matter and form, that is. These are also the components of the uniform parts by which we say that growth occurs in animals, so that also have this dual aspect. Now each uniform part is described variously by its matter and by its form. When we describe flesh as fluid and in a state of continual dispersal and assimilation, we say that the flesh is affected in this way with respect to its matter. Again when we describe flesh as remaining the same, we base this attribution on the form itself and the flesh *qua* form. Indeed every-

γινομένην ἄλλοτε ἄλλην, κατὰ δὲ τὸ εἶδος ἐν αὐτῷ κατ' ἀριθμὸν ἕκαστον μένει. κατὰ γὰρ τὸ εἶδος ἡ σὰρξ ἡ αὐτὴ κατ' ἀριθμὸν μένει. κἂν τὸ μὲν ἀπορρέῃ τῆς ὑποβεβλημένης ὕλης αὐτῇ, τὸ δὲ ἐπιρρέῃ, ἔστ' ἄν τι μένῃ αὐτῆς ἐν αὐτῷ, φυλάσσει τὸ τῆς σαρκὸς εἶδος, τῇ κατὰ διάδοσιν μονῇ κωλῦον αὐτοῦ τὴν παντελῆ φθοράν. οὐ γὰρ τὸ εἶναι τῇ σαρκὶ ἐν τῷ τόσῳδε μεγέθει, ὃ οὐ ταὐτὸ μένει διὰ τὴν τῆς ὕλης ῥύσιν, ἀλλ' ἐν τῷ εἴδει τῷ τοιῷδε, ὃ ταὐτὸν μένει, ἔστ' ἂν σώζηταί τι τῆς σαρκός.

οὕτως τοίνυν
236 τούτων ἐχόντων, κειμένου δὲ καὶ τοῦ τὸ αὐξανόμενον δεῖν ὑπομένειν, ἂν μὲν κατὰ τὴν ὕλην τις λέγῃ γίνεσθαι τὴν τῆς σαρκὸς αὔξησιν, ἀναιρήσει τὸ τὸ αὐξόμενον τὸ αὐτὸ σωζόμενον κατ' ἀριθμὸν αὔξεσθαι (αὕτη γὰρ ἄλλοτε ἄλλη), εἰ δέ τις κατὰ τὸ εἶδος λέγοι τὴν αὔξησιν γίνεσθαι, οὐκέτι ἂν ὁ λόγος ἐμποδίζοιτο. τοῦτο γὰρ ὑπομένει, τοῦ δὲ ὑπομένοντος ἦν.

κειμένου δὲ τοῦ τὰ αὐξανόμενα κατὰ τὸ εἶδος αὔξεσθαι, ἐπὶ τούτοις ἄξιον ἰδεῖν, πῶς τῆς τοῦ αὔξοντος αὐτὰ προσκρίσεως αὐτοῖς γινομένης αὔξεται. εἰ μὲν οὖν κατὰ τὴν ὕλην ἡ αὔξησίς τε καὶ ἡ πρόσκρισις ἐγίνετο, ἀναγκαῖον ἂν ἦν λέγειν τῷ κατὰ πᾶν μόριον αὔξεσθαι τὰ αὐξανόμενα λέγοντι, κατὰ πᾶσαν τὴν ὕλην γενέσθαι τὴν προσθήκην τὴν ἀπὸ τῆς τροφῆς, ἥτις οὐκ ἄλλως ἐδύνατο γίνεσθαι (τῆς γε ὕλης τῆς αὐτῆς ὑπομενούσης), ‹ἢ› σώματος χωροῦντος διὰ σώματος, εἴ γε σῶμα ἡ τροφή· ἐπεὶ δὲ κατὰ τὸ εἶδος ἡ αὔξησις, οὐκέτι ἀναγκαῖον, εἰ καὶ κατὰ πᾶν τὸ εἶδος αὔξεται, τοῦτο γίνεσθαι σώματος χωροῦντος διὰ σώματος. οὐδὲ γὰρ τὴν ἀρχὴν προϋπάρχον ὑπομένει, ὥστε γίνεσθαι τῷ αὔξοντι σώματι δι' αὐτοῦ τὴν ὁδόν, ἀλλὰ γίνεται μὲν ἡ τροφὴ τῆς σαρκὸς ὕλη, προσκρινομένη δὲ κατὰ μέρη τινὰ τῇ τῆς προϋπαρχούσης ἔτι, μενούσης τε καὶ τὸ εἶδος φερούσης, εἰ μὲν εἴη ἐλάττων ἢ ἴση τῆς ἀπορρεούσης, σώζει [οὐ] μόνον τό,

---

235. 30 ἔστ' ἄν τι μένῃ coni. Br., cf. 235. 34, Alex. *Quaest.* 13. 32, Aristot. *de gen. et corr.* 322a 24: ἔστι μένον codd., obelo not. Br.
235. 32 αὐτοῦ Apelt: αὐτὸ codd.
235. 32-33 ἐν τῷ τοσῷδε F, Montanari p. 37, Apelt Br. Rodier: ἐντὸς τόσῳ δὲ codd.
235. 33 ὃ οὐ Apelt: οὐ codd.
235. 34 ὃ Apelt Rodier: οὐ codd.
236. 9 τῷ Apelt: τὸ codd.
236. 12 ἢ suppl. Ideler
236. 13 εἰ καὶ Ideler: εἶναι codd.
236. 17 τε Rex: δὲ codd. Br.
236. 18 σώζει Ideler: σώζειν codd. οὐ om. F, Montanari p. 37, Rex, cf. μόνον σώζει, 236. 25

thing that has its being in matter undergoes various changes in its matter because it does not remain one in number, but continually changes its parts and assumes varying states, while in form everything is stable in number; for as regards the form flesh remains the same in number, and if one part of its supporting matter disperses while another accedes then, as long as some matter remains in it, it preserves the form of the flesh and prevents its total destruction by the stability of its distribution. For the being of flesh does not lie in the extent of its magnitude which does not remain the same on account of the fluidity of the matter, but in the sort of form it has, which remains the same as long as any flesh is preserved.

35. 34 If this is so, [236] and if it is established that the body that grows must be stable, then to say that the growth of flesh occurred with respect to the matter would abolish the principle that what grows will grow while remaining the same in number (for matter is different at different times); but our theory would no longer be hindered by the claim that growth occurred with respect to the form, for form is stable and [growth], as we saw, indeed involves that which is stable.

36. 6 Since it has been established that things that grow grow in their form it is worth seeing how in this case they grow through assimilating the body that causes their growth. Now if growth and assimilation occurred with respect to matter one would have to say to the view that things that grow grow in every part, that addition from nutriment occurs in the whole of matter, and at least while the matter remains stable, it could only occur by body going through body, if nutriment is body. But since growth occurs with respect to form it cannot occur by body going through body, even if body grows with respect to the whole form.

36. 14 For matter is not at all stable in a preexistent state so that the body causing growth passes through it, but nutriment becomes matter for flesh and by being mixed in specific parts with the [matter] of the still preexistent flesh which is both stable and contains the form, then (1): when the nutriment is less than or equal to the deteriorating flesh it preserves only the thing by which it is assimi-

Sem. Griekse Letterkunde
Blandijnberg 2
Gent

ᾧ προσκρίνεται, τρέφουσα αὐτὸ καὶ κωλύουσα πᾶν διαφορηθῆναί τε καὶ φθαρῆναι διὰ τὴν συνεχῆ ῥύσιν, ὅταν δὲ πλείων ἡ ὑπὸ τῆς θρεπτικῆς δυνάμεως κατεργαζομένη τε καὶ προσκρινομένη ὕλη τῆς ἐπαναλισκομένης τε καὶ ἀπορρεούσης, τότε πρὸς τῷ τρέφεσθαι τὸ ὑποκείμενον καὶ αὔξεται, τῆς μὲν οὐσίας τοῦ σῴζοντος εἰς τὸ εἶναί τε καὶ σῴζεσθαι συντελούσης τῷ τρεφομένῳ, τῆς δὲ ποσότητος εἰς αὔξησίν τε καὶ μέγεθος. διὸ ὅταν μὲν ἐνεργῇ καθὸ τροφή, τότε μόνον σῴζει τὸ ὑποκείμενον, ὅταν δὲ καὶ ὡς ποσή, πρὸς τῷ σῴζειν καὶ συναύξει.

[μὴ] πάντη δὲ τῆς προσκρίσεως γινομένης, αὔξεται πάντως, ὅτι τῆς θρεπτικῆς δυνάμεως τῆς ἀλλοιούσης τε τὴν τροφὴν καὶ ἐξομοιούσης καὶ προσκρινούσης τῷ τρεφομένῳ ἴδιον ὥσπερ τὸ σῴζειν τὸ ὑπ' αὐτοῦ τρεφόμενον, οὕτω δὲ καὶ μετὰ τοῦ ἰδίου αὐτὸ τηρεῖν σχήματος· μᾶλλον δὲ ἕπεται καὶ τούτῳ τὸ σῴζειν. τότε γὰρ κυρίως σῴζεταί τι, ὅταν τηρῆται μετὰ παντὸς τοῦ κατὰ φύσιν ὑπάρχοντος αὐτῷ, ὧν ἐστι καὶ τὸ σχῆμα. πᾶν γὰρ ἔμψυχον αὐτό τε μετά τινός ἐστι οἰκείου σχήματος καὶ τῶν μερῶν ἕκαστον αὐτοῦ, ὃ φυλάσσεται τότε σχῆμα, ὅταν προσκρινόμενον ἐκ τῆς τροφῆς μὴ μένῃ τὴν καθὸ προσετέθη
237 χώραν φυλάσσον καὶ κατὰ τοῦτο προσκείμενον, ἀλλὰ προωθῇ τὸ πρὸ αὐτοῦ, κἀκεῖνο τὸ πρὸ αὐτοῦ πάλιν, μέχρι τοσούτου, ἕως ἂν ἕκαστον τῶν μορίων αὐτοῦ κατὰ ἀναλογίαν προτιθεμένου τοῦ τρέφοντός τε καὶ αὔξοντος ἴσην ἐπίδοσιν λαβὸν ταὐτὸν ἔχον τὸ σχῆμα μείνῃ, ὃ εἶχε καὶ πρὸ τῆς προσκρίσεως τοῦ τρέφοντος.

ὡς γὰρ τῆς γῆς βαρέος ὄντος σώματος καὶ κάτω τε καὶ ἐπὶ τὸ μέσον φερομένου καὶ κατὰ τὴν αὐτοῦ φύσιν καὶ τῷ ἰδίῳ μέσῳ τὸ τοῦ παντὸς κατέχοντος μέσον τε καὶ κέντρον, ἂν βάρος ἐπ' αὐτῇ τι μεῖζον τεθῇ κατὰ θάτερον ἡμισφαίριον, οὐ διὰ τῆς προϋπαρχούσης γῆς διαδύεται τοῦτο τῆς ἐν τῷ μέσῳ μενούσης, ὡς τῆς ἰδίας ῥοπῆς τὸ κέντρον ἔχειν κατὰ τὸ τοῦ παντὸς μέσον, ἀλλ' ἄνωθεν ‹ἐν› τῇ προϋπαρ-

---

236. 19 διαφορηθῆναι F, Montanari p. 38: διαφθαρῆναι codd. Br.
236. 26 μὴ secl. Ideler
237. 1-2 πρὸ αὐτοῦ bis scripsi, cf. Aristot. *de caelo* 297a 27, v. infra Comm. 237. 5-15: πρὸ αὐτοῦ codd.
237. 4 ἔχον τὸ σχῆμα coni. Ideler Rex: ἔχοντα σχῆμα codd. Br. εἶχε coni. Ideler, cf. 213. 6, 237. 5: ἔχει codd. Br.
237. 8 τι Diels: ἢ codd.
237. 9 τοῦτο Apelt: τὸ ὑπὸ codd.
237. 10 ‹ἐν› coni. Schwartz

lated, nourishing it and preventing it from being totally dispersed and corrupted through continual deterioration; but (2): if the matter which is digested and assimilated by the nutritive faculty is greater than the matter which is used up and deteriorates, then in addition to being nourished the substrate also grows, while the substance of the body that preserves [i.e. the nutriment] produces being and stability for the body being nourished, as its quantity achieves growth and magnitude. Nutriment, then, only preserves the substrate when it acts *qua* nutriment, but when it also acts as quantity it contributes to growth as well as preservation.

36. 26 When assimilation occurs in all parts of a body so does growth,
because it is peculiar to the nutritive faculty which transforms and
assimilates food and absorbs it for the body being nourished that
just as it preserves the body that is nourished by it so it maintains
it along with its own shape; indeed the preservation results from
this. In a strict sense the body that is nourished is only preserved
when it is maintained along with the whole of its natural endowment
36. 32 which also includes shape. Indeed every living creature is in itself
endowed with its own particular shape as is each of its parts, and it
protects this shape when on the assimilation of nutriment it is not
stable in retaining the area where the addition is made [237] and in
being added to at this point, but rather thrusts forward the part in
front of it, and again that part the part in front of it, and this just
until each part of the body to which a proportional addition is made
retains the same shape that it had before assimilating nutriment by
taking an equal addition from what nourishes and causes its growth.

37. 5 Take the earth, a heavy body that moves downward and towards the middle, and both in its own nature and by its own middle occupies the middle and centre of the universe—if a weight greater [than it] were added to it in one hemisphere this would not sink through the preexisting earth that remains in the middle so that it has the centre of its own impulse aligned with the centre of the universe, but rather in the preexisting earth it thrusts forward from above

χούσῃ γῇ μέχρι τοσούτου τὸ πρὸ αὐτοῦ μόριον τῆς γῆς προωθεῖ, ἕως ἂν τὸ γενόμενον τοῦ τῆς γῆς βάρους κέντρον διὰ τὴν προσθήκην τοῦ βάρους γένηται κατὰ <τὸ> τοῦ παντὸς μέσον (τοῦτο γὰρ κατὰ φύσιν τοῖς βαρέσι τε καὶ κάτω φερομένοις σώμασιν· οὗ γινομένου καὶ τὸ σχῆμα ἅμα ἡ γῆ φυλάσσει τὸ σφαιρικόν, ὃ εἶχε καὶ πρὸ τοῦ), οὕτως ὑποληπτέον καὶ ἐπὶ τῶν ὁμοιομερῶν [γίνεσθαι], τῶν αὐξομένων διὰ τὴν ἀπὸ τῆς τροφῆς πρόσκρισιν, εὐθὺ τῇ προσκρίσει τε καὶ ἐξομοιώσει πρὸς τὸ τρεφόμενον τῆς τροφῆς παρ᾽ αὐτῶν τῶν μερῶν τοῦ τρεφομένου πρόσοδόν τινα γίνεσθαι καὶ μετάστασιν ἄλλου προωθοῦντος ἄλλο, ἕως ἂν τὸ πᾶν, ἐν ᾧ πρότερον ἦν σχήματι, ἐν τούτῳ καὶ μετὰ τὴν ἀπὸ τῆς τροφῆς προσθήκην γένηται.

ὡς

γὰρ αὔξεταί τε καὶ φυλάσσεται τὰ τῶν ὁμοιομερῶν τοῦ τρεφομένου σχήματα ἐν τῇ τῶν ὁμοιομερῶν αὐξήσει, οὕτως καὶ τὰ τῶν ἀνομοιομερῶν ὑποληπτέον φυλάσσεσθαι σχήματα τῆς κατὰ μόρια προσκρινομένης αὐτοῖς τροφῆς ὑπὸ τῆς ἐν αὐτοῖς θρεπτικῆς δυνάμεως εἰς πάντα αὐτὰ ἀγομένης ἀναλόγως. καθ᾽ ἣν ἀναλογίαν μετὰ τῆς τοῦ σώματος σωτηρίας ἕκαστον αὐτῶν πᾶν αὔξεται, καὶ δεῖ τὸ γινόμενον ὑπὸ τῆς φύσεώς τε καὶ θρεπτικῆς δυνάμεως τοῖς διὰ τῆς τροφῆς αὐξανομένοις ὑπολαβεῖν ὅμοιον εἶναι, ὡς ἂν εἰ <ὕδωρ ἢ> οἶνόν τις ἐπινοήσαι διά τινος σωλῆνος φερόμενον, φυλάσσοντος μὲν τὸ σχῆμα ταὐτόν, δι᾽ ὑγρότητα δὲ καὶ μαλακότητα, ὅταν μὲν ἔλαττον ᾖ τὸ δι᾽ αὐτοῦ φερόμενον ὑγρόν, συστελλομένου τε καὶ τὸ σχῆμα ἐπὶ ἐλάττονι σῴζοντος ὄγκῳ, ὅταν δὲ πλεῖον ᾖ τοῦτο, ἀνευρυνομένου τε πάντῃ καὶ μείζονα τὸν ὄγκον λαμβάνοντος. ὡς γὰρ ἐπὶ τοῦ σωλῆνος τοῦ τοιούτου οὐ τὸ ὕδωρ τὸ αὐξανόμενόν ἐστιν, ὅ γε τὴν ἀρχὴν οὐδὲ ὑπομένει,
238 ἀλλὰ ἄλλο τι γίνεται, καὶ ποτὲ μὲν πλέον, ποτὲ δὲ ἔλαττον, τὸ δ᾽ ἐπὶ τῷ ὕδατι σχῆμά ἐστι τὸ μένον καὶ τὴν ἐπίδοσίν τε καὶ τὴν συστολὴν λαμβάνον, οὕτως ὑποληπτέον γίνεσθαι καὶ ἐπὶ τῶν αὐξανομένων φύσει· τὴν μὲν ὕλην, ἐφ᾽ ἧς τὸ αὐξόμενον εἶδος, ἄλλοτε ἄλλην εἶναι δεῖν διὰ τὰς

237. 11 πρὸ αὐτοῦ scripsi, cf. 237. 1-2: πρὸ αὑτοῦ Apelt: πρὸς αὐτὸ codd. Br.
237. 13 <τὸ> coni. Br.
237. 16 γίνεσθαι secl. Br.
237. 19 ἄλλου προωθοῦντος F, Montanari p. 37, Apelt Br.: ἀλλ᾽ οὐ προοδοῦντος codd.
237. 21 ὁμοιομερῶν Apelt: ἀνομοιομερῶν codd.
237. 24 αὐτοῖς [sc. μορίοις] scripsi, cf. 234. 14-15: αὐτῇ [sc. τροφῇ] codd. Br.
237. 26 δεῖ Apelt Rodier: διὰ codd.
237. 28 <ὕδωρ ἢ> coni. Diels, cf. ὑγρόν 237. 30
237. 31 τοῦτο Rex: τούτων codd. Br.
238. 4 δεῖν coni. Br.: δεῖ codd.

the part in front of it just until the centre of the weight of the earth produced by the additional weight is aligned with the middle of the universe; (for this is natural with heavy bodies that move downward, and when it occurs the earth preserves the spherical shape
37. 15 that it also had previously). Now just the same must be understood to apply to the uniform parts that grow through the assimilation of food, where the absorption and assimilation of nutriment by the body that is nourished immediately causes a forward movement from the actual parts of the body that is nourished and a change of position, with various forward motions until after the addition of nutriment the whole body has the same shape that it formerly had.

37. 20 Indeed just as the shapes of the uniform parts of the body that is nourished grow and are maintained by growth in these parts, so must it be understood that the shapes of the non-uniform parts are maintained when nutriment is assimilated part by part and distributed to all of them proportionately by their nutritive faculty. Each of them grows as a whole by this proportion, while they preserve the body that they are in, and we must understand that the case of bodies that grow through nutriment by the agency of nature and
37. 28 the nutritive faculty is like conceiving of <water or> wine moving along a channel which maintains the same shape, but because of its elasticity and pliancy is compressed and maintains its shape in a smaller volume when the water moving along it is less than its volume, but when this is greater, it expands in every direction and assumes a larger volume. As with such a channel the water is not what grows, as it does not remain at all stable [238] but assumes varying volumes, but the shape containing the liquid is what is stable and what assumes expansion and contraction—so too must the process involved in natural growth be understood: that while the matter to which the growing form belongs must vary at different

συνεχεῖς ἀποκρίσεις, τὸ δὲ εἶδος τὸ μένον ἐπὶ τῇ ὕλῃ τῇ ῥεούσῃ ἀναλόγως τῷ σωλῆνος σχήματι μειοῦσθαι μέν, ὅταν ἐλάττων ἡ ὕλη γένηται, αὔξεσθαι δέ, πλείονος τῆς κατὰ τὴν ὕλην προσκρίσεως γινομένης, μετὰ τῆς τοῦ οἰκείου σχήματος σωτηρίας. τὸ γὰρ ἐπὶ πλείονι ὕλῃ ταὐτὸν εἶδος μεῖζον τοῦ ἐπ' ἐλάττονι, ὡς καὶ τὸ σχῆμα τοῦ σωλῆνος, τοῦ δι' ὑγρότητα συμμεταπίπτοντος τῷ δι' αὐτοῦ ῥέοντι.

ἡ δὲ πρόσκρισις ἡ τῆς τροφῆς τοῖς τρεφομένοις γίνεται μὲν κατὰ μεταβολὴν ἐξομοιουμένης αὐτῆς τῷ τρεφομένῳ, οὐ γενομένη δὲ πρῶτον ἡ αὐτὴ τῷ τρεφομένῳ, ἔπειτ' αὐτῷ προσκρίνεται. γένεσις γὰρ τοῦτο σαρκός, ἀλλ' οὐ τροφή. ἀλλ' ἡ ἐσχάτη τροφή, ὅπερ ἐστὶν ἐπὶ τῶν ἐναίμων τὸ αἷμα, ἐπὶ δὲ τῶν ἄλλων τὸ ἀνάλογον, ἑκάστῳ τῷ τρεφομένῳ διὰ τῶν ἀγγείων φερομένη τῶν καθηκόντων ἐπ' αὐτὰ καὶ ἐπιρρέουσα ὑπὸ τῆς ἐν ἑκάστῳ δυνάμεως, μεταβάλλεταί τε καὶ ἐξομοιοῦται τῷ τρεφομένῳ. καὶ δεῖ τὸ γινόμενον ἐπινοῆσαι, ὡς εἰ οἴνῳ ἐπιχεομένου ὕδατος ἡ ἐν τῷ οἴνῳ δύναμις ἀεὶ τὸ ἐμπῖπτον εἰς αὐτὸν ὕδωρ οἶνον ποιοῦσα, σώζει τε καὶ αὔξει τὸν οἶνον. ὅπερ ὁρᾶται καὶ τὸ πῦρ ποιοῦν, ἐπειδὰν αὐτῷ γειτνιάσῃ τι τῶν καυστῶν.

οὕτως δὲ γινομένης τῆς αὐξήσεως οἶμαι φανερὸν εἶναι τὸ μηδὲ τὴν αὔξησιν συντελεῖν τοῖς σῶμα διὰ σώματος χωρεῖν λέγουσιν πρὸς τὴν κατασκευὴν τοῦ παραδόξου τε καὶ θαυμαστοῦ δόγματος.

---

238. 5 τὸ μένον Rex: τὸ μὲν codd. Br.
238. 13 ἡ Ideler: ἢ codd.
238. 14 τροφή] τροπή Montanari pp. 46-47
238. 18 οἴνῳ ἐπιχεομένου Apelt: οἶνον, ᾧ ἐπιχέον οὗ codd.
238. 19 σώζει τε Apelt: σώζεται codd.

times on account of its continual deterioration, the form, on the other hand, that remains proportionately with the fluid matter like the shape of the channel, diminishes when the matter decreases, but grows when the assimilation of matter increases, maintaining all along the body's individual shape. For an identical form belonging to a larger amount of matter is larger than one which belongs to a smaller amount, just like the shape of the channel that because of its elasticity adjusts to what flows through it.

38. 10 The absorption of nutriment by bodies that are nourished occurs by a process of alteration involving the assimilation of nutriment by the body that is nourished; the nutriment does not first become the same as the thing that is nourished and then absorbed by it. That would be the coming-to-be of flesh, not nutrition. But the ultimate nutriment, blood in redblooded creatures, and its analogue in other things, by passing through the vessels that belong to them in each body that is being nourished and flowing in, is altered and assimilated by the body that is nourished through the [nutritive] faculty that each possesses.

38. 17 And we must conceive of the process as like that of water being poured into wine, where the power in the wine continually makes wine of the water poured in and so preserves and increases the wine. This [process] can also be seen in the activity of fire when something combustible is in its vicinity.

38. 20 Since growth occurs in this way I think that it is clear that it does not help to support the remarkable and outlandish theory of those who say that body goes through body.

GLOSSARY OF TERMS

actualisation, actuality	τελειότης
assimilation	πρόσκρισις, προσκρίνεσθαι
blend (total blending)	κιρνᾶσθαι (κρᾶσις δι' ὅλων)
body going through body	σῶμα διὰ σώματος χωροῦν
coming-to-be	γένεσις
complete mutual coextension	ὅλα δι' ὅλων ἀλλήλοις ἀντιπαρέκτασις (ἀντιπαρεκτείνεσθαι)
composition	σύνθεσις
constituents	τὰ μιγνύμενα, κιρνάμενα
contrariety (opposites)	ἐναντίωσις (τὰ ἐναντία)
dissolution	διάκρισις (διακρίνεσθαι)
be divided through and through	πάντῃ διῃρῆσθαι
elasticity	ὑγρότης
enmattered form	ἔνυλον εἶδος
exist independently (individually)	ὑφεστάναι καθ' αὑτά (κατ' ἰδίαν)
fusion	σύγχυσις
growth	αὔξησις
joint-destruction	σύμφθαρσις
by juncture	καθ' ἁρμήν
juxtaposition	παράθεσις, παρατίθεσθαι
magnitude	μέγεθος
malleable	εὐπαθής
matter/form	ὕλη/εἶδος
mixture	μῖξις, μίγνυσθαι
nature	φύσις
[process of] nourishment, (thing that is nourished)	τὸ τρέφεσθαι (τὸ τρεφόμενον)
notion (common, natural)	ἔννοια (κοινή, φυσική)
nutriment, nutrition	τροφή
nutritive faculty	θρεπτικὴ δύναμις

passing-away	φθορά
preconception	πρόληψις
preexist	προϋπάρχειν
sense-presentation (after Sandbach, in Long ed. *Problems in Stoicism*, p. 10)	φαντασία
reduction-to-the-elements	ἀναστοιχείωσις
standards of truth	κριτήρια τῆς ἀληθείας
State	ἕξις
substantiality	ὑπόστασις
tension	τόνος
(reciprocal) interaction, (be acted on)	ποιεῖν τε καὶ πάσχειν (ὑπ' ἀλλήλων), (πάσχειν)
uncompounded bodies	ἁπλᾶ σώματα
uniform (non-uniform) parts (after Peck; see *Aristotle, Parts of Animals* [Loeb ed.], p.28)	τὰ ὁμοιομερῆ (ἀνομοιομερῆ)
whole	τὸ πᾶν

NOTES ON THE TEXT

213. 7-8. I have deleted τὸ so that δύνασθαι depends directly on ὅσον. Cf. the common ὅσον μ' εἰδέναι. The τὸ could have originated in a repetition of the last two letters of the preceding τούτῳ as appears to have happened in the Aldine edition.

215. 22. ὁμοίως δὲ καὶ is convincing here, for why should Alexander be tentative? At 214. 27-28 he readily enough associates the theory of matter as ὁμοιομερίαι with the Democritean theory of mixture, and he now completes his programme by marshalling the theory that matter is composed of planes (cf. 214. 5) behind the Epicurean theory.

218. 12. The claim that the Stoics use every notion but the common notions is certainly borne out elsewhere in the work; cf. 220. 23-29, 222. 26-29.

218. 19-20. I have translated Diels' σαφήνειαν ἔχει δι' *faute de mieux*. Clearly some expression similar to διὰ τῆς αἰσθήσεως γνώριμον (218. 30-31) is intended. Ideler, for example, suggested ὀφθαλμοῖς ἔστιν ἰδεῖν. ἣν . . .

220. 27. I read αὑτοῖς since two or more bodies must be extended "through one another" (cf. 216. 16, 217. 29) or as "whole through whole" (cf. 218. 16 and *Mant.* 140. 10). For αὑτοῖς = ἀλλήλοις cf. 220. 2 above, and cf. 220. 1 (αὑτὰ) and 220. 11 (αὑτοῖς) where by emendation the same sense is achieved in a context similarly dealing with bodily interpenetration.

221. 33. This independent use of ἀλλήλων is suspect and δι' ὅλου is unparalleled in Alexander in this sense. κράσεις διόλου at Diog. Laert. VII. 151 is also questionable, as we have seen (p. 31 n. 45).

222. 5-6. For διαιρεῖσθαι in 222. 4 and 5 I read διῃρῆσθαι since a *process* of division cannot be in actuality, only its product. Hence we need a present perfect tense; cf. 222. 12-13, 15-17 and Comm. on 222. 4-13. Similarly at 222. 9 the change to διῃρῆσθαι is necessary since as 9-10 show the division in question must be complete. Bruns' attempt at 222. 5 to establish a disjunction with a second

ῇ falls foul of the ἐνεργείᾳ in both clauses which indicates that they must be parallel. This parallelism moreover begins at οὕτως δὲ καὶ κ.τ.λ., which should not be bracketed as it is by Bruns, and lies between the actual division of any body (τι σῶμα) and the total mixture of bodies (σώματα) so divided. This connection is reasserted at 222. 17 when the inquiry into this aspect of the relation between mixture and division is resumed after the only alternative to it, that of an uncompleted infinite division, has been dismissed at 222. 6-13.

222. 15. As with the argument discussed in the preceding note the present argument can only be valid if we read διῃρῆσθαι for διαιρεῖσθαι. The reading in the manuscripts says that constituents form an infinite number of completely divided bodies by being able to undergo a process of division (διαιρεῖσθαι) rather than, as we would expect from the conclusion (κατ' αὐτοὺς etc.) and from διῃρημένον (line 15), by being able to reach a state of complete division. With this reading εἰ δὲ λέγοιεν (222. 14) introduces exactly the opposite hypothesis to that of an uncompleted infinite division furnished by εἰ μὲν γὰρ λέγουσιν κ.τ.λ. at 222. 6ff.

The palaeographical possibility of the change proposed in this and the preceding note rests on the standard confusion of H and EI. This could have resulted in a corruption to διηρεισθαι and its simplifying emendation to διαιρεισθαι.

222. 25. Aristotle, *de gen. et corr.* A2 316a 29-30 suggests that it follows from points being parts of the line that a magnitude is composed of non-magnitudes, and not vice-versa.

224. 11. ὁμοίως ἀλλήλοις needs to be qualified as at 224. 13 by τοῖς συνεχέσι κ.τ.λ. Here it is gratuitous. For the word order of my emended τὰ ἡνωμένα πάντα cf. 233. 31.

224. 13. It would be contradictory in this argument to say that pneuma separated bodies and hence I add συνέχεσθαι. For the conjunction of συνέχεσθαι and ἡνῶσθαι cf. 223. 31.

224. 24-25. Cf. Von Arnim (*SVF* II, p. 146. 10) where the change to αὑτοῦ and αὑτὸ is made tacitly. ὄν too must agree with pneuma rather than being followed by a comma as Rodier suggested. The

whole expression may well be modelled after Arius Didymus Fr. 28 (Stob. *Ecl.* I. xvii. 4, p. 154. 1W = *SVF* II, p. 152. 32-33). As we saw (pp. 54ff. above) this text was probably Alexander's source in *de mixtione* 3.

226. 33a-34. Bruns was the first to recognise that 227. 23-24 was in some sense the continuation of 226. 34, but it was Brinkmann (art. cit.) who demonstrated on palaeographical grounds the link between the passages. It is his reconstruction that I have followed.

230. 8. Montanari (p. 37), who mistakenly refers to this locus as 238. 8, endorses the reading in F. He says that this coincides partly with Bruns' emendation which in fact is αὐτό. He recommends καὶ ἰσχύον as maintaining a symmetry with πλεονάσαν. This may be right although it involves postponing the one remaining finite verb, γίνεται, very late in the sentence.

230. 15. I agree with Montanari (p. 39) that the contrast in 230. 14-19 is between blending and passing-away, whether the latter is the standard reciprocal process of coming to be and passing away, or the hypothetical case of complete destruction. Alexander's point is that blending is a unique type of passing-away in which constituents are preserved *in a sense*. φθορὰ σωζομένων τινων is a more precise formulation of the definition at 13. 228. 34-35 that ἡ ὡς κρᾶσις μῖξις occurs οὐ σωζομένων ἔτι τῶν μιγνυμένων. Its clarification in the course of ch. 14 allows Alexander to say at the end (231. 10-12) that blending is unification without any of the initially juxtaposed constituents passing away. The passing away excluded here is described at 230. 15-16 (cf. 13. 230. 5-13). Thus a specific theory is developed from the general intuition that constituents are preserved; see 215. 13-14 and cf. Aristotle *de gen. et corr.* A10 327b 30-31. The reading of φθορά for μῖξις at 230. 15 and 19 is, I think, justified.

231. 20-21. The reconstruction of these lines that I have offered may be compared with Bruns' suggestion in his apparatus: διὸ καὶ ὀλίγης δεῖται βοηθείας [τῆς] πρὸς <τὴν> αὐτῶν εἰς ταύτην (scil. δύναμιν) μεταβολήν, ὅσα γε αὐτῶν τοιοῦτο χωρεῖ, <ἐπεὶ> μὴ πάντῃ ἀχώριστα τὰ κεκραμένα. Rodier altered πρὸς αὐτὴν εἰς ταύτην μεταβολήν (codd.) to πρὸς αὐτοῦ εἰς ταῦτα μεταβολήν. Rex offers the following: διὸ καὶ ὀλίγης τινος (Apelt) δεῖται βοηθείας τῆς πρὸς αὐτὴν εἰς ταύτην (scil.

ἐνέργειαν) μεταβολήν, ὅσα γε αὐτῶν (Br.) τοιαῦτα χωρεῖ μὴ πάντῃ χωριστά τε καὶ πεπερασμένα (Schramm).

233. 17. τὸ ὑπάρχον σῶμα is unusual. The normal equivalents for τὸ αὐξόμενον are τὸ ὑποκείμενον (e.g. 236. 22) or προϋπάρχειν in the present participle qualifying a relevant noun (e.g. ὕλη at 236. 17, or μέγεθος at Philop. *de gen. et corr.* 87. 6) or in the form τὸ προϋπαρχον (Philop. 118. 23). It may be best to follow A^2 and Ra.

COMMENTARY

Chapters 1 and 2. After a statement of the central paradoxes in the Stoic theory of blending (213. 2-13), the opening chapter describes the main differences in theories of blending resulting from differences in theories of matter (213. 13-214. 16). Those based on a corpuscular theory of matter—principally the Atomist and Epicurean theories—are then expounded and criticised in chapter 2. Although the Stoic theory of blending is described in chapter 3 with some reference to the theory that matter is unified, none of the subsequent criticisms take this into account, nor is the Aristotelian theory of blending closely related to a theory of matter as unified, nor, finally, are these two theories critically contrasted (cf. note on 3. 216. 1-4). The polemical remarks on total blending at 1. 213. 2-13 are taken up and elaborated in chapter 7, while the reference to "body going through body" at 213. 13 surfaces again in chapters 5 and 6.

Chapter 1

213. 2. The opening of the treatise is missing. It probably contained some general discussion of blending and possibly some programmatic statement of the superiority of the Aristotelian theory. At 13. 228. 5 Alexander says that he is returning to "the original statement" (τὸν ἐξ ἀρχῆς λόγον) which 228. 5-7 suggests involved some decision to demonstrate the absurdity of other theories before expounding Aristotle's. There is nothing in the *de mixtione* really corresponding to this and it may well have been conveyed by the lost portion of this chapter.

213. 2-13. Right at the outset (2-6) Alexander asserts a bald contradiction in the Stoic theory between (a) the totality of blending, and (b) the preservation of their surfaces by the constituents. The former requires that constituents lose their identity, the latter that they retain it, so that they can be separated (7-8). This contradictory theory is then contrasted (10-13) with the Aristotelian theory (8-10) that constituents are separable from a blend in which they are not completely mixed.

This contradiction at the heart of the notion of "body going through body" (13-14) is not exploited until ch. 7 when total blending is shown to be either a case of juxtaposition (where constituents are not totally blended) or fusion (where they are completely blended but inseparable); see on ch. 7, 220. 23-29. The topic surfaces again in ch. 15 where Alexander shows that constituents are separable from a homoiomerous blend in such a way that they can be regarded as not being "wholly blended," i.e. being in potentiality; see on 15. 231. 12-29. Alexander could have derived this contradiction from the Stoic classification of mixtures where both the constituents of juxtapositions and blends are said to preserve their own substances (cf. 3. 216. 19-20 and 216. 30-31) while in the former case this is "with respect to their surface" (κατὰ τὴν περιγραφήν, 3. 216. 20). From this it would follow that what constituents preserve in a blend also are their surfaces (213. 3, 5, 12), while the blend would only be total if they lost this individuality (213. 4, 11-12); yet the Stoics seem to want both of these to occur in total blending. Whether or not the passage was known to Alexander (cf. on 4. 217. 36), Hierocles' description of the relation between soul and body (IV. 4-10 Von Arnim) could certainly also encourage such criticism: "... [the soul] is totally blended with it, so that not even the smallest part of the mixture does not share in their mutual participation (ὡς μηδὲ τοὐλάχιστον τοῦ μίγματος μέρος τῆς ὁποτέρου αὐτῶν ἀμοιρεῖν μετοχῆς [cf. 3. 217. 10-11]); for the blending is very similar to what occurs with heated iron; for in each of these cases alike the juxtaposition is complete (δι' ὅλων ἐστὶν ἡ παράθεσις; [cf. on 4. 218. 8])." A Stoic might argue that such complete juxtaposition was not paradoxical because it only occurred between bodies of complementary physical properties of activity and passivity, such as pneuma and matter (cf. pp. 35-36 above). Even so, as long as the passive principle is defined as body it is difficult to see any reason in physics why it could be occupied by another body with which it is *also* juxtaposed. This is certainly what the Stoic doctrine of pneuma's presence in matter has to mean (cf. p. 35 n. 67 above), but it is difficult to accept it as a satisfactory conception, and Alexander's response seems quite justified.

213. 4-5. "in a uniform body" (ὁμοῦ μὲν ... ὁμοῦ δέ). Cf. 213. 10-12, Plutarch *de comm. not.* 37. 1078B (p. 104. 29 Teubner), and note on 7. 221. 9-10 where Alexander reduces total blending to a

case in which bodies fuse to form a "uniform" (ὁμοιομερές) compound.

213. 7-8. "bodies that have been blended can be separated again" (τὸ δύνασθαι τὰ κεκραμένα χωρίζεσθαι πάλιν). Cf. Aristotle *de gen. et corr.* A10 327b 29, and here at 213. 8-9, 216. 32 (again with reference to the Stoics), 222. 30, 224. 12-13, and 228. 14-15.

213. 10. "universally held notions" (τὰς ἁπάντων ἐννοίας); an alternative expression for "common notions" (κοιναὶ ἔννοιαι). On Alexander's use of this terminology see on 5. 218. 11.

213. 13. "the notion of body <going> through body" (τό τε οὖν σῶμα διὰ σώματος <χωρεῖν>). This important expression first enters the *de mixtione* as a general characterisation of total blending. It is anticipated by 213. 11 (δι' ὅλων ἀλλήλων χωρεῖν τινα σώματα; cf. 217. 10). Alexander seems to regard it as an accepted redescription of total blending, as he does not justify its introduction here; presumably an earlier polemical tradition had ensured its close association with the Stoic doctrine, so that this redescription could be taken for granted (cf. Pt. II-iii, *passim*). Cf. on 7. 220. 23-29.

213. 14-15. "philosophical principles" (τὰ κατὰ φιλοσοφίαν δόγματα). These are the theories of matter on which different theories of mixture are founded; see next note. Cf. 12. 227. 3 where the same expression occurs; by that stage however Alexander's analysis of bodily interpenetration has led him to claim that the doctrines making up the Stoic theory of a unified cosmos are derived from such a form of mixture rather than total blending being derived from a theory of continuous matter, as claimed here and in ch. 3. As I argue above (p. 38ff.) this accords more closely with Stoic thinking than the present systematisation.

213. 15-18. For a categorisation of theories of matter in these terms see also Galen, *Quod. mor. an.* 5 (IV. 785K *Script. Min.* II p. 46. 13-15) and *De Nat. Fac.* I. 12 (II. 27K). While the grouping of corpuscular theories is legitimate enough since mixture involves the combination of such primary bodies, no indication is given here or later as to precisely why the theory of a single unified matter should be relevant to any theory of mixture. This is a particularly

ahistorical presentation, for Aristotle's theory of mixture was clearly a reaction against the Empedoclean theory of the mixture of four elements *in a corpuscular form* (see *de gen. et corr.* B7 334a 27-30; cf. Aetius I. 17. 3), yet the latter is not referred to here, perhaps in order to establish a more definite polarisation between the two theories of matter. On the Presocratic background to Aristotle's theory of mixture see F. Solmsen, *Aristotle's System of the Physical World*, ch. 19, and on its relation to the Stoic theory cf. pp. 49-50 above.

213. 20. "...other things come to be by the composition and particular interlinking..." "Composition (σύνθεσις) here describes the formation of secondary bodies, and even with the standard ad ditions of περιπλοκή (interlinking), τάξις (order), and θέσις (position) the Atomist-Epicurean theory is distinguished only slightly from that of Anaxagoras where bodies are formed κατὰ σύγκρισιν καὶ σύνθεσιν ("by compounding and composition"). This accords with the doxographical account (at Aetius I. 17. 2) that Anaxagoras and Democritus thought that mixtures occurred by juxtaposition (παράθεσις), but as we see in ch. 2 Alexander requires separate doctrines of mixture and coming-to-be of these thinkers, and obtains them by rather forced means (see on 214. 18-28, 214. 26-28, and 214. 28-215. 8).

214. 3-6. "Anaxagoras and Archelaus..." Archelaus is just as closely bracketed with Anaxagoras at *DK* 60A 4, 5. The precise formulation of this theory of matter is of course Aristotelian. Anaxagoras probably did not use the word "uniform" (ὁμοιομερής) to describe his principles; see W. K. C. Guthrie, *A History of Greek Philosophy*, II, 325-236.

The theory of partless bodies was held by Diodorus Cronus (fl. c. 300 B.C.), to whom Alexander attributes it explicitly at *de sensu* 122. 23 and 172. 29, and presumably by some of his followers. On this theory and its relation to Epicureanism see D. J. Furley, *Two Studies in Greek Atomism*, I. ch. 9.

On the Platonic doctrine of planes see Aristotle *de caelo* Γ1 299a 3-300a 19. Alexander *Quaest.* II. 13 attacks the doctrine, as does Alex. ap. Philop. *de gen. et corr.* 81. 28-31. Cf. on 2. 214. 27-28 below.

On the Pythagorean doctrine of numbers see Aristotle *Met.* A5 *passim* and cf. *de caelo* 300a 14-19.

214. 10-16. Cf. *de fato* 165. 25-27 for a similar formulaic description of prevailing dissension at the opening of a work.

Chapter 2

214. 18-28. (214. 18-25 = *DK* 68A 64). This "Democritean" theory would seem to be based on the unattributed theory that blending is relative to perception mentioned by Aristotle at *de gen. et corr.* A10 328a 8-16. Cf. especially 214. 19-20, 22-23 with 328a 7-8, and see Joachim on 328a 1-2. There is no record of it being propounded by Democritus, and since its presence here is simply to fulfill Alexander's programme (213. 15-17) that theories of matter have corresponding theories of mixture, this is not good evidence for attributing it to him.

214. 23. "...its own nature, which they had even before the mixture" (φύσιν, ἣν εἶχον καὶ πρὸ τῆς μίξεως). Cf. exactly the same expression used of constituents' surfaces at 213. 6, 13 in the description of the presence of constituents in a "Stoic" blend. Here the preservation of "natures" is an aspect of juxtaposition; the earlier text would seem, as we have said (on 213. 2-6), to be setting up the Stoic theory of blending for the argument reducing it to juxtaposition that we eventually find at 7. 220. 35-221. 7.

214. 26. "the consequence" (τὸ ἑπόμενον) simply being that mixture is relative to perception. Epicurus tried to evade it (28-29) by making mixture the juxtaposition of atoms which, presumably, necessarily escape perception (cf. Alex. *Quaest.* III. 12, 105. 13) whereas the Democritean corpuscles just happen to be imperceptible, but might be accessible to the proverbial Lynceus (*de gen. et corr.* A10 328a 15).

214. 27-28. It is not clear whether the juxtaposition of homoiomerous bodies is blending, or, as at 214. 2, coming-to-be. If the former, then we would need evidence that Anaxagoras believed that corpuscles were juxtaposed in blending in some different way from the composition (σύνθεσις) by which they formed compounds. Similarly with the attempt to line up the theory that matter is composed of partless bodies behind the "Epicurean" theory of blending at 215. 22-23. The whole distinction between coming-to-be and blending is, as we have seen (on 1. 213. 20), unprecedented for the proponents of this theory of matter.

214. 28-215. 8. (= Usener, *Epicurea*, No. 290). This, as Joachim (on *de gen. et corr.* 328a 1-2) saw, is an attribution to Epicurus of the theory of the mixture of indivisibles described at *de gen. et corr.* A10 328a 16-17. Sambursky's elaborate speculations on the meaning of this text (*Archiv. Int. d'Hist. des Sciences* 2 [1958] 254-255) therefore seem unnecessary. Philoponus (*de gen. et corr.* 193. 20-24) thought that this theory had to be Democritean, and if Alexander did not perversely distinguish blending from the combination that created compounds, his description at 213. 18-23 would have entailed that both the Atomists and Epicurus held the same theory of blending *qua* coming-to-be described here and would have made the attempt to distinguish between them unnecessary.

This interpretation also makes redundant speculation such as Kerferd's (*Phronesis* 16 [1971] 90) on whether or not the reference (at 215. 5-6) to "wine-producing" (οἰνοποιός) and "water-producing" (ὑδροποιός) atoms implies an Epicurean molecular theory. For a complementary example of Alexander's confusion over the differences between Epicurus and the Atomists see *Met.* 36. 25-27 where he attributes the theory of partless parts of atoms to both when it is, as the doctrine of minimal parts, only Epicurean; see Furley, *Two Studies*, pp. 98-99.

215. 1. "in the division" (ἐν τῇ διαιρέσει). Juxtaposition involves the division of initially continuous bodies into indivisible particles.

215. 10. "preconception" (πρόληψις). Alexander uses this in the same sense as "common notion" (κοινὴ ἔννοια) or "natural notion" (φυσικὴ ἔννοια), and often uses the expression "common preconceptions" (κοιναὶ προλήψεις, 215. 32, 220. 25). Further on Alexander's use of this Stoic terminology see my discussion at *Symb. Os.* 48 (1973) 61-62 and on 5. 218. 11.

215. 11. "in the unification of the bodies that have been blended" (ἐν ἑνώσει τῶν κεκραμένων). "Unification" (ἕνωσις) terminates blending for Aristotle (*de gen. et corr.* A10 328b 22) as is constantly emphasised by Alexander in his restatement of the Aristotelian theory (228. 35, 229. 1, 230. 33, 230. 36, 231. 4, 12, 15-16, 26-27). Cf. Alex. ap. Simplic. *de caelo* 659. 23-26 for an argument that neither planes nor atoms can form a continuous body.

215. 13. "Reduction-to-the-elements" (ἀναστοιχείωσις) (also at 215. 18, 23) is a rare word. *LSJ* does not record these instances nor

Philop. *de gen. et corr.* 193. 28-29 (on 327b 31ff., and probably derived from Alex. *ad loc.*—possibly itself a source of the present passage). Philo (at *SVF* II, p. 188. 6) reports Chrysippus as using the expression "fire that reduces the order of the universe to itself" (τὸ ἀναστοιχειῶσαν τὴν διακόσμησιν εἰς αὑτὸ πῦρ), and it may well be a Stoic technical term.

215. 13-14. "for mixture should occur between bodies that are preserved" (ἡ γὰρ μῖξις σωζομένων γενέσθαι δοκεῖ). In an Aristotelian blend constituents were present in potentiality but "their power is preserved" (σώζεται . . . ἡ δύναμις αὐτῶν, *de gen. et corr.* A10 327b 30-31). Cf. also on 230. 14-15.

215. 16-17. "by their particular composition and interlinking (τῇ ποιᾷ συνθέσει τε καὶ περιπλοκῇ); cf. 1. 213. 20-21 where this expression describes, in part, the coming-to-be of secondary bodies from atoms for *both* the Atomists and Epicurus. It shows again how forced is the attempt to distinguish between their respective theories of mixture in this chapter.

215. 26-27. "By juxtaposition of parts" (κατὰ παράθεσιν μερῶν) refers to the "Democritean" theory at 214. 18-28, and "by mixture of atoms" (μῖξιν ἀτόμων) to the "Epicurean" (214. 28-215. 8). There is no evidence that there was also a separate theory of mixture *qua* juxtaposition of planes, or partless bodies, to complement the explanation of coming-to-be (γένεσις) in terms of these principles at 214. 1-5. Alexander would have had to be as arbitrary as he is in the cases of Democritus and Epicurus (cf. on 214. 18-28 and 214. 28-215. 8) if this were to be established.

215. 30. "not a demonstration of great learning" (ἐπίδειξις ἱστορίας). This perhaps means that the erudition of a commentary is not to be expected; but cf. *de fato* 165. 5-7 and p. 20 n. 95.

Chapters 3 and 4. The commentary on these chapters has been largely preempted by the discussion in Pt. II-ii. On 3. 216. 14-217. 2 see especially pp. 54-65, and on chapter 4, pp. 68-71; for a general analysis of this whole text see pp. 54-65.

Chapter 3

216. 1-4. "matter is entirely unified" (κοινῶς ἡνῶσθαι τὴν ὕλην). That is "matter is unified as common to [or shared in by] all the bodies

for which it is a substrate" (cf. ὑποκεῖσθαι at 1. 213. 16; 13. 229. 16). This language may derive from Aristotle *de gen. et corr.* B5 332a 18 where the elements are said to have a ὕλη κοινή ("common matter"). At *Quaest.* II. 28 there is an interesting discussion of the sense in which matter is "common" to forms as compared to the sense in which genus is common to species. For ὕλη described as κοινὸν ὑπόκειται πᾶσιν in an Aristotelian context see Ps.-Ocellus 16. 23 Harder.

In the *de mixtione* Alexander initially attempts to group the Stoic and Aristotelian theories of blending as both being dependent on a theory of unified matter; 1. 213. 15-18 and 214. 8-10 refer back to the general description of the main difference between these two theories at 213. 2-15. In the present chapter the description of matter being unified by pneuma at 216. 14-15 picks up 216. 1 and 4-5. However this glosses over the important difference in the sense of "unified" (ἡνῶσθαι) in each of these cases. In the former unification is the result of a physical force pervading matter, whereas in the more general description unification is an inherent property of matter. It is the latter sense which emerges clearly in Alexander's description of matter's relation to the constituents of blends in ch. 13 (229. 9, 10, 16; cf. 14. 230. 20-21, 231. 6) where bodies are said to share in (κοινωνεῖν) the same matter; matter's unity is simply its being the identical substrate for different forms. Hence at 10. 223. 32-34 Alexander can suggest that pneuma's function of unifying matter (he mentions its creation of sympathy between bodies—cf. 216. 16-17 below) is best explained by "the sharing of bodies in matter" (διὰ τὴν τῆς ὕλης κοινωνίαν); here the Aristotelian sense of matter as a common substrate is opposed to the Stoic notion of matter being unified by something. This establishes the distinction implicit in the present text, and may explain why in ch. 13-15 in the exposition of the Aristotelian theory of blending no mention is made of the Stoics, and of the conflict between those who held the same theory of unified matter referred to at 1. 214. 8-9.

Marcus Aurelius does refer to "one common substance" (μία οὐσία κοινή) at XII. 30, and κοινός is widely used by the Stoics to describe the λόγος or principle of universal order (*SVF* IV, Index s.v.). However it is always implied that this common factor is a result of a force creating a relation between things rather than, as in Alexander's Aristotelian usage, a common substrate for bodies.

216. 6-14. This is not a very convincing description of dissent among the Stoics. In physics Antipater of Tarsus was a conventional Stoic (cf. *SVF* III, pp. 249-251) and Sosigenes is only referred to in this text. Alexander must be speculating that there was Aristotelian influence on the Stoic theory of mixture; for short of forcing a total reconstruction, this could only have been minimal. The passage is best regarded as further elaborating the conventional description of the dispute in ch. 1; cf. on 1. 214. 10-16.

216. 15-16. "through which the whole is held together and is stable and sympathetic with itself" (ὑφ' οὗ συνέχεταί τε καὶ συμμένει καὶ σύμπαθές ἐστιν αὑτῷ τὸ πᾶν). This use of τὸ πᾶν overlooks the Stoic distinction between the whole (τὸ ὅλον) and the all (τὸ πᾶν) where the latter includes the cosmos and the infinite external void; cf. p. 56 n. 142 above, and Alex. *Quaest.* III. 12, 101. 22-102. 19, an argument that turns on an equation of ὅλον and πᾶν. The triad is typical of Alexander's characterisations of Stoic doctrines: cf. 4. 218. 2-6, 10. 223. 8-9 = the present text, 11. 225. 2; and *de fato* 178. 8-9 (προκαταβεβλημέναις καὶ ὡρισμέναις καὶ προϋπαρχούσαις τισὶν αἰτίαις). He also employs it to describe Peripatetic doctrines—cf. below, 10. 223. 10-12, 16. 236. 27-28, and *Met.* 358. 36—as well as in non-philosophical contexts (e.g. *de fato* 204. 20-21). Though a somewhat rhetorical device, in the present case all the components are significant, though elsewhere terms seem to be added for effect (see notes on 4. 218. 2-6 and 16. 236. 27-28). συνέχεται refers to the "force creating continuity", the συνεκτικὴ δύναμις of pneuma (cf. above p. 36); for συμμονή ("stability") as an effect of pneuma cf. *SVF* II 550. As for συμπάθεια, there is no reason to deprive early Stoicism of this doctrine as K. Reinhardt (*Kosmos und Sympathie*, p. 54 n. 2) wished; see the criticisms of M. Pohlenz, *GGA* (1926) 277, followed by many since.

216. 18. "two or more" (δύο ἢ καὶ πλείω). Cf. 216. 28-29; 220. 27; and Stob. *Ecl.* I. 17, p. 154. 15, 22 and p. 155. 11 (= *SVF* II, p. 153. 6, 12).

216. 19. " 'by juncture' as he says." Cf. 221. 19 (ὑπὸ μιᾶς, ὥς φασιν, ἕξεως), and *Mant.* 115. 10, 131. 4, for similar disclaimers when a Stoic technical term is employed.

217. 2-13. This text links the description of the tripartite classification of mixtures (216. 14-217. 2) and the account of the proof for

total blending in ch. 4. There is, as we have seen, a conflict between the notions of total blending expressed in these two texts which can only be reconciled by recognising their distinct purposes (pp. 65-71 above *passim*). But for Alexander they almost certainly represented simply the evidence of two different sources; 216. 14-217. 2 probably being based on a doxographical source (see pp. 54-65), and ch. 4 perhaps emanating from a Stoic scholastic source (see on 217. 36). In the text under consideration he is attempting to provide a logical transition from the one to the other. He was evidently aware that the argument in ch. 4 was based on the common notions; at 217. 13 he describes the πίστεις ("suasions") that prove that body completely extends through body, picking up the claim at 217. 2-4 that the differences between the three kinds of mixture are "established" (πιστοῦσθαι) by the common notions. However he only has available the analogical argument for total blending given in ch. 4 and therefore has to correlate the use of the common notions there with the existence of three types of mixture in the classification in ch. 3. This he partly achieves by claiming that Chrysippus said that we "at least" (γοῦν, 217. 4) have a different "presentation" (φαντασία) of the *product* of each type of mixture (note the perfects συγκειμένων, συγκεχυμένων, κεκραμένων at 217. 5-6) which would not be possible if all the *processes* of mixture (παντὰ τὰ ὁπωσοῦν μιγνύμενα 217. 8, cf. 216. 17) were alike He then emphasises this by repeating (217. 9-13) the distinction between the processes of juxtaposition and blending from the earlier tripartite classification. However the fact that such different processes result in different presentations (φαντασίαι) does not constitute proof of how these processes occur, and Alexander has certainly not explained why in ch. 4 only total blending can be proved from the common notions. However at the same time he has introduced the Stoic concept of presentation (φαντασία) in addition to that of the common notions and has to establish their relation. The standard Chrysippean doctrine is that the "apprehensive presentation" (καταληπτικὴ φαντασία) is the "standard of truth" (κριτήριον τῆς ἀληθείας); cf. Diog. Laert. VII. 54 = *SVF* II 105, and see F. Sandbach in A. A. Long, ed., *Problems of Stoicism*, pp. 9-21, for further discussion of the doctrine. Alexander was presumably aware of this; at *de an.* 71. 10-15 he assimilates into his account of the Aristotelian doctrine of the imagination (φαντασία) the Stoic distinction between apprehensive and non-apprehensive (ἀκατάληπτος) presentations (this text is at *SVF* II 70, but is clearly a vestigium). However since the presentations of the products of the mixtures only provide, as we have seen, a limited proof that the corresponding processes occur, they cannot be the highest standards of truth; this title must be reserved for the common notions which in ch. 4 show how at least one of these processes, total blending, occurs. Hence, the description at 217. 2-4 (cf. 218. 12, 220. 25-26) of these notions as μάλιστα δὲ κριτήρια τῆς ἀληθείας ("above all the standards of truth"). Apart from the fact that this is unparalleled, we could doubt its veracity simply because independent proofs from the common notions for all three types of mixture are not provided, and, if we read chapter 4 as I have suggested (see pp. 68-71 above), need not be provided since only total blending is of any significance to Stoic physics. Hence 217. 2-4 must be regarded as Alexander's own rather weak rationalisation of the relation between the tripartite classification of mixtures and the proof in ch. 4 of only total blending; this text can therefore scarcely serve as evidence that the common notions are products of similar presentations which act as a second standard of truth after individual presentations, as Gould (*Philosophy of Chrysippus*, pp. 61-62) has argued. Cf. also K. Abel, *Gnomon* 44 (1972) 650 with n. 25, and my criticisms at *Symb. Os.* 48 (1973) 56-57. There however I did argue (pp. 57-59) that the common notions are principal standards of truth in that they serve, as chapter 4 shows, as a basis from which a central theory of Stoic physics can be derived (see also above, pp. 41-44 with n. 106). This suggestion was an attempt to make sense of *de mixt.* 3. 217. 2-4 on the assumption that it represented Chrysippean doctrine. Since it now appears to be Alexander's independent rationalisation, that interpretation must be modified. I still think that the common notions are, as ch. 4 shows, general concepts from which central doctrines of Stoic philosophy can be inferred (cf. p. 41 with n. 99 above), but now doubt that this view can be grounded in the actual expression "highest standards of truth," taken as authentic Stoic terminology. At the same time I do not think that Alexander can be acquitted of having tried to impose a hierarchical structure on Stoic epistemology by our translating μάλιστα κριτήρια (217. 3) as "excellent standards" with Sandbach (*Problems of Stoicism*, p. 23); the context shows that he obviously regards them as "excellent" vis à vis φαντασίαι.

217. 2 "their own natures" (φύσεις); cf. 217. 7. This use of φύσις in the sense of form is Aristotelian; cf. *Met.* Z6 1031a 30, 1031b 1; also 16. 234. 3 where a body undergoing growth is said to preserve its οἰκεία φύσις.

217. 11-12. "in such a product of mixture through blending" (ἐν τῷ τοιούτῳ κεκραμένῳ μίγματι). An odd expression, but logical if we recall that mixture (μῖξις) is presented in this chapter as a generic notion so that the result of a blend can be termed a blended mixture.

Chapter 4

217. 13. "suasions" (πίστεις); cf. 217. 2. That is, these are not formal proofs, but like the non-demonstrative rhetorical arguments to which the term applies very often in Aristotle, e.g. *Rhet.* 1355b 35. It is not a Stoic term and has almost certainly been imposed by their Peripatetic critic. Cf. its use at *de fato* 202. 2. This is a slight revision of my suggestion at *Symb. Os.* 48 (1973) 50 n. 18.

217. 14. "the proponents of this doctrine" (οἱ προϊστάμενοι τῆσδε τῆς δόξης). At 3. 217. 2-13 Chrysippus was the subject carried over from 3. 216. 14-15, where he was, arbitrarily I believe (see p. 64 n. 175), made the author of the whole tripartite classification of mixtures. The switch here is almost certainly due to Alexander's use of a new source—see on 217. 36—the later reference to Chrysippus at 218. 8-9 notwithstanding (cf. note *ad loc.*).

217. 24. "the part that befalls us" (τὸ ἐπιβάλλον ἡμῖν μέρος). Farquharson (II, p. 722) compares this with Marcus Aurelius VII. 7 where τὸ ἐπιβάλλον is used more generally than here. Like Marcus II. 1 and VI. 14 this seems to be an ethical version of the notion of cooperative causation that we have located in this argument; cf. above, pp. 39ff.

217. 33. "clear evidence" (ἐναργῆ μαρτύρια) is probably Alexander's expression; cf. *de fato* 186. 23. It is vague in meaning and conceals the precise logical relation of 217. 34-218. 10 to the preceding analogical argument. As I suggest above (pp. 41-44), this evidence is, for the most part, to be regarded as a set of inferences from the case of a wine drop's extension through water.

217. 33-218. 1. "soul ... Nature ... State" (ψυχή, φύσις, ἕξις). On this tripartite classification see above, p. 41 n. 94. It is criticised at 9. 222. 26-34.

217. 33-34. "its own substantiality" (ἰδία ὑπόστασις). This expression is related to κατ' ἰδίαν [ὑπόστασιν understood] ὑφεστάναι that we find at 9. 222. 29 and 13. 228. 24 (cf. καθ' αὑτὰ ὑφεστάναι at 228. 13-14) used in an Aristotelian sense to mean "exist individually or independently." Cf. also on 13. 228. 13-14. This shows that the present text reflects Alexander's assimilation of this originally Stoic terminology rather than its use in an authentically Stoic fashion. I translate the noun "substantiality" since it is only used in the singular as an abstract noun to mean "property of independent existence." The evidence for the use of ὑπόστασις by Stoics is complicated: see the discussion by H. Dörrie, *Nachr. Akad. Gött.* (1955) 49-58, and for the debate about the relative meanings of ὑπάρχειν and ὑφεστάναι in Stoicism, the articles by P. Hadot, *Arch. für Begriffsgesch.* 13 (1969) 115-127, A. Graeser, *ibid.* 15 (1971) 299-305, and V. Goldschmidt, *REG* 85 (1972) 331-344. I would only note here that there is a very extensive usage of ὑπόστασις/ὑφεστάναι and ὕπαρξις/ὑπάρχειν in Alexander and the later commentators that deserves to be analysed in detail; conceivably it would throw some light on the vexed question of the relation of these terms in Stoicism. H. Dörrie, *op. cit.*, pp. 59-60 gives only a brief sketch of Alexander's usage of ὑπόστασις. For other examples of his use of καθ' αὑτὸ ὑφεστάναι see *de an.* 19. 7, 20, and for κατ' ἰδίαν ὑφεστώς see *Met.* 84. 28. Two texts which use ὑπόστασις and ὕπαρξις in single contexts are *de an.* 90. 1-9 and *Top.* 355. 11-16, where there seems very little difference between them, the former specifying something's existence as a substance, while the latter refers to its existence in general. Further on ὕπαρξις see C. Kahn in *Islamic Philosophy and the Classical Tradition*, pp. 151-155.

217. 36. "for there is nothing in the body possessing the soul that does not partake of the soul" (οὐδὲν γὰρ ψυχῆς ἄμοιρον τοῦ τὴν ψυχὴν ἔχοντος σώματος). Cf. Hierocles IV. 6-8 Von Arnim on the relation between soul and body: ὡς μηδὲ τοὐλάχιστον τοῦ μίγματος μέρος τῆς ὁποτέρου αὐτῶν ἀμοιρεῖν μετοχῆς (trans. above, on 1. 213. 2-13). In view of this general similarity of expression (cf. 217. 10-11 and note on 218. 8 below; also cf. Hier. IV. 4-5 with Alex. *de an.* 13. 19, 14. 23, and IV. 10-11 with Alex. *Mant.* 117. 9-11), and of the general relevance of Hierocles' description of total blending at IV. 5-10 Von Arnim to Alexander's critique (see on 213. 2-13), it is quite likely that Alexander's source for ch. 4 was a Stoic scholastic

of the type represented by Hierocles. Possibly this is why the whole argument in ch. 4 is attributed to "proponents" of the doctrine of total blending (217. 13-14) rather than Chrysippus who is not referred to until 218. 8-10. Cf. also the reference to Chrysippus' "followers" (οἱ κατ' αὐτὸν φιλοσοφοῦντες) at 218. 10.

218. 1-2. "that fire passes completely through the iron." Cf. above, p. 40 n. 93 and p. 56 n. 141 on the reasons for the presence of this example here. It is used as an illustration of the blending of soul with body at Hier. IV. 8-9 Von Arnim, and is adapted to the case of the incarnation at Origen *Contr. Cels.* III. 41 and *De Princ.* II. 6. 6. It is subjected to a detailed empirical analysis at 12. 227. 13-228. 4.

218. 2-6. Two triads of adjectives describing the two pairs of elements: "rare"/"dense" (λεπτομερής/παχυμερής) are Aristotelian (e.g. *de caelo* 304a 30-31); "having/lacking tension" (εὔτονα/ἄτονα) are Stoic (cf. *SVF* II 444); "light"/"heavy" (κοῦφα/βαρῆ) seem to make up a redundant completion of the triad (cf. on 3. 216. 15-16). This whole text serves as an alternative description of the relation between pneuma and matter in terms of the active and passive elements (cf. p. 35 above), yet Alexander does not discuss it in his critique of the theory of pneuma in chapter 10. It would be particularly relevant to his comments on the relations between pneuma and the elements; see notes on 10. 224. 14a-22, and 11. 225. 5-18. It is hard to believe that Alexander did not see the relation between this description of the two pairs of elements and the description of pneuma at 3. 216. 15-16. He could not however have acknowledged it without adopting the critical approach to source material totally absent from ancient philosophical polemics.

218. 6. "continuity" (συνέχεια). Literally: "the state of being held together." This is not entirely accurate. Pneuma, *qua* the two active elements Fire and Air, has an internal coherency in its own tension (see *SVF* II 447), while the passive elements are held together by it. Each pair does not have its own separate "continuity" (cf. *SVF* II 439, 440, 444).

218. 6-8. "poisons that are destructive..." There is no parallel, as far as I know, to this rather vague example of total blending. It does not self-evidently illustrate the blend of a small volume with a larger. Perhaps it was displaced here by a process similar to

that probably involved in the location of the case of fire pervading iron at 218. 1-2.

218. 8. "in a total juxtaposition" (ὅλα δι' ὅλων παρατιθέμενα) seems to contradict "blended" (κιρνᾶσθαι) in 218. 7, and in view of the tripartite classification is completely contradictory. We do find δι' ὅλων ... ἡ παράθεσις ("complete juxtaposition") at Hier. IV. 9-10 Von Arnim (cf. on 1. 213. 2-13) and Alexander may have slipped into this authentic Stoic terminology; cf. on 3. 217. 36.

218. 8-9. The thesis that light is a body is attacked as a case of body going through body at *Mant.* 139. 1-17; cf. above, p. 85 n. 246 and also 9. 222. 33-34. The reference to Chrysippus may indicate that Alexander has added this additional example of total blending from a source other than that from which bulk of chapter 4 is derived.

Chapters 5 to 9. These chapters constitute Alexander's critique of total blending taken in isolation from the Stoic physical system, while 10 and 11, as we shall see, deal with its ramifications in the theories of pneuma and God. Despite Alexander's acknowledgement at the opening of chapter 12 (227. 1-11) that total blending is the basis for these and related doctrines (fate, providence, sympathy, the soul, etc.) he never criticises this relationship as such, but takes each of the terms in it separately: total blending in chapters 5 to 9, pneuma and God in chapters 10 and 11. See further on this the introductory note to chapters 10 and 11.

The general Peripatetic background to the arguments against body going through body in chapters 5 and 6 (to 220. 3) has been traced at Pt. III-iii above. Chapters 7 and 8 are independent attacks on the notion of total blending; in the first case in the light of the tripartite classification of mixtures, and in the second, that of the requirement that constituents be divisible. Chapter 9 considers total blending with reference to the doctrine of form and matter, a critique anticipated at 6. 220. 6-11, and later included in the polemics against the doctrines of pneuma (see 10. 223. 30-34) and God (see 11. 224. 27-32 and 226. 10-24). After the digression in these two chapters, chapter 12 returns to the theme of chapters 5-9 by offering a final detailed critique of one of the illustrations of body going through body (227. 10-228. 4)—a notion which 10 and 11 have shown to be at the root of two major doctrines of Stoic physics (cf. 10. 223. 18, 11. 225. 4-5, and 12. 226. 34-227. 10).

Chapter 5

218. 11. "the common notions." Alexander's implication (by our reading of the text) is that the notions employed by the Stoics are not in fact "common," since bodily interpenetration is not conceivable (προείληπται, 218. 17). For a notion to be common or natural for Alexander it must be a direct reflection of experience; he demands, in effect, that the common notions be principles of common sense, and uses the Stoic terminology in this way in several other places (e.g. *de fato* 165. 15—see further *Symb. Os.* 48 [1973] 62 nn. 83, 84). This completely rejects the analogical argument in chapter 4 in which the notion of a body extending over another body was not formed directly from experience, but by analogy from certain causal relations manifest in experience. If my interpretation is correct (see p. 41 n. 99 above) the common notions serve as a synthesis of experience (in this sense they are natural) from which the theory of extension of body through body can be derived. It is in principle impossible for such analogical argumentation to directly reflect experience as Alexander demands. As we see in ch. 15, Alexander himself has the problem of explaining the sense in which constituents are in potentiality in a blend by analogy with the fact that they are easily separable; he leaves exactly the same gap between experience and theory that here he refuses to allow the Stoics in insisting that bodily interpenetration be based on naïve common sense (see notes on 15. 231. 12-29, and 232. 32-233. 2). Also important is Alexander's use of the terms "common notions," "preconception" (πρόληψις), "common preconception," "natural notion" as precise equivalents—cf. 2. 215. 10, 32; here at 218. 14-15, 17; 7. 220. 25; 9. 222. 26; 11. 227. 11; 13. 233. 1. I have argued (*Symb. Os.* 48 [1973] 51-54) that the common notions are distinct from the two equivalent terms "preconception" and "natural notion," and that this difference can be gleaned in part from ch. 4 (cf. p. 45 n. 107 above).

218. 15-16. "...bodily interpenetration with full coextension" (τό τε γὰρ σῶμα διὰ σώματος χωρεῖν ὅλον ὅλῳ παρεκτεινόμενον...). A clear illustration of how the theoretically rich notion in the verb describing extension is simplified by the general verb of motion; cf. 217. 9-10 for a similar juxtaposition.

218. 16. "fails to strike us" (<οὐ> προσπίπτει). For προσπίπτειν used particularly of sense-presentations (φαντασίαι) of states of affairs

being encountered see Epictetus II. 28. 8 and Marcus Aurelius III. 2. This Stoic usage can be found in Alexander's discussion of the relation between φαντασία and action in the *de fato*; e.g. 184. 3 or 186. 6-7. Cf. also *Quaest.* III. 12, 101. 12. I have only found the usage earlier at Philo *De Ebr.* 47 (II. 208. 7-8 Cohn-Wendland) apart from the doxographic definition of τὸ εἰδικόν as προσπῖπτον ἤδη αἰσθητόν (Aet. IV. 9. 13 = *SVF* II. 81); it is difficult therefore to determine whether it is an early Stoic term. At any rate for Alexander its association with φαντασία could be taken for granted and implicit in this sentence therefore is a continuation of the critique of the Stoic employment of their doctrine of common notions, for they, as we have seen (on 217. 2-13) are derived from φαντασίαι. The reference to preconception at 218. 17 (ἀλλὰ καὶ ...) shows that Alexander saw that it was a stage beyond presentation although he did not distinguish between it and the formation of common notions; see preceding note.

218. 17-24. "It is, indeed, a natural notion..." The general claim is that if body cannot be received by body then it is received by place, which is defined as "receptive of body" (σώματος δεκτικόν, 218. 21). It might appear that Alexander is somehow endorsing the interval theory of place, especially when he refers in 218. 23 to "empty interval" (κενὸν διάστημα). This cannot be so; this theory is refuted by Aristotle at *Phys.* Δ4 211b 18-212a 4 before he establishes his theory that place is the limit of the surrounding body, and we find the argument at 211b 19-25 repeated at Alex. *Quaest.* III. 12. 106. 10-13 (anti-Epicurean in this context rather than anti-Stoic as implied by *SVF* II 536). It is rather that the logic of the present argument leads him to suggest that if a body cannot go through a body then it must enter spaces or intervals which in the opinion of some (δοκεῖ τισιν, 218. 20) can be defined as place *qua* receptacle. This then demands an analysis of what possible receptacles could explain bodily interpenetration which 218. 24-219. 9 provides. Alexander is simply following a general notion to its logical limits and can therefore introduce ideas required by the argument rather than ones he, or for that matter his opponents, would endorse. Thus while the Stoics defined place as occupied interval (e.g. *SVF* II 503, 504) this is irrelevant to the present argument since they could not have considered the possibility of bodily interpenetration through pores. It is probably

because the present text is so influenced by Aristotelian texts in which Atomism was criticised (see next note), that Alexander is led to introduce the Atomists' definition of place, and they are the likeliest "some" referred to in 218. 20.

218. 17-18. "...a natural notion that what is full can no longer receive anything in itself" (φυσικὴ γὰρ ἔννοια τὸ τὸ πλῆρες μηκέτ' ἐν αὑτῷ δύνασθαι δέχεσθαί τι). This very closely reflects Aristot. *Phys.* Δ6 213b 6-7, which is part of an argument for the void which goes on (213b 7-12) to claim that without the existence of the void two solids will occupy the same place, and *a fortiori* a large body will occupy the place of a small one. On the significance of this for the cluster of arguments surrounding the notion of body going through body, see above pp. 80, 87-88. The use of "receive" (δέχεσθαι) reduces the relation of bodily interpenetration to a case of two bodies simply being in the same place. Any hint of the dynamic physics involved in the relation in the context of the theory of pneuma is removed. Cf. also 218. 28; 219. 9, 30, 33; 220. 4, 5, 11, 12, and *de an.* 20. 9-15 for a similar usage.

218. 21-24. "full of itself" (πλῆρες αὑτοῦ). Cf. *de an.* 20. 10, 11; and for the equivalent μεστὸν αὑτοῦ, *de an.* 20. 9, 10 and *Mant.* 139. 34-35.

218. 24-219. 9. This is one of three sets of arguments in *utramque partem* employed in the *de mixtione*; cf. chapters 7 and 8 *in toto*. I have argued elsewhere (*Philologus* 116 [1972] 300-305) that 218. 33-219. 5 is an outgrowth of Aristotle's argument against the "Empedoclean" theory of pores at *de gen. et corr.* A8. Cf. 326b 6-10 with 218. 33-219. 1, and 325b 5-9 with 219. 1-5. This seems confirmed by Philoponus on *de gen. et corr.* 326b 6-10 (178. 5-20) which basically parallels 218. 24-219. 5. Cf. further pp. 78-79 above on the background to these arguments and for further parallels. In the article cited I have also discussed the extent to which this and related texts could be directed against Strato of Lampsacus' theory of interstitial pores. It may be this passage which led Plotinus to say that those who believe in total blending give their opponents the chance to suggest "exits of air" (ἀέρων ἐξόδους) as an explanation; see II. vii. 1. 34 (cf. 20-21).

218. 31. "the narrowest vessels (ἀγγεῖα) of the body." Alexander may have had in mind a case such as that described by Aristotle

Sem. Griekse Letterkunde
Blandijnberg 2
Gent

at *de gen. an.* B4 738a 10-16 where nutriment blocks "very fine blood-vessels" (λεπτοτάται φλέβες) and leads to a haemorrhage. For ἀγγεῖα = φλέβες see Bekker, *Index*, 5a 22-24.

219. 3-9. Cf. 16. 234. 29-32 and see note on 16. 234. 23-32.

Chapter 6

219. 9-22. This argument is the corollary of 5. 218. 17-18. If a solid cannot receive a solid in itself, then when compounded with one an increase must occur. Alexander now proves this in detail in the face of the possibility, presented by the notion of bodily interpenetration, that there is no such increase. The cases in which this happens are then seen not to involve relations between bodies; see 220. 6-11.

219. 11-12. "why any given body does not contribute to an increase in the size of a similar body in all dimensions" (τί δήποτε οὐ τὸ τυχὸν σῶμα [τὸ τυχὸν] πάντῃ συναυξητικόν τοῦ ὁμοιοῦ). This is the general form of an argument we find particularly applied to the theory that light is a body to yield the paradox that light should increase the size of its medium: see *Mant.* 124. 13-14, 16-20, 123. 12-20 and cf. Ps.-Galen *de qual. incorp.* (XIX. 473 K = Westenberger, p. 9. 10-16); see also *Mant.* 116. 1-2 (with reference to the thesis that the soul is a body), and 116. 13-15 (regarding corporeal virtues).

The definition of body as a three-dimensional solid (here at 219. 16-17) on which the arguments of this chapter depend is assumed in these other texts, and occasionally made explicit (cf. Nemes. *de nat. hom.* 2. 71 Matth., and Plot. VI. 1. 26 [= *SVF* II 315]). It was, of course, standard: see Aristot. *Top.* 142b 25, *Phys.* 204b 20, and *de caelo* 268a 2, 268b 14, and 298b 3 where body and magnitude are equated; cf. also Aetius I. 11. 1, Ar. Did. at *Dox. Graec.*, 449. 6-11, and the Stoic definitions at *SVF* II 357, and III, p. 259. 24-26 (Apollodorus). As we have argued (above, pp. 47-48) the Stoics could accommodate this definition insofar as it applies to intracosmic bodies; indeed, they must have, or else there would have been no need to prove the extension of body through body by analogy with actual phenomena, as in *de mixt.* 4; it would have had actual exemplifications, as Alexander demanded (see note on 5. 218. 11). An alternative Stoic definition of body was only necessary, we have argued, for the relation between pneuma and matter, and this was provided by appropriate characterisations in terms of a qualitative physics.

219. 15. "they will necessarily have something beyond contact" (ἐξ ἀνάγκης ἕξουσί τι παρὰ τὴν ἁφήν). Beyond contact is continuity; the two lines will therefore be continuous (συνεχές). Cf. Aristot. *Phys.* 231a 21-23. Plotinus (II. 7. 1. 44-48) may have borrowed and developed this text in this comparison of total blending to a case where lines, instead of being in contact and causing an increase, as here, would be superimposed on one another and produce no increase (ὡς ἂν γραμμὴ γραμμῇ ἐφαρμοσθείη, ὥστε αὔξην μὴ γίνεσθαι). II. 7. 1. 55-56, a few lines later, is almost certainly Alexandrian; Theiler and Beutler rightly refer to *de mixt.* 16. 234. 7-8.

219. 19. "makes the compound both less somehow and equal..." (ἔλαττόν ποτε καὶ ἴσον τὸ ἐξ ἀμφοῖν ποιεῖν). Less than and equal to the constituents, that is. If A goes through B, and occupies its place, then the compound AB is less than A + B, but equal to B in this case, and to A if the process is reciprocated, and if we assume A and B to be of different size. If they are of equal size, AB would anyway be equal to A or B taken separately. Alexander develops paradoxes of such reciprocal equalisation in ridiculing the Stoic view that in total mixture a small body is blended with a large one; see *Mant.* 140. 10-25, 141. 6-22. Particularly relevant to the present passage is 140. 15-18.

219. 20-22. "for the subject ... is abolished along with its abolished proprium." This applies the principle of "joint abolition" (συναναίρεσις): when something definitionally prior to something else is abolished it leads to the accompanying abolition of that thing; see Aristot. *Top.* 123a 14, 141b 28-29, and cf. Alex. *Top.* 127, 23-26 and 236. 20-21. The *de fato* contains a number of uses of the principle; e.g. 180. 3-4, 184. 20, 196. 3-7, 206. 30, 209. 8-9, 209. 30-31, 211. 22-28.

219. 23. "which comes to be in some body." γίνεσθαι ἐν ("comes to be in"), like δέχεσθαι ("receive") (see on 218. 17-18), is a term that eliminates the dynamic element involved in the Stoic account of body extending through body by means of the tensional motion (cf. p. 37 nn. 85, 86 above). It occurs also here at 219. 24, 27;

12. 227. 30; and at *de an.* 20. 14-15; cf. also Plut. *de comm. not.* 37. 1078B.

219. 24. "occupies no additional place..." (μηδένα προσεπιλαμβάνειν ἄλλον τόπον). Cf. Simplic. *Phys.* 591. 21-22 for a similar expression used of a body in motion. There is no parallel in Aristotle for this or the ἐπιλαμβάνειν at 16. 234. 21.

219. 22-28. Cf. Philop. *de gen. et corr.* 90. 17-20 for an identical argument. At *Mant.* 141. 4-8 Alexander argues that if soul and body occupy the same place, and soul is separated from body, then it will either occupy void (κενόν) or always be in a body; the first alternative implies that it has left a place empty.

219. 28-32. This argument seems to rest on a paradox expressed in rather compressed terms. I think it can be rationalised as follows. If A and B are totally blended then they occupy the same place; but if, *per impossibile,* an increase occurs when they are so blended then B cannot be in A's place but must occupy another place. However by the definition of total blending it can only occupy A's place. Therefore nothing (no body) occupies this other place. This amounts to an elaborate way of saying that the total blending of bodies is incompatible with their being aggregated into a larger compound (cf. 219. 19-20). If that occurred then the increase in size would not be evident; the body causing it would not be body.

219. 32-220. 3. This argument may be clarified by this diagram:

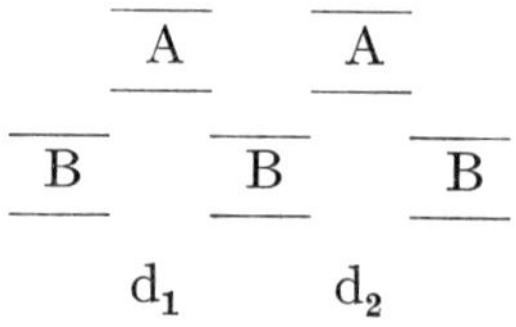

If A is being blended with B and is received at d_1 and d_2, then insofar as B receives A it does not do so *qua* body. For a division is not a body; cf. 8. 222. 3 for this point in a different context.

220. 3-11. This anticipates the argument in ch. 9 that the total mixture of bodies is in fact explaining what must be seen as the relation between form and matter; cf. 222. 35-223. 6. For the example of the ash (220. 10) see Aristot. *Phys.* Δ6 213b 21. Proponents of the void argued that growth occurs as water sinks into ash because it contains interstitial void, and it is impossible for two

bodies to be in the same place (213b 18-22). Simplic. *ad loc.* (*Phys.* 651. 4-7) uses the expression "body going through body," and it is presumably because of the example's association with this topic that Alexander finds it necessary to stress that what occurs here is change and not the passage of body through body.

220. 12. "mutual replacement." Alexander considers mutual replacement (ἀντιπερίστασις) as an alternative to bodily interpenetration also at *Mant.* 129. 1-2 and *de sensu* 29. 12-30. 1. He is undoubtedly thinking of Aristotle's reference to it as an explanation of the motion of projectiles at *Phys.* Θ10 267a 15-20; on this see Solmsen, *Aristotle's System of the Physical World*, pp. 136-137. The Stoic alternative is not simple bodily interpenetration but the motion of pneuma, the "tensional motion" (ἡ τονικὴ κίνησις); see note on 10. 224. 23-27. Conceivably the term ἀντιπαρέκτασις ("extension-through"), which we have seen (pp. 37-38) describes this motion in its most significant aspect, was formulated by Chrysippus as a deliberate response to ἀντιπερίστασις.

220. 13-23. This is a rather offhand treatment of the blending of a small body with a larger one (cf. 217. 31-32), the central feature of the argument in ch. 4. Except for this text Alexander offers no justification for ignoring it and only discussing the general notion of body going through body. At *Mant.* 140. 10-25 and 141. 9-25 there are more extensive arguments against the blending of unequal volumes (see pp. 87-88 above). Cf. also Plot. II. 7. 1. 49-56 (especially 53-55 with *Mant.* 141. 16-17), and IV. 7. 8^2. 13-15. On the relation between this aspect of blending and the theory of pneuma, see pp. 37-38 above.

220. 17-18. "the fact that incense and bodies that are similarly incinerated." This strictly empirical analysis of the example of heated incense cited from the Stoic argument at ch. 4 (217. 15-17) tries to establish its irrelevance as a foundation for the notion that a body can preserve its qualities when extended. This is an *ad hominem* point; several other illustrations of a less questionable nature were provided in ch. 4 (217. 19-26), and only if Alexander showed that the notion of a body in some way extending itself is inconceivable under any conditions would the basis for the whole analogy drawn in that argument be undermined.

220. 18. "over a considerable expanse" (ἐπὶ πλεῖον). Cf. ἐπὶ πλέον at 220. 20, and ἐπὶ τοσοῦτον at 4. 217. 30; see p. 31 n. 47 above.

220. 23. "association" (παράθεσις). For παρατίθεσθαι in this sense cf. Plato *Phaedo* 65e 8, and Plut. *de comm. not.* 1081C.

Chapter 7

220. 23-29. This reintroduces the notion, ridiculed in ch. 1 (213. 2-8), that constituents can be totally blended yet maintain their original qualities, and be separated from the blend. It was later repeated without critical comment in the definition of total blending at 3. 216. 25-217. 2 (cf. 217. 9-12, 27-31); now Alexander returns to it in the light of the tripartite classification of mixtures offered in ch. 3 and shows that blending must in fact be one of its limiting cases, juxtaposition (220. 37-221. 7), or fusion (221. 7-25). Only the former ensures that constituents preserve their original character, while only in the latter process are they totally mixed; yet juxtaposition is not total blending, and bodies cannot be separated from a fusion as must be the case with constituents of a blend.

This sentence shows very clearly how the polemic against "body going through body" is separate from that directed against total blending, as defined in chs. 3 and 4. 220. 23-26 summarises the successful critique in chs. 5 and 6, while 26-29 begins an attack on the specific details of the theory of total blending. Again (cf. on 1. 213. 13) we see the use of "body going through body" as an independent notion entailed by total blending rather than an integral part of the theory. See also on 227. 2.

220. 37-221. 7. The earlier emphasis (cf. on 1. 213. 2-13) on the preservation by constituents of their surfaces now pays dividends; for when juxtaposition is defined as surface contact (216. 20), and total blending is said to involve the preservation of surfaces (213. 5, 12), it necessarily follows that such blending is juxtaposition. If the particular expression "preserving surfaces" were Stoic it is strange that it is omitted in the formal exposition of the classification of mixtures in ch. 3, and only used in these two polemical contexts.

221. 7. "...that mixture and blending are distinct" (ἄλλο φασὶ μῖξιν καὶ ἄλλο κρᾶσιν εἶναι). "Mixture" here is used where we would expect "juxtaposition," and we can only assume that Alexander

has slipped into the usage we find in both Menon (Anon. Lond. XIV. 20-23 Diels), and Philo (at *SVF* II, p. 153. 29). Elsewhere (at 228. 27-28) μῖξις is a generic term for Alexander; cf. p. 56 n. 142 above. This text probably refers back in particular to 3. 217. 7-13 where emphasis is laid on the difference in the Stoic theory between blending and juxtaposition.

221. 7-20. This argument reduces total blending to fusion; for since the constituents of a total blend have now been shown to be preserved only if they are juxtaposed (i.e. not blending) then they are blended only if they are destroyed. Both Plutarch (*de comm. not.* 37. 1078B-C [cf. p. 69] and 40. 1080F-1081A) and Ps.-Galen (*de qual. incorp.* 471 K [pp. 7. 20-8. 7 Westenberger]), though with much less explicit argumentation, regard the only form of total blending beyond juxtaposition as fusion. They may have been drawing on an argument, such as the present one, in which the stages of this reduction were made clear.

221. 9-10. "totally uniform" (πᾶν ὁμοιομερές). Alexander slips into Aristotelian language (cf. 231. 15-16, 26-27) to emphasise that blending is not juxtaposition. When he goes on to say that such a homoiomerous compound will not be in juxtaposition but a total blend (κρᾶσις δι' ὅλων) he has therefore already altered the sense of "total" to mean a process that is complete and results in a unified body, rather than, as the Stoics I believe held (see pp. 32-33 above), one which is total because one body pervades the whole volume of another. Hence elsewhere Alexander could use δι' ὅλων μίγνυσθαι etc. to describe the completeness of a blend in Aristotelian context: cf. *de sensu* 63. 23, 64. 15, 25, and similarly Galen *de tempr.* I. 9, and *passim*.

221. 17. "then not even their States would be preserved." When bodies are totally blended and fused they lose their "states" (ἕξεις), and the product of this blend is a unified body with a specific "state" (ἕξις). Alexander's *ad hominem* point here rests on a standard definition of ἕξις as that by which a body is held together, e.g. *SVF* II, p. 124. 19-21. It is paradoxical, he is saying, for a body to lose this in a blend, but necessary that a blend with a "state" be indissoluble and its constituents inseparable.

The argument shows very clearly the effect of taking the tripartite classification of mixtures in ch. 3 as a description of the behaviour of intracosmic bodies. If Alexander had taken account of his own description of ἕξις at 217. 37-218. 1 he would have seen that it itself was a constituent that extended through the bodies held together by it, and that the theory of total blending was designed to explain *this* relation and not that between bodies with ἕξεις, as the present argument assumes. 9. 222. 27 and 31-32 does in fact argue against ἕξις in this capacity, but that critique is dissociated from the present *ad hominem* argument—piecemeal criticism being Alexander's method throughout—cf. on 220. 23-29 above, and on 6. 220. 13-23 and 220. 17-18.

221. 23-24. "we do see that in some cases they are separated." This looks ahead to ch. 15, especially 231. 30-232. 31 where the separability of constituents is illustrated and made the basis on which the nature of their presence in potentiality in a blend is explained.

Chapter 8

221. 25-28. This chapter is derived from Aristotle's claim (*de gen. et corr.* A10 328a 24, 328b 17) that only moist and easily divided bodies (εὐδιαίρετα) can be blended; cf. below 13. 228. 36 and 14. 231. 12-13. As a result, before bodies are fully blended they are juxtaposed as corpuscles (328a 33-328b 2; cf. 14. 231. 13-14), and their divisibility facilitates the blending in which a single unified body is formed. Here this causal relation between bodily division and blending is applied to the Stoic theory of total blending, and an exhaustive survey is made of all possible types of division in order to show that this theory is not satisfactorily explained by any of them. The discussion assumes, as was argued in the second part of ch. 7 (221. 7-25), that a total blend for the Stoics must be a unity, much as it was for Aristotle (hence the significance of 221. 9; cf. note *ad loc.*); for the whole chapter neglects the Stoic claim (216. 3-31) that the constituents are preserved in a blend, presumably on the basis of the argument at 7. 220. 37-221. 7.

221. 26. "particularly in liquids" (ἐν τοῖς ὑγροῖς ... μάλιστα). Cf. *de gen. et corr.* A10 328b 3, and 13. 230. 34-36 below.

221. 32-33. "thus blending would again occur by juxtaposition." Cf. the earlier arguments leading to this conclusion at 7. 221. 5-6 and 5. 219. 1-3.

221. 34-222. 3. This idea that bodies can be divided into incor-

poreal divisions is part of an argument for indivisible bodies quoted in *de gen. et corr.* A2 by Aristotle; see 316a 23-30 and 316b 12-14, cf. note on 222. 22-26 below. "Division" (διαίρεσις) seems to be a more generic term than "point" (στιγμή, 222. 25-26): cf. Aristot. *de an.* 430b 20-21, *Met.* 1060b 18-19. Its incorporeality is also emphasised in Alexander's argument at 6. 219. 32-220. 3. On division as an attribute (πάθος) of body see Alex. *Met.* 230. 29-30. For this notion of exhaustive division also note Plot. II. 7. 1. 31-32, and IV. 7. 8². 15-18 where division into points is mentioned and contrasted with juxtaposition.

222. 4-13. The general reasoning of this passage I take as follows; since bodies are totally blended insofar as they are divided, and since the previous argument has shown that such division can neither be limited nor exhaustive, then the only alternative is that constituents be divided to infinity. This must involve division into an actual infinite set of parts, since infinite division as an incompleted process would not explain total blending (6-13). However once one admits that total blending can only be explained by division into actual infinite sets (14-17), and says that these are composed of bodies, various paradoxes ensue (17-22), which, given that these bodies are not atoms, cannot be evaded by saying that the infinite sets are composed of incorporeal entities which cannot be the parts of a body (22-26).

222. 4-6 is thus a bald assertion that the one notion of infinite division open to the Stoics is impossible. But why this notion? First, division has to be complete (πάντῃ [222. 5] "in all places" or "through and through") if mixture is to be total (δι' ὅλων). Secondly, it has to be "in actuality" (ἐνεργείᾳ), since again the mixture has to be complete and a division which was only potentially infinite could not achieve this. By these criteria the division can therefore only be "into actual infinite sets" (εἰς ἄπειρα ἐνεργείᾳ [cf. Plot. IV. 7. 8². 18-20]); that is, since the division is complete, actual, and infinite it must necessarily terminate in an actual infinite set of parts. This is why I believe we should read διῃρῆσθαι at 222. 5-6 which is the perfect passive infinitive with a present sense—lit. "is in a state of having been divided." This is the only conclusion to which the Stoics can be led if they are to explain total blending in terms of infinite division, given that there is a causal relation between division and blending (221. 31-32).

There is an illuminating discussion of the broad distinction between the process of infinite division and a complete division into infinites (which, as 6-17 now shows, the Stoics are not allowed to draw) at Simplic. *Phys.* 141. 16-142. 3.

222. 6-13, then, simply reemphasises that the concept of infinite division applicable in this case must be that just described, and not the alternative raised here—that division be an unterminated process. This is, in an unrefined form, the theory held by both Aristotle and the Stoics (see on 222. 7), but Alexander only reaches it as an alternative to be excluded once it is established that only an actual complete infinite division can explain total mixture. In effect this argument is that at 221. 29-34, where it was said that if there was any remainder to a division then bodies would be unmixed (221. 30-31 and 222. 13), placed in the context of infinite division.

222. 7. "through the division never letting up..." (τῷ μηδέποτε ἐπιλείπειν τὴν τομήν). The Stoic and Aristotelian theories of infinite division were very close, and 222. 6-8 expresses a general view compatible with both. For the Stoics see for example *SVF* II 482. The only difference was that for Aristotle (*Phys.* Γ6 *passim*) the infinite was in potentiality, a concept not invoked by the Stoics (see my discussion at *Apeiron* 7 [1973] 23 nn. 13, 14). If Alexander were arguing explicitly against the Stoics at this point rather than against one of the theoretically possible forms of infinite division he would presumably reject it for this reason; this appears to happen in some of the commentators' arguments against other theories of infinite division (see on 222. 17-22 below).

222. 14-17. By reading διῃρῆσθαι (the present perfect) at 222. 15 (see note on the text) this sentence spells out the consequences of infinite division which 222. 4-6 has implied that the Stoics would have to accept if they wished to explain the totality of blending; it also introduces the arguments (222. 17-22) which sustain the earlier claim that this is impossible (ἀδύνατον, 222. 4).

222. 17-26. For the broad alternative of infinite division into magnitudes or non-magnitudes posed here, see also Simplic. *Phys.* 142. 3-15 (continuing the passage referred to on 222. 3-13 above, and so suggesting that Alexander's critique here may have originated

in a similar discussion in his lost commentary on the *Physics*). Cf. also Philop. *de gen. et corr.* 35. 13-36. 30.

222. 17-22. If the constituents are divided to infinity so that they each have a set of actual infinite parts, then, by this argument, they will each be infinitely large, since the sum of an infinite set of parts with size is an infinite body; for they must be composed of the things into which they are divided. The general principle that the sum of an infinite number of bodies is an infinite body seems to have been clearly formulated first by Eudemus (ap. Simplic. *Phys.* 459. 22-26) in commenting on Aristotle's theory of the infinite. Certainly elsewhere (*Phys.* 492. 14-21) Simplicius makes it clear that Aristotle's theory of the infinite as being in potentiality evaded this paradox. For Epicurus (*Ep. ad Hdt.* 57) it was a reason for positing minimal parts in the atom (see Furley, *Two Studies*, pp. 14-15), and is, I have argued, an influence on Chrysippus' theory of infinite divisibility (see *Apeiron* 7 [1973] 21-29). It may also have been a presupposition of Zeno's major argument against plurality (DK frs. 1 and 2); see Vlastos, *Encyclopaedia of Philosophy*, VIII, 370-371. It occurs in a variety of contexts in Simplicius' commentaries (*Phys.* 168. 34-169. 1, 172. 28-31, 459. 22-26, and *de caelo* 608. 12-17 and 635. 16-17) but was also known to Alexander (ap. Simplic. *Phys.* 173. 22-23 and *Quaest.* III. 12. 103. 34-36) and Philoponus (see on 222. 21-22 below). Cf. also Sext. Emp. *P.H.* III. 44.

As the present passage shows, this principle rests on the assumption that there is a complete infinite set of parts existing in actuality. This is required by the Stoic notion of total blending, but elsewhere we find it imposed on the notion of an infinite number of atoms in Epicureanism (Alex. *Quaest.* 103. 34-36) and on Anaxagoras' theory of an infinite number of homoiomerous bodies (Simplic. *Phys.* 172. 28-29). In none of these cases does the opponent assert that an infinite set can be complete, and this principle seems to refute their theories of the infinite only because they do not like Aristotle allow for a potential infinite, and can therefore be assumed to be positing an actual or complete infinite. The paradox thus offends our intuitive conception of the infinite as incomplete, and serves as a *reductio* of these doctrines. Further on the general logic of this principle, see Vlastos, *loc. cit.*

222. 20-21. "if something must be composed of just the things into

which it is divided" (εἴ γε εἰς ἃ τέμνεταί τι καὶ συγκεῖσθαι αὐτὸ ἐκ τούτων ἀναγκαῖον). Cf. 222. 1-2, and 222. 24-25. This principle of simple addition is thus applied to an infinite series just as it might be to a finite series. Cf. Jackob Bernoulli (quoted by Vlastos, *Encyclopaedia of Philosophy,* VIII, 371): "If 10 members are present [in an infinite progression] the 10th necessarily exists, if 100 then necessarily the 100th, ... if therefore their number is infinite then the infinitieth member must exist." So by the principle that Alexander cites, a body is as large as its constitutive units; if these are infinitely numerous, it is infinitely large.

222. 21-22. "and this would result in there being still further additional infinites" (συμβαίνοιτο <δ'> ἂν οὕτω καὶ πλείω ἂν εἶναι σώματα ἄλλα ἄπειρα). If the sum of an actual infinite set of bodies is an infinite body, why should there be further infinite bodies, when the original set of which the first body is the sum is, by definition, complete? For Alexander must mean not that these additional infinite bodies will simply be further constituents in a blend, but that they are the logical consequence of there being a single infinite body.

A possible Aristotelian source is *Phys.* Γ5 204a 20-28 where it is argued that an infinite body will be divided into infinites, so that the same body will be many infinites—which is impossible (πολλὰ δ' ἄπειρα εἶναι τὸ αὐτὸ ἀδύνατον, 204a 25-26). The logic behind this is not developed; Aristotle simply says that if air is part of air, so infinite is part of infinite. As far as the present text goes, nothing more profound may be implied. There are however arguments against Anaxagoras by Simplicius (*Phys.* 172. 28-31) and Alexander (ap. Simplic. *Phys.* 173. 20-27) that there will be additional infinites (and indeed an infinite number) if each of the infinitely numerous homoiomerous bodies is itself infinitely divisible. Thus a theory of infinite divisibility is criticised because it entails both that a body is infinitely large (as the sum of an infinite number of parts), and that these can also be infinitely divided (*ex hypothesi*) to produce further infinites, and eventually an infinite number of infinites.

This reasoning would be inappropriate to the present text without attributing to the Stoics the theory of infinite division as an unending process ruled out at 222. 6-13. For only this could serve as the premise for the conclusion that there are additional infinites. Yet if Alexander is not adding this further paradox because it was conventionally attached to that of infinite body (perhaps by the naive argument at Aristot. *Phys.* Γ5) he must be relying on such logic. For in general this argument could only succeed if someone made the contradictory assertions that (1) there is infinite division of body, and (2) that body is divided into a complete infinite set. (2) would entail that there was an infinite body, and (1) that there were further infinites. No one ever made these assertions (though, according to Vlastos, *Encyl. Philos.*, VIII, 370-371 Zeno did, at least tacitly), and so the arguments derived from them in the commentators are an illicit imposition, and the major flaw is in the critics' paradox itself. This is a limitation that can perhaps also be found in Lucretius' argument at I. 615-622.

Elsewhere we find this argument used against the Epicureans whose (alleged) belief in an infinite number of worlds each being composed of an infinite number of atoms is said to entail that there are an infinite number of infinites (ἀπειράκις ἄπειρα), since separately each world is an infinitely large body: see Alex. *Quaest.* III. 12. 104. 4-12 and cf. a related polemic of his ap. Philop. *de gen. et corr.* 12. 9-10, 18-22 against the theory that each atomic shape has an infinite number of instantiations. Also note *Quaest.* III. 12. 103. 12-34 where Alexander has a very general argument that if there are numerous infinites then one infinite must be greater than another.

In view of the use of these arguments by Alexander and Simplicius from an Aristotelian standpoint, it is interesting to see Philoponus' rather crude attempts to turn them back on Aristotle by claiming, without any justification, that the infinite cannot be in potentiality, but must be actualised. So, in the context of a universe that has existed from infinite time—a notion Philoponus is concerned to deny by these polemics—it will follow that any infinite series that is actualised at a given point in time will be exceeded as that series continues; there could therefore be up to an infinite times infinite number of human beings created (*Phys.* 428. 25-429. 20). For other examples see S. Sambursky in *Islamic Philosophy and the Classical Tradition*, pp. 351-353, who seems unaware of their long history in the exegetical tradition.

222. 22-26. This returns to the earlier argument (221. 34-223. 3) that an exhaustive division will terminate in incorporeal divisions. The reference to points connects it closely with *de gen. et corr.* A2

316a 30-34 (cf. Simplic. *Phys.* 142. 6-10), as probably does the reference to indivisible bodies (222. 23-24); for the whole argument quoted at *de gen. et corr.* A2 316a 14-316b 14 is designed to demonstrate the necessity of such bodies (see Furley, *Two Studies*, pp. 90-94).

Chapter 9

222. 27-28. "State ... Nature ... soul." On this triad see p. 41 with n. 94 above. Cf. its description at 4. 217. 33-218. 1, the text against which the first part of this chapter (26-35) is directed in the light of the general principle that constituents are separable from a blend (cf. 1. 213. 7-8; 3. 216. 32-217. 2; 7. 220. 34ff.).

222. 31-32. "But no State is separable from what has it." Cf. Dexippus *Cat.* 50. 31-32 (= *SVF* II 461) for the same point. In Alexander's case it may simply reflect *de gen. et corr.* A10 327b 16-17: "nor in general are qualities and states capable of being mixed with things" (οὐδ' ὅλως τὰ πάθη καὶ τὰς ἕξεις οἷόν τε μεμῖχθαι τοῖς πράγμασιν).

222. 32. "capable of independent existence" (καθ' αὑτὴν εἶναι δύνασθαι). Here used synonymously with κατ' ἰδίαν ὑφεστάναι (222. 29), just as 13. 228. 18 (καθ' αὑτὰ εἶναι δύνασθαι) is with 228. 14-15 (καθ' αὑτὰ ὑφεστάναι). This shows clearly the sense in which ὑφεστάναι refers to the being of substances. Cf. on 4. 217. 33-34.

222. 35. "enmattered form" (ἔνυλον εἶδος). This expression is widely used by Alexander to characterise form in its individuating aspect. Cf. *de an.* 16. 2 where soul is described as here; also *Met.* 373. 22-24. (On form, on the other hand, as universal, and hence immaterial [ἄϋλος] see *de an.* 87. 5-23, where this universality is grounded in the activity of the intellect; cf. J. Owens, *Amer. Philos. Quart.* 3 [1966] 167-169, who discusses this topic in Aristotle). Enmattered form complements Alexander's notion of matter as a common factor shared by all individual bodies: see on 10. 223. 30-34, and cf. *Quaest.* II. 28. 78. 1-3. There are useful discussions of it by C. Baeumker, *Problem der Materie*, pp. 296-297, and by P. Merlan, *Monopsychism, Mysticism, Metaconsciousness*, pp. 16-17.

222. 36-223. 6. This is a theoretical analysis of the illustration of fire totally pervading iron at 4. 218. 1-2, complementing the strictly empirical analysis at 12. 227. 13-228. 4. It applies the concepts of

form and matter to the proof for total blending that in ch. 4 immediately succeeds those discussed in the same terms in the first part of the present chapter (26-35).

The train of thought in this passage is slightly difficult to follow. After saying that fire is not mixed with bodies, Alexander states that in general matter is not mixed with form (cf. also 13. 228. 21-22). Now the relation of fire to iron is not that of form to matter but of quality to body, and for *that* reason cannot be mixed with it as 6. 220. 8-9 shows. 223. 2-3 is then rather perplexing. Fire is said to have matter in the sense that bodies that are burnt are its matter, but what does the further claim ἀλλ' ἡ μὲν ἄφθαρτος, ἡ δ' οὔ ("the one is indestructible, the other not") mean? It must, I think, refer to two kinds of matter, both matter in general and the matter of fire just described to us. Alexander must then be saying that matter in its relation to form cannot be destroyed except where, as when being burnt, it is a material body. That is, he is drawing the simple "classical" distinction between primary and proximate matter, which we find more explicitly established elsewhere; see *de an.* 2. 25-7. 8, and cf. especially the distinction between ἡ κυρίως ὕλη (matter in the strict sense) and ἡ προσεχὴς ὕλη (proximate matter) at *de an.* 4. 22-27. 223. 3-6 then simply describes how such proximate matter is destroyed and eventually loses its form, supporting 223. 1-3. If this happens fire cannot be mixed with a body; it can simply destroy proximate matter, without affecting the fundamental metaphysical relation between matter (*qua* primary matter) and form.

223. 6. "...and expelled from their own form" (τοῦ οἰκείου εἴδους ἐξίσταται). Cf. Alex. *Quaest.* 53. 5. This is probably derived from expressions such as ἐξίστασθαι ἐκ τῆς φύσεως (e.g. Aristot. *Phys.* Θ7 261a 20) or τῆς οὐσίας (*Top.* 145a 4, 10), though it has no precise Aristotelian parallel. For a general survey of such language in Aristotle see G. Else, *Aristotle's Poetics—The Argument*, p. 496 n. 30.

Chapters 10 and 11. The discussion of the Stoic doctrines of pneuma and God in these chapters seems to constitute a digression from the main object of the work, the search for a theory of blending that accords with our natural intuitions (2. 215. 30-32). It is not clearly justified until the beginning of chapter 12 (226. 34-227. 1) where Alexander says that it was provoked by denials of Aristotle's doctrine of aether, the substance of the heavenly bodies, which is referred to at 223. 10-14 and 223. 33 as an alternative to pneuma, and at 225. 9-10 and 225. 30-34 in relation to the Stoic theory of God; see further on 223. 6-14 below. Alexander then (227. 2-10) attempts to integrate these two chapters with the critique of "body going through body" by saying that the Stoic doctrines of soul, fate, providence, the principles (ἀρχαί), God, and the whole concept of the unification and sympathy of the universe are dependent on this single notion; for all these doctrines, he claims, are simply cases of God going through matter. This enlarges the two very brief arguments that both the theories of pneuma and God are cases of body going through body at 10. 223. 18 and 11. 225. 4-5. This is a revealing explanation; for if these central Stoic doctrines are indeed dependent on the notion of body going through body, and this notion is at the core of the theory of total blending, as ch. 5 suggests, and as 12. 227. 4-5 states, then total blending, as we have argued (at pp. 34-49), must be the foundation of the theory of pneuma and its associated doctrines (sympathy, fate, providence, etc.). That is, Alexander here allows, at least implicitly, that the Stoic argument presented in ch. 4 shows the theory of pneuma to be dependent on the notion of total blending, a point rather obscured in that text (cf. on 4. 217. 33 and see also on 12. 227. 9-10); he does not however actually criticise the structure of that argument, but only the notion of blending itself, either as a case of body going through body (chs. 5, 6, and 12) or as total blending (chs. 7-9).

Chapter 10

223. 6-14. The critical contrast drawn here between the motion of pneuma (or God) through matter and the action of the fifth body on the sublunar elements is repeated with variations at 223. 30-34 below, at 11. 225. 18-226. 10 (cf. esp. 225. 30-34) and summed up at 12. 226. 34-227. 1, where Alexander tries to justify the whole argument of chs. 10 and 11 as an attempt to ward off a Stoic challenge to the doctrine of the fifth body.

Now the general conception of the heavenly bodies as preserving a process of continuous coming-to-be (γένεσις) in the sublunar world, as a result of their continuous motion, is an orthodox Aristotelian doctrine: see especially *de gen. et corr.* B10. For a description of the general doctrine of the relation between the heavenly bodies and the interchange of the four elements, see A. L. Peck,

Aristotle: Generation of Animals (Loeb ed.), App. A, and H. Happ, *Hyle*, pp. 503-512.

Alexander we know elaborated this predominantly physical theory into a doctrine of providence in a special treatise, the Περὶ Προνοίας (*De Providentia*), almost certainly as a response to the Stoic theory of "total" providence, where a material deity totally pervades the universe. Our main sources are unfortunately very disparate: three texts in the *Quaestiones*—I. 25, II. 21, II. 3 (on this see Moraux, *Hermes* 95 [1967] 159-169)—which in the case of the latter two deal with problems in the theory of providence rather than systematically expound the doctrine; some six short dogmatic quotations in Cyril of Alexandria's *Adversus Julianum* (*PG* 76, 596B, 621C, 625C, 628D-629A, 704B, 741AB), translated and briefly discussed by R. M. Grant, *JTS* 15 (1964) 265-269; and an Arabic text (see A. Dietrich, *Nachr. Gött. Akad.* [1964] 97-98 on the manuscripts) discussed and paraphrased (being too fragmentary to be translated) by P. Thillet, *Actes du Premier Congrès International de Philosophie Médiévale* (1960) 313-324; there is a briefer summary of a different ms. by S. Pines at *Archives d'histoire doctrinale et littéraire du moyen âge* 34 (1959) 295-299. P. Moraux has given two general accounts of the doctrine, largely on the basis of the texts from the *Quaestiones* at *Alexandre d'Aphrodise*, App. II, and, more briefly, at *D'Aristote à Bessarion*, pp. 58-64. See also H. Happ, *Antike und Abendland* 14 (1968) 82-83, who comments on the Arabic text and sets the doctrine in historical perspective.

Briefly, Alexander held that the providential influence of the heavenly bodies extended into the sublunar world, but not to the level of the individual, as the Stoics believed (cf. 226. 24-30), but only to the level of the *infimae species*; the heavenly bodies themselves were not in need of providence. This excluded an earlier Peripatetic theory (see e.g. Diog. Laert. V. 32, and cf. Festugière, *L'idéal religieux*, Excursus C) which held that God's providence had only extended to these bodies (cf. Moraux, *Rev. Phil. de Louvain* 47 (1949) 33-34).

An attempt to reconstruct the *de providentia* would be most welcome. Here I shall only try to show that at several places in *de mixt.* 10 and 11 we can detect traces of Alexander's theory of providence, and that these previously neglected texts should be taken into account in future work. They are certainly the best evidence in Greek of Alexander's thoughts on providence in the light of the alternative Stoic doctrine; see further notes to 223. 30-34, and II. 224. 30-32, 225. 18-226. 10, 226. 24-30, 226. 30-34. En route I shall also try to indicate the significance of the *de mundo* for Alexander's conception of providence. He seems to have known this work, and probably regarded it as authentic—see Moraux, *Hermes* 95 (1967) 160 n. 2.

223. 7-9. "that the whole of substance is unified by a pneuma..." cf. 3. 216. 14-17. A genuinely historical critique of pneuma's relation to matter would perhaps deal with the sense in which it creates a unified substance "in which" bodies can be mixed together (see 3. 216. 17), instead of simply comparing it with the fifth body.

223. 10-14. The broad duality established here between the fifth body and matter seems to reflect the doxographic presentation of Aristotelian principles (ἀρχαί). Cf. Aet. I. 3. 22, Epiphanus (*Dox. Graec.*, p. 592. 10-14), and Hermias (*Dox. Graec.*, p. 653. 31); in the latter the aether and the four elements are contrasted, in Stoic terms, as active and passive principles. See also Sext. Emp. *P.H.* III. 31 and *A.M.* X. 316 where the four elements and the κυκλοφορητικὸν σῶμα (circular-moving body) are described as principles. Although Alexander was aware of the more orthodox descriptions of the relation of the heavenly bodies to the four elements (see II. 225. 30-35 with note) he might find it convenient in a popular treatise such as the *de mixtione* to present Aristotelian doctrine in these simplified terms, particularly in order to point up a contrast with the Stoics (see on 223. 11-12). Cf. his use of the sculptor-statue example at *de fato* 3 to illustrate the four causes; although misleading as an exposition of Aristotle (see R. Sprague, *The Monist* 52 [1968] 298-300) it was a convenient illustrative device in an admittedly pedagogical work. Probably drawn from some doxographical source (cf. Sen. *Ep.* 65. 2-10, Clem. Alex. *Strom.* VIII. 9 [= Stählin-Freuchtel, III, p. 98. 3-5]), it no more represented his considered analysis of Aristotle's text than the present description.

223. 11-12. "by surrounding the whole enmattered, affectable and alterable substance..." This almost parrots the description of pneuma's action. "Affectable" and "alterable" (παθητός, μετάβλητος) are indeed elsewhere applied to matter *qua* passive principle in accounts of Stoic doctrine: see *SVF* II, pp. 158. 21, 115. 7,

116. 21. ἔνυλος ("enmattered") is an authentically Peripatetic term most frequently applied to form (see note on 9. 222. 35), but also used, as in this text, in a general characterisation of the sub-lunar world at *Quaest.* I. 25. 41. 10-11. Cf. also *Met.* 223. 5 where Alexander describes matter as παθητή and Aet. *Plac.* II. 4. 12 where Aristotle is said to have regarded the sublunar world as παθητόν. Note also the Stoicising language at *de mundo* 2 392a 34 where sublunar matter is termed δι' ὅλων παθητή τε καὶ τρεπτή.

223. 13-14. "holds together and preserves the whole . . . by causing in a fixed order the interchange of bodies" (μεταβολὰς ἐν ὡρισμένῃ τάξει ποιουμένη συνέχει καὶ σώζει τὸ πᾶν). For τάξις ὡρισμένη used in a description of seasonal order caused by the heavenly bodies see *Quaest.* 89. 9. Cf. a more general use at *de fato* 168. 3-5 where things that have nature as the cause of their coming-to-be are said to achieve their end κατά τινας ἀριθμοὺς καὶ τάξιν ὡρισμένην. συνέχει seems to parrot συνέχεται (223. 8) in the description of pneuma's action on matter; such assimilation could have been encouraged by the Stoic expression συνεκτικὴ αἰτία at *de mundo* 6 397b 9.

For the providential import of σώζει here see note on 11. 224. 30-32.

223. 16. "to certain bonds" (δεσμοῖς τισι); see 223. 34 below. Philo (at *SVF* I 106, p. 30. 35) uses the term "bond" to describe the tension of pneuma, and one must assume that it was a technical term for the Stoics.

223. 19-27. This very literal-minded argument takes pneuma in its basic sense of "breath" and points out that, as such, it cannot be actually present in every body; how then can it unify the universe (τὸ πᾶν, lit. "the all")? I have translated the word "pneuma" in inverted commas where it is used in the literal sense to clarify the argument. For a closely related critique of the soul as air see *Mant.* 115. 16-20.

223. 27-30. "for some bodies are continuous, others discrete." This is another common-sense argument which could be provoked by a literal reading of the distinction Stoics drew between total blending and juxtaposition, or between unified bodies and those composed of divided parts (see above, p. 69 n. 189). It is included with other arguments against the notion that pneuma holds the cosmos together by Alex. ap. Simplic. *de caelo* 286. 15-21 (not in

SVF). In the present chapter it is complemented by the arguments at 224. 6-14 below. For a general argument against pneuma's capacity to hold the cosmos together, see Alex. ap. Simplic. *Phys.* 671. 8-12 (*SVF* II 552).

The Stoic answer to this literal-minded critique would presumably be that the unity caused by pneuma's motion through matter is not reflected directly in phenomena, any more than that motion itself which, as the argument in ch. 4 shows (see pp. 40-43 above), has to be demonstrated analogically. Such a position is implied by the general doctrine of pneuma, though we have to turn to later authors for a clear statement of it—e.g. Marcus Aurelius IV. 40, VI. 25, XII. 30.

223. 30-34. Alexander divides pneuma's effects into two classes: (1) the internal unification and containment of a body, and (2) its sympathy with other bodies. (1) he says can be better explained by form (cf. *Mant.* 115. 9-10 where pneuma's "tension" is compared to form), and (2) by matter, in which all bodies share, and by the action of the heavenly body. He does not attribute any independent significance to matter in the Stoic theory, although the passive elements are the complementary physical field of pneuma's motion (cf. p. 35 above). This contrasts somewhat with the arguments at 6. 220. 6-11 and ch. 9 *passim* where the two elements in a blend (e.g. soul and body) were compared with form and matter. Some of the arguments in ch. 11 show why Alexander would have difficulty in acknowledging that matter was independently significant in Stoic physics—see on 225. 5-18 and 226. 10-16. Several modern scholars, it should be noted, regard the Stoic principles as a materialistic version of the form-matter distinction: see Christensen, *Essay on the Unity of Stoic Philosophy*, p. 12; Hunt, *Apeiron* 1 (1967) 9; and Furley, *BICS* 13 (1966) 32. This view is closer to Alexander's in ch. 9 than here or at 11. 224. 27-32.

By οἰκεῖον εἶδος ("its own form") Alexander must mean ἔνυλον εἶδος ("enmattered form," cf. on 9. 222. 35), where form individuates matter. For matter as that in which bodies share (κοινωνεῖν) see on 3. 216. 1-4. Alexander does not elsewhere describe matter as causing sympathy between bodies; cf. however, Philop. *de gen. et corr.* 41. 25-26 who says that if atoms exist "continuity and sympathy [presumably in matter] will be destroyed" (ἡ συνέχεια καὶ ἡ συμπάθεια ἀναιρεθήσεται). Like much else in this commentary (see ch. 16, introductory note) this could be Alexandrian. Neither, as far as we know, does Alexander use such language elsewhere to describe the effect of the heavenly bodies on the sublunar world. We know that for Alexander (see note on 223. 6-14) these bodies influenced individual form, insofar as they ensured the preservation of the species, and unless this can be thought of as creating sympathy, we should not assume that Alexander used the Stoic term in his own doctrine of providence. Still the contrast between its Stoic sense here, and its sense in the Peripatetic theory described at Diog. Laert. V. 32 remains notable. There terrestial bodies are said to be governed "by their sympathy with the [heavenly bodies]" (κατὰ τὴν πρὸς ταῦτα συμπάθειαν). Conceivably Alexander could have revised this relational sense of sympathy to establish a concept of a sympathetic interrelation between bodies caused by the aether.

Whether or not Philoponus' description of a sympathy between natural bodies (the elements) in blending (*Phys.* 113. 8-9) is to be connected with this text is unclear. Alexander certainly does not connect blending with sympathy in the later discussion in chs. 13 and 14 except perhaps by a very general implication at 14. 231. 3 (see note *ad loc.*).

223. 34-36. "the tension of pneuma" (ὁ τόνος τοῦ πνεύματος) is in itself an ambiguous expression. It may mean pneuma's own internal principle of coherency (the πνευματικὸς τόνος: see *SVF* II 447), or it may refer to the tension which pneuma transmits to other bodies (see e.g. *SVF* II 444). In its former aspect it is criticised at *Mant.* 114. 28-29 (cf. Galen at *SVF* II, p. 144. 37-39) in its role as soul for being a "self cohering" agent (αὐτὸ συνέχον), regarded as an absurd notion since whatever "binds together" has itself to be "bound together" and so on *ad infinitum.* In this chapter no attention is paid to this aspect of pneuma, and in this text Alexander is only thinking of the tension transmitted to bodies internally (πρὸς τὰ οἰκεῖα μέρη) and in their relation to other bodies, i.e. in their universal sympathy.

223. 36. "pneuma is forced by something" (βιαζόμενον ... ὑπό τινος τὸ πνεῦμα). This is a Peripatetic rationalisation of pneuma's motion. Pneuma, like any body, must be either self-moving or moved by something else (see Aristot. *Phys.* Θ4). For the Stoics it was by definition self-moving, but since this would be unintelligible to a Peripatetic a speculative version of the second alternative had to

be provided. Cf. the comparison between pneuma and aether at Ar. Did. fr. 28 (*SVF* II 471, p. 152. 33-35—quoted at p. 52 above). Also cf. 224. 2 ("through motion in a mass it acquires strength" [ὑπὸ τῆς ἀθρόας κινήσεως ἰσχύν τινα λαμβάνει]) with Aristot. *Meteor.* 367a 30-31 where wind is said to make earthquakes more severe when moving in a mass (ἀθρόως φερόμενον). The terms "moist" and "easily-divisible" (224. 3) recall the qualities that Aristotle claimed necessary for the constituents of a blend: cf. note on 8. 221. 25-28. They are not essential to the Stoic theory.

224. 5-6. "something void and of an intangible nature." This speculation would offend the basic conception of pneuma as an inherently active force; it was presumably forwarded by opponents of the Stoics.

224. 6-14a. There are, I think, three separate points here, developing the argument at 223. 27-30. (1) (6-9) Pneuma, by definition, causes the stability of bodies (συμμένειν, cf. 223. 8, 216. 16), but as a matter of fact some collapse and therefore cannot be held together by it. (2) (9-11) Unified bodies are in fact divided, but this would be impossible if they were all similarly unified. (3) (11-14) If all bodies are similarly unified, then it would be impossible to distinguish those that are, in fact, divisible (juxtaposed bodies) from those that are not. Conceivably (3) was derived from the Stoic distinction between juxtaposed bodies formed ἐκ διεστώτων, and ἡνωμένα (see pp. 69-70 above), but could have been arrived at simply on general grounds. Like 223. 27-30 this whole argument criticises pneuma in the same general common-sense manner as the arguments against body going through body in chs. 5 and 6.

224. 14a-22. If the Stoics asserted (a) that pneuma is composed of fire and air, and (b) that fire and air are created by pneuma, they would fall victim to this argument. Where, I think, Alexander's argument breaks down is that, for the Stoics, the elements referred to in each case differ. That is, the active elements in (a) in virtue of which pneuma acts on the passive elements (see above on 4. 218. 2-6) cannot be the same as the four elements generated in this action and from which the cosmos is formed. Such a distinction is not explicitly drawn in our texts, but is implied by the general distinctions between principles and elements drawn at *SVF* II 299, 408. If pneuma is the active principle, its qualities cannot be those

of the elements evolving from fire and engaged in cosmogony (cf. *SVF* II 413). Only by neglecting this distinction does it become reasonable to ask—as Alexander does—how pneuma can be a compound formed from two elements, while at the same time functioning as the force that generates these same elements.

This is a special case of a wider range of arguments against the whole relation of God to matter in ch. 11 (see note on 11. 225. 5-18 below) where this problem of a principle's apparent dependency on the matter supposedly dependent on it is further exploited. Cf. also Alex. *de an.* 19. 21-30 for an argument that if the soul is a single element generated from other elements it is not uncompounded. Possible Aristotelian sources are *Metaph.* Λ10 1075b 1-6 and *de gen. et corr.* A1 315a 19-25. Cf. also Plot. II. 9. 5. 21 on the dependency of soul on its constituent elements.

224. 22-23. If pneuma is a compound of Fire and Air, and these constituents are present in actuality (as they would be in a Stoic total blend), then Hot and Cold are actually present together. A possible Aristotelian source for this argument is *de gen. et corr.* B5 332a 12-17 where it is argued that if Air is directly produced from Fire, then it would have to be both hot and cold; hence they must share in a common matter. Also cf. Alex. *de an.* 19. 34-36 where the impossibility of coexistent opposites in a blend is said to show that one could not come to be from any of the others. In Stoic physics it would seem that the property of being active which Hot and Cold share outweighs their contrariety, and so makes it plausible that they be compounded.

224. 23-27. There is a more extensive critique of the concept of pneuma's motion at *Mant.* 131. 5-22 in a discussion of the extension of the air in perception (συνένταcις); see my article on this concept, *Grazer Beiträge* II (1974) 251-261. Here the thrust of the argument seems to be that it is not one of the five regular motions, nor, Alexander might have added, a case of mutual-replacement, which at 6. 220. 11-13 he stressed was ruled out by body going through body.

Chapter 11

224. 27. "as I said." I.e. at 10. 223. 30-34.

224. 30-32. "and in that these things are preserved and stable, as they are, [they seem] to attribute the preservation, being, and

stability of each of them (τό τε εἶναι ἑκάστου καὶ τὸ σώζεσθαί τε καὶ συμμένειν) to a pneuma which allegedly pervades everything." The language used here to describe individual identity is Alexander's own. Although συμμένειν ("stability") perhaps has a technical sense for the Stoics (see on 216. 15-16) Alexander uses it himself (at *de an.* 15. 21, 21. 20) to describe individual substantiality (Aristot. *Met.* M2 1077a 24 is only loosely comparable). For the combination of σώζεσθαι and εἶναι ("stability" and "being") see *de an.* 75. 31, and for σώζεσθαι and συμμένειν Alex. ap. Simplic. *de caelo* 83. 30-32. The notion of preservation strongly suggests the idea of providence, and its usage here may be intended to reflect such texts as *de mundo* (genuine for Alexander) 399b 9-11 or 400a 4 where God is said to cause the preservation (σωτηρία) of the universe. See the fragment of Alexander's *de providentia* at Cyril *Adv. Jul.* (*PG* 76 625c) where divine providence is said to be responsible for the preservation and coming-to-be (σωτηρία and γένεσις) of terrestial bodies, and their eternal continuation in form (κατ' εἶδος ἀίδιον διαμονήν). Preservation and stability are elsewhere connected in contexts where some form of providential control of the universe is being described: see Ps.-Ocell., p. 13. 11, 16 Harder; Cleomedes I. 3, p. 28. 20 Ziegler; Clem. Alex. *Strom.* VII. 11. 12-13; and cf. Alex. *Meteor.* 6. 15-17 where the heavenly bodies are said to be the cause of the universe's being and eternal stability (τοῦ τε εἶναι καὶ τῆς ἀιδίου διαμονῆς).

The present text would therefore seem to continue the preoccupation with the issue of the providentially guaranteed status of bodies clearly raised in the earlier comments on matter and form as alternatives to pneuma at 10. 223. 30-34; cf. also on 16. 235. 12-14.

224. 33. "reasonably." This because the Stoic doctrine of active and passive principles, God and matter, described here at 224. 33-225. 3, precisely embodies the error expressed in Aristotelian terms in the preceding sentence (224. 27-32). Two of the subsequent arguments in this chapter (226. 10-24) take up the critique of God's relation to matter in its terms.

224. 34. "two universal principles matter and God" (τὸ δύο ἀρχὰς τῶν πάντων ... εἶναι ὕλην τε καὶ θεόν). For similar testimonia see *SVF* II 300, 301, 304, and cf. p. 35 nn. 65-67 above. For a recent discussion of the whole doctrine of principles which argues that it originates in the physical theory of the early Academy, see H.-J.

Krämer, *Platonismus und Hellenistische Philosophie,* pp. 108-130. Although all the arguments of this chapter are directed against various aspects of this theory of principles it is not criticised directly until the very end, at 226. 30-33b (+ 227. 23-25).

225. 3. "intelligent pneuma" (πνεῦμα νοερόν). Cf. *SVF* II, p. 113. 17 and II, p. 146. 17-18 for similar expressions. This identification is particularly important for the argument at 225. 5-18 where the relation between God and matter *qua* four elements is discussed; it justifies treating God's relation to matter as problematic in just the same way as pneuma's relation to the elements in the argument at 10. 224. 14-22.

225. 5-18. Gibbon (*Decline and Fall of the Roman Empire,* ch. 2) remarked that "as it was impossible for [the Stoics] to conceive the creation of matter, the workman in the Stoic philosophy was not sufficiently distinguished from the work." He was echoing the criticism of several Christian fathers—see H. Wolfson, in *The Classical Tradition, Studies in Honour of H. Caplan,* pp. 409-410, and *Philosophy of the Church Fathers,* I, 87-88. Alexander's criticisms in these terms probably spring more directly from the Platonist tradition in which God is said to be identical with matter, often as either one of the elements or a compound (see Plut. *de comm. not.* 1085B [= *SVF* II 313]; Albin. *Didask* 10.8; Sext. Emp. *A.M.* IX. 180-181; with reference to the theory of soul as pneuma, Alex. *Mant.* 115. 6-9, *Top.* 173. 14-16; cf. Ps.-Galen *de qual. incorp.* XIX. 478-479 K [= *SVF* II 323a]; and for Plotinus see Graeser, *Plotinus and the Stoics,* p. 37). An ancestor of this may be Aristotle's critique of Empedocles' four elements (*de gen. et corr.* A1 315a 19-25), or of the theory of soul as harmony (*de an.* 408a 18-28, cf. *Phaedo* 93a 6-7) where an epiphenomenon is claimed to be identical with its formative parts; cf. also Plotinus II. 9. 5. 19-23.

This argument could be applied to the Stoics for two reasons. First critics found it difficult to grasp the general relation between the active and passive principles, where matter was qualityless and passive before an entirely active principle, yet neither was distinguishable except in theory. It was all too easy to argue, as did Sextus Empiricus (*A.M.* IX. 240) that these were two names for the same thing, or that God was identical with matter (as here), or that matter was not qualityless and independent (as below at 226. 10-16; cf. Plut. *de comm. not.* 1085B).

Secondly the relation between the four elements and the two principles was not always clear. In the doxography, for example, they are listed in parallel (Aet. I. 3. 25), and Sextus Empiricus (*P.H.* III. 31) merely lists the elements as principles and dismisses "the mysterious qualityless matter." Though elsewhere the elements are said to be generated from matter when it is acted on (e.g. Sext. Emp. *A.M.* X. 312 [= *SVF* II 309], cf. *SVF* II 316) it would be easy to confuse matter itself with the elements (as Alexander does) and to argue that God is simply composed of them. This problem would be compounded by the doctrine of pneuma in which a distinction was drawn between active and passive elements. In the note on 10. 224. 14-22 we argue that these were not the elements generated from matter and serving as constituents in the cosmogony. But it would be easy and natural for a critic to take the elements in only one sense, and thus see pneuma as composed of matter like any other compound.

Our text is the product of both kinds of misunderstanding. At 225. 5-10 God's relation to the elements is seen much as pneuma's at 224. 14-22 (cf. 225. 6-8 for the comparison with pneuma). At 225. 11-17 the general relation between God and matter is criticised (as later at 226. 10-16) as it is conveyed by the description of the principles at 224. 34-225. 1; the references to the elements at 225. 10-11 and 17-18 are not important for this main point, and merely repeat 225. 5-6.

225. 8-10. "the divine body will be a fifth substance for them." An ironic consequence in view of the claim that the Stoics in fact ought to invoke this concept—see 10. 223. 30-34. Cf. Plot. VI. 1. 26. 20-22 for a general argument that if God is a body, composite in substance but without matter, the Stoics will have to introduce another matter, that of God. Since the overall context employs the concepts of form and matter, this particular argument may have emanated from Peripatetic sources. See further Graeser, *Plotinus and the Stoics*, pp. 91-93.

225. 10. "with appropriate support" (μετὰ τῶν οἰκείων). Von Arnim's addition of ἀποδείξεων (*SVF* II, p. 112. 38) is unnecessary. Rex's "im Zusammenhang mit seinen eigenen System" is loose, but conveys the essential meaning.

225. 16-18. "he would be eternal for them in name alone." That is, God would not be independently eternal (cf. 225. 4), but only

as eternal as matter or its compounds on which he is dependent for his being. The next sentence explains this by again (see 225. 5-6. 10-11) referring to the fact that God must be an element or a compound if he is material. For basically the same argument see Cic. *de nat. deor.* III. 14. 34, where it is attributed to Carneades, and Sext. Emp. *A.M.* IX. 180. Cicero stresses the fact that matter will decompose and therefore a material being cannot be eternal. Perhaps this is implicit in Alexander's argument. These parallels may also help to justify Schwartz's reading ἀΐδιος for αἰδοῖος in line 16.

225. 18-226. 10. It is clear that early Stoic physics developed Aristotle's concept of nature as exhibiting an immanent teleology like that of an art (τέχνη) without an artist (*Phys.* B8—cf. esp. 199b 26-32) in its doctrine of "artistic fire" (πῦρ τεχνικόν); cf. F. Solmsen, *Aristotle's System of the Physical World*, p. 115 n. 91. At 225. 20-27 we see another use of the φύσις/τέχνη (Nature/Art) contrast to show that nature acts as an internal force in matter and is therefore a superior craftsman to one who works only on the exterior of his material. That is, nature is like art in its rationality, but differs in its total effect on matter; cf. also Ps.-Gal. *de qual. incorp.* XIX. 478 K (p. 14. 6-8 Westenberger = *SVF* II, p. 116. 11-13) for the same contrast between divine and human craftsmanship.

Alexander's response to this is to simply state the basic principles of Aristotelian cosmology. Natural change occurs in the sublunar world as a result of the varying motions of the heavenly bodies, which are dependent on an external mover (225. 30-34). 226. 6-10 (esp. 9-10) implicitly refers to God as the first mover of these bodies. 225. 34-226. 3 simply adds a series of natural changes also dependent on the motion of the heavenly bodies, while 226. 4-6 adds some examples from intracosmic processes.

Now this response must be seen in the context of Alexander's general theory of nature. For him nature had two dimensions. (1) It was subject to providence by the action of the heavenly bodies in ensuring the continuity of species. Cf. note on 10. 223. 6-14, and see *de fato* 169. 23-26 (cf. 10. 223. 12-13) where the motion of the heavenly bodies is described as the first cause of natural change as here it is described at 225. 29-30 as the "cause of the first change in matter." (2) Within this matrix Alexander saw the processes of nature as self-propelled motions towards an end, but not as closely resembling the action of an art (τέχνη) as Aristotle had argued in *Phys.* B8. In fact he drew a sharp distinction between the rational processes of art which involve choice (προαίρεσις), and therefore the capacity equally to do or not do something, and those of nature in which barring any impediment an end will be reached (*de fato* 168. 3-18; cf. *Mant.* 153. 19-20). For this reason he could identify fate (εἱμαρμένη) and nature (*de fato* 5 *passim*) without compromising freedom, a difficulty entailed, he thought, by the "Stoic" view that everything occurred by fate (attacked at length in the *de fato,* esp. chs. 11-20). This is complemented by Alex. ap. Simplic. *Phys.* 310. 31-311. 37, where it is said that because nature does not possess the faculty of choice it is irrational (ἄλογος—311. 1). It is also said to be a mechanical process towards an end proceeding from "the first source established in receptive matter" (ἀρχῆς καταβληθείσης τῆς πρώτης ἐν ὕλῃ τῇ δεκτικῇ—311. 1-2, cf. 311. 7, and here 225. 29-30). This source is not specified but presumably is the motion of the heavenly bodies described here at 225. 30-34. Thus Alexander's view of nature as irrational *vis-à-vis* art (and hence definable as fate) dovetails into his view of nature as a providentially governed process; cf. Moraux, *Alexandre d'Aphrodise,* p. 198. Fate and nature, on the one hand, and providence on the other, are related without resulting in the pan-providential and also completely determined cosmos of Stoicism; cf. G. Verbeke, *AGPh* 50 (1968) 80 n. 1. Alexander's critique of the Stoic God as an internal creative force in the present text is, I believe, an offshoot of this more systematic reaction to Stoicism; cf. also pp. 25-26 above. I have only given the briefest indication of its character. It merits further study both in Greek and Arabic sources; for some indications of the latter see H. V. B. Brown in *Islamic Philosophy and the Classical Tradition,* p. 41.

225. 22. "are completely formed" (δι' ὅλων εἰδοποιεῖται). Elsewhere Alexander uses εἰδοποιεῖσθαι in an Aristotelian sense (e.g. *de an.* 20. 23, *de sensu* 73. 24) and this must be regarded as his rephrasing of the Stoic argument. Cf. also the reference to statues at 225. 24, a standard Aristotelian illustration (cf. Alex. *de an.* 3. 7-13).

225. 31. "cooling and heating" (ψῦξις, θερμότης). Cf. Aristot. *de gen. an.* Δ10 777b 29-33 where these processes are linked with coming-to-be and passing-away and said to be dependent on the motion of the stars. Cf. also *Meteor.* A2 339a 21-33 for a similar program-

matic statement which is paraphrased in language very similar to that of the present passage by Alexander, *Meteor.* 6. 9-20 (cf. also 7. 9-14).

226. 2-3. "bodies that are frozen on the earth and compounded in the upper air have their creator separate." See *Meteor.* B5 361b 36-362a 11 where the causative power of the sun on frozen earth is described, and see *ibid.* A9 on the sun's influence on rain and winds.

226. 8-9. "there is another cause prior to what creates in this way." For πρῶτος = πρότερος see Alex. *Met.* 105. 8. If God is totally blended with matter he is part of the process of natural change which by definition (see 225. 28-30) must have an external cause. The consequences of this close identification of God with matter have been explored at 225. 5-18.

226. 10-16. The comparison of the relation between God and matter to that between form and matter develops the similar critical comparisons made with reference to pneuma at 223. 30-34 and 224. 27-32 above. The difficulty now emphasised is how matter can be in itself qualityless if it is dependent on form (*qua* God) for its individuality (συμμένειν, εἶναι; cf. note on 224. 30-32 above); cf. *de an.* 17. 21-18. 3 for this point in more general terms. This, in effect, complements the whole argument at 225. 10-18 where God was shown to be dependent for his existence on matter; for such criticism of the Stoic God explicitly in terms of the form-matter relation see Alex. *Met.* 178. 17-19 (*SVF* II 306). This argument again raises the problem of how the Stoic principles can be described as two inseparable bodies that can each be conceived of independently. Only if, like matter and form, they were not bodies that had interpenetrated (cf. 220. 6-8) would their separability in theory (νοήσει or ἐπινοίᾳ as Alexander says of form and matter at *de an.* 4. 9-10, 6. 17-19) be acceptable to a Peripatetic. See also my note on this text at *Hermes* 101 (1973) 278-282.

226. 12. "...and God is the power of matter" (καὶ ἡ δύναμις τῆς ὕλης ἐστὶ θεός). The "divine power" (θεῖα δύναμις) is a Stoic term (cf. 225. 27 above) that Alexander used at *Quaest.* II. 3. 49. 30-34 in an account of the providential relation of the heavenly bodies to matter; see Moraux, *Hermes* 95 (1967) 160 n. 2.

226. 16-24. Alexander assumes (εἴ γε, 226. 17) that in the primal Fire God and matter are copresent. This cannot be regarded, as it

is by Krämer (*Platonismus und Hellenistische Philosophie*, p. 109 n. 9), as evidence that the two Stoic principles were maintained in the conflagration. It is simply a premise expressed in Aristotelian terms from which the consequence of a form changing itself can be asserted as a paradoxical redescription of the evolution of bodies from Fire. The paradox arises in the light of the principle that if a body loses its form it is destroyed (cf. 9. 226. 21-22, and 13. 230. 7-12).

For a more general criticism of God as fire undergoing change see Plut. *de def. or.* 426B (*SVF* II 1055), and cf. Ps.-Gal. *de qual. incorp.* XIX. 479 K (*SVF* II 1056). Also cf. Philo *de aet. mund.* 16, on the difficulties inherent in the theory of conflagration.

There is no clear evidence on whether or not the active and passive principles were maintained in the primal fire. This would be unlikely, since the very purpose of this dualism is to explain the activity of the active principle (God) on matter from which everything comes to be. Such a process is necessarily not envisaged in the conflagration.

226. 24-30. It was an old criticism of the Stoics that they made God a meddler in details; see e.g. Cic. *de nat. deor.* I. 52-56. Christian authors emphasised the indignity that this imposed: Justin (at *SVF* II 926) and Clem. Alex. (at *SVF* II 1040); cf. also *Mant.* 113. 12-15, Ps.-Galen *de qual. incorp.* XIX. 479-480 K who argues that for the Stoics God becomes the disgusting things he makes, and Sext. Emp. *P.H.* III. 218. Almost certainly the present argument was connected with Alexander's critique of the Stoic theory of providence. The use of προηγούμενον ("pre-meditated") at 26 suggests this in itself; on such language see Moraux, *Alexandre d'Aphrodise*, p. 196 n. 2. Again the Arabic fragment of the *de providentia* refers to a critique of the view that God is occupied with particulars (Thillet, p. 319). Cf. also *de fato* 203. 10-12 where in a reaction to the behaviour of the oracle in the story of Oedipus Alexander remarks that the Epicurean theory of "absence of providence" (ἀπρονοησία) would be more pious (εὐσεβέστερον) than to have God responsible for such evil. For the connection between piety (εὐσέβεια) and providence cf. also Alex. *de prov.* ap. Cyril *Adv. Jul.*, *PG* 76 628D, 741AB, and cf. *de mundo* 398b 3-7 where it is said to be unfitting for God (ἀπρεπὲς θεῷ) to administer details in the world.

226. 24-25. "... demeans our preconception of the deity" (ἀνάξια

τῆς θείας προλήψεως). Cf. *de fato* 202. 25-26 where, with reference to the Oedipus story, Alexander asks who, by telling this, περὶ θεῶν εὐσεβεῖς διδάσκει προλήψεις. Cf. also *Quaest.* II. 21. 69. 30-31. πρόληψις is undoubtedly used here in the sense of common notion; cf. above on 2. 215. 10.

226. 29. "of grubs and gnats" (σκωλήκων τε καὶ ἐμπίδων). These are the lowest form of animal life, spontaneously generated from matter (e.g. Aristot. *Hist. An.* A1 487b 5). Cf. their employment in an argument against Platonic ideas by Alexander at *Met.* 104. 16-18. Like κορόπλαθος (lit. "maker of small-puppets" [κόροι]) this symbolises the extreme level of detail to which God's influence extends in the Stoic doctrine.

226. 30-33b (+ 227. 23-25; see the apparatus). There must (ἀνάγκη, 226. 31) be interaction between constituents of a blend only because for Alexander, like Aristotle, "action" and "being acted on" (ποιεῖν, πάσχειν) are reciprocal relations; cf. further 229. 3-6 where this is described as part of the Aristotelian theory of mixture. In Alexander's own description of the active and passive principles at 224. 34-225. 3 it was evident that this relation was non-reciprocal (cf. also above p. 35 with n. 66), so that the whole argument here is an illegitimate restatement of the Stoic doctrine in Aristotelian terms. Cf. also Sext. Emp. *A.M.* IX. 254 for a general argument that the copresence of the active and passive principles means that they will interact. This argument could well be one form of a more general critique of the Stoic theory of providence, that it would involve God being dependent on mortals: see *Quaest.* II. 21. 68. 19-35, and cf. the Arabic ms., Thillet, p. 319. Note *de prov.* ap. Cyril *Adv. Jul., PG* 76 596B for a general statement that God will be providential only because he can be so. Presumably in this way Alexander's final theory evaded the danger of a reciprocity between God and the lower world leading to his dependence on it, as envisaged by the present argument.

Chapter 12

226. 35-227. 1. "the only theories worthy of divine things." Cf. Alex. *de providentia* at Cyril *Adv. Jul., PG* 76 628D-629A, and 741AB for expressions of the value of revering "the divine" (τὰ θεῖα, τὸ θεῖον), and attaining happiness by contemplation of "divine things."

227. 1. "ambitious attempts" (πειρωμένους διὰ φιλοτιμίαν). Cf. *de fato* 189. 9.

227. 2. "unaware of even the source of the stupidity of their statements." If "body going through body" is at the root of these various Stoic doctrines, but their proponents are unaware of it, then it must be, as we have argued at length (see Pt. II-iii *passim*), an independent notion imposed on them rather than an actual doctrine of theirs; cf. also introductory note to chs. 10 and 11.

227. 9-10. "for the God that pervades matter is all of these things for them." This suggests that the theory of total blending referred to at 227. 5 is that implied by the description of the principles at 11. 225. 1-2. Alexander has not related it closely to the discussion of total blending in ch. 4 where its relation to the theories of pneuma and soul is somewhat obscure (see pp. 40-43 above). Even so 227. 5-10 provides good evidence that that earlier argument is to be interpreted as showing the derivation of these theories from total blending, as we argue above. Cf. note on 213. 14-15.

227. 13-17. "it gains conviction (πίστις) from it being allegedly clear..." At 4. 218. 1-2 the case of fire pervading iron is listed as part of the "clear evidence" (ἐναργῆ μαρτύρια) that body goes through body. Cf. note *ad loc.*

227. 17. "other reasons aside" (πρὸς τοῖς ἄλλοις). This must refer to 9. 222. 35-223. 6 where the case of "fire in iron" (222. 36) is said not to be a case of mixture since this involves matter being unmixed with form (see note *ad loc.*). Alexander is now going to deal with this phenomenon in strictly empirical terms; cf. 6. 220. 16-22 for a similar kind of analysis of the case of heated incense which was one of the proofs that bodies could extend themselves.

227. 29. "nor any other matter supporting the fire" (οὔτε τις ἄλλη τῶν ὑποκειμένων ὕλη τῷ πυρί). For a demonstration that fire cannot be in such bodies as wood or coal "as in a substrate" (ὡς ἐν ὑποκειμένῳ) see *Mant.* 120. 17-33.

227. 33. "to the extent that there is moisture in it." On the relation between the moisture in a body and the different effects produced by fire see Aristot. *Meteor.* 387b 23-388a 9, and cf. *Mant.* 149. 10-13. For the general principle that moisture is nutriment for fire see Aristot. *de an.* B4 416a 26 with Hicks' note *ad loc.*

Chapters 13 to 15. These three chapters are a restatement of Aristotle's theory of blending, based mainly on the account given in *de gen. et corr.* A10, but also from a standpoint dictated by some of the themes of the *de mixtione*. Thus although this whole text may have been excerpted from Alexander's lost commentary on the *de generatione et corruptione* (on this see note at the end of Comm. on ch. 16), as parallels with Philoponus' commentary suggest, it has almost certainly been restructured to fit the present treatise. It is this aspect that the following commentary will particularly emphasise. Aristotle's theory of mixture can best be approached independently through Joachim's commentary on the *de generatione et corruptione* and his still valuable article at *JP* 29 (1904) 72-86. There are more recent accounts in Solmsen, *Aristotle's System of the Physical World*, ch. 19, G. A. Seeck, *Über die Elemente in der Kosmologie des Aristoteles*, pp. 50-59, and R. A. Horne, *Chymia* 11 (1960) 21-27; see also the notes in W. J. Verdenius and J. H. Waszink, *Aristotle on Coming-to-be and Passing-away* (2nd ed., 1966).

Chapter 13 is a straightforward account of the types of mixture, the candidates for blends, the conditions under which these can occur, and finally a description of the interaction of the four elements. In the remainder of his discussion Alexander does not go on to examine the blending of the elements, but rather discusses the essential differences between blending and coming-to-be (in ch. 14), and the separability of constituents from a blend (in ch. 15). The latter topic picks up a recurring theme in the earlier critique of the Stoics (see 1. 213. 2-13, cf. 7. 221. 14-15, 9. 222. 29-35) and is Alexander's main answer to their theory. For by showing how constituents are separable he hopes to show in what sense they are present in a blend; in this way he offers the Aristotelian response to the notion of body going through body, which had seemed so absurdly linked to the idea that bodies were present in and separable from a blend (see on 1. 213. 2-13).

Chapter 13

228. 5. "our original statement." See on 1. 213. 2.

228. 10-25. This general account of the constituents of blending is derived from *de gen. et corr.* A10 327b 12-23. The constituents of mixtures must be substances, Alexander sees, if they are to be separable from one another (228. 14-15). There is almost a play

on words (reflecting *de gen. et corr.* 327b 27-29) in the claim that only what is separable (χωριστόν, 228. 15) can be separated from a blend. Cf. also Philop. *de gen. et corr.* 192. 2-4 for a similar statement.

In theory this requirement is reasonable enough. If blending is to involve the unification of bodies without their complete destruction (see *de gen. et corr.* A10 327a 35-327b 10) it must occur between bodies that have some independent status. However this creates difficulties for Aristotle's account in three areas. (1) If substances are not destroyed but exist in a unified blend in potentiality, what exactly is their ontological status? Can a form be diminished in this way? Alexander simply assumes that there is no difficulty here; he regards the arguments in ch. 15 as proving that such a diminished existence is possible, and in ch. 14 sees no difficulties in distinguishing this form of change from coming-to-be and passing-away. See further on 14. 230. 14-24, 15. 231. 16-19, and 232. 29. (2) If the theory of blending is intended to explain the combination of the four elements to form homoiomerous bodies (as *de gen. et corr.* B7 334b 8-30 indicates), then the elements must be substances. But if blending is primarily the interaction of opposites (see 229. 17-18), are not these "powers" rather than the elements themselves the basic constituents of blends; and if this is the case, is it really substances that are interacting? See also on 229. 18-30. (3) The specific examples of blending given by Aristotle and Alexander are not of substances but of "masses," i.e. random volumes of matter: tin and copper (*de gen. et corr.* A10 328b 9), milk (= cheese and whey) (15. 231. 29), wine and water (232. 2), mead (232. 5), and some unspecified liquids (233. 2-14). Such masses (σωροί) are not substances at least according to one statement of Aristotle (*Met.* Z16 1040b 5-10; cf. p. 58 n. 150 above), and if they are merely meant to be illustrations of a fundamental form of blending occurring between the elements (or powers) they do not directly establish that theory; for in the examples given they are not opposites, as constituents of blends must be. See on 229. 21-25.

228. 13-14. "now mixture and blending occur among things that are naturally independent substances" (ἐν τοῖς καθ' αὑτὰ ὑφεστάναι φύσιν ἔχουσι). Cf. 228. 24-25 and on 4. 217. 33-34 above. The present translation is clearly justified by the conclusion of this section, 228. 22-25.

228. 17. "Anaxagoras." Cf. *de gen. et corr.* A10 327b 19-20 where there is a general reference to the doctrine of everything being mixed with everything.

228. 20-21. "For even if they are also substances, they do not exist with equal independence" (καὶ γὰρ εἰ οὐσίαι καὶ ταῦτ', ἀλλ' οὐ καθ' αὑτὰ συνυφίστανται). Here, I think, Alexander is pointing out that although substance may be defined as form (cf. Aristot. *Met.* Z17) this does not mean that forms are independent of substances, for there is still, as the next sentence (228. 21-22) makes clear, a dependency of form on matter. Thus to say that form is substance is not to say that there is "an independent substance *along with*" (συν-) substance itself. This usage differs from the simple sense of "coexist" that we find, for example, at Philop. *de gen. et corr.* 301. 27-28 where place is said to coexist with the body occupying it (τῷ σώματι συνυφίσταται) and not to be independent (ἔχει καθ' ἑαυτὸν ὕπαρξιν); for in this case coexistence is exactly the sort of interdependency that Alexander claims for form and substance by denying their joint independence in the present text.

228. 27-28. "mixture is the more generic, blending the more specific." Cf. Aristot. *Top.* 122b 25ff. for this distinction. On Aristotle's terminology in general see Joachim, *JP* (1904) 73-74, and on his use of classifications of mixtures in other contexts see M. J. Aubonnet, *Actes du VIIe Congrès de l'Assoc. G. Budé*, Paris (1964), p. 300. The distinction between composition and mixture (30-34) corresponds to that drawn between composition and juxtaposition by Galen *de const. art. med.* I. 241-242 K. Alexander also uses μῖξις (mixture) for juxtaposition at 7. 221. 7, though if it is a generic term (as here and in ch. 3) it should not have been used in such a subordinate classification. Cf. p. 56 above.

228. 36. "...among easily-bounded and moist bodies." In addition to being easily-divided (see note on 8. 221. 25-28) constituents of a mixture must have fluid surfaces. Cf. Aristot. *de gen. et corr.* 328b 3 where this type of substance is regarded as a species of divisible bodies; cf. also 328a 17, and 14. 231. 12-13 below.

229. 3-21. For the main theme of this paragraph see Aristot. *de gen. et corr.* A10 328a 18-28.

229. 3. "because of its viscosity" (διὰ γλισχρότητα). Cf. 231. 1, and Aristot. *de gen. et corr.* A10 328b 4-5 where however no examples

of viscous substances are given. On oil as viscous see *Meteor.* Δ6-7 383b 34-384a 2.

229. 6-8. "the whole divine body, which is active, is unmixed." This special emphasis on the fifth element, the aether, being free from blending seems to pick up the argument at 226. 30-33b where the Stoic God, it was claimed, was involved in the same reciprocal interaction as is now attributed to constituents of blends; see note *ad loc.* For aether as "unaffected" see *de caelo* A3 270b 2. An aporia as to whether or not the divine body is unaffected can be found at Alex. *Meteor.* 18. 8-19. 19 (= Philop. *Meteor.* 47. 27-48. 13). See E. Évrard, *REG* 78 (1965) 593-594 on both texts.

229. 8-11. "of bodies that do share in (κοινωνεῖν) the same matter." On this language see on 3. 216. 1-4.

229. 15. "sound ... cannot be acted on by line." Simplicius (*Phys.* 516. 27-28) remarks that sound cannot be in contact with line apropos *Phys.* 208a 11-14 where Aristotle points out that random bodies cannot be in contact. This provides another example of a simple illustration applied by commentators to only broadly similar points; cf. above, p. 80 and p. 87 n. 253.

229. 17. "... a common underlying nature that can receive the qualities of each" (κοινὴν ὑποκεῖσθαί τινα φύσιν τῶν ἑκατέρων παθῶν ἐπιδεκτικήν). Aristotle uses δεκτικός ("receptive") of matter at *Met.* 1055a 29-30, and *de gen. et corr.* 320a 2-4, and the same term analogically at *ibid.* 328b 11 and *de an.* 414a 10. For Alexander's derivative use of ἐπιδεκτικός see *de an.* 5. 7-8, *Met.* 157. 23-24.

229. 18-30. "interaction, then, occurs among bodies with ... a mutual contrariety." Blending occurs between opposites, and the only genuine opposites in Aristotelian physics are the four primary qualities considered here. However at *de gen. et corr.* B7 334b 8-31 blending is discussed only in terms of these qualities and not their pairings as elemental qualities with forms described by Alexander at 229. 30-230. 5. While it is questionable whether or not Alexander's criterion that only substances blend (228. 10-25) is in fact fulfilled by Aristotle's account (on this see Sokolowski, *JHP* 7 [1970] 271-272; also cf. 229. 29-30 and on 15. 232. 29), Alexander himself seems to have no difficulty in seeing elements as substances with forms: cf. esp. *de an.* 7. 9-11. 13 where the forms of the elements

are conceived of as the primary forms of all bodies compounded from them. This systematisation roughly corresponds with Joachim's claim (*JP* 29 [1904] 75-76) that for Aristotle all chemical compounds are *au fond* "quaternary"—i.e. the product of the four elements. It is, however, not significant in the present discussion where the role of elements in blending is for various reasons (see on 229. 30-230. 14) neglected.

229. 21a-25. "and also their intermediaries." Cf. Aristot. *de gen. et corr.* B7 334b 8-30 where the blend of pure opposites is rejected and allowance is made for varying degrees of hot and cold in blends. The mean (τὸ μέσον) along which such degrees can be found is "extensive and not divisible" (πολὺ καὶ οὐκ ἀδιαίρετον, 334b 28-29). This view assumes that only the powers and not the elements are significant constituents of blends. It is also irrelevant, as Seeck (*Die Elemente*, pp. 56-57) notes, to the discussion in *de gen. et corr.* A10 of the blending of substances such as wine and water (328a 27-28), or metals (328b 10-14).

229. 29-30. "they are differentiated by forms and accidental qualities." At *de an.* 7. 13-14 Alexander makes it clearer that the elements interact to form compounds by their incorporeal qualities (κατὰ τὰ ἀσώματα ποιεῖν τε καὶ πάσχειν). Cf. however *de sensu* 73. 19-21 where this view is termed Platonic—a description confirmed by Albinus *Didask.* 11 (p. 22 Hermann = p. 65 Louis). This only shows the extent to which he is trying to systematise Aristotle's often confusing comments on the interaction of the elements, where regularly the greatest emphasis is laid on the "powers" rather than the elements themselves—cf. esp. *de part. an.* B1 646a 13-21 and see F. Solmsen, *Aristotle's System of the Physical World*, pp. 345-352. Alexander may also have been provoked by the doctrine that qualities are bodies, attributed to the Stoics (cf. p. 85 n. 247 above); Alex. *Mant.* 122. 16-125. 4, like Albinus *Didask.* 11 and Ps.-Galen *de qual. incorp.*, attacks this doctrine at length.

229. 30-230. 14. The principal Aristotelian source of this description of elemental qualities and interchange is *de gen. et corr.* B4. Alexander is justified in introducing the elements into his account since they are the primary constituents of blends for Aristotle, as *de gen. et corr.* B7 makes clear. However their significance is not really explained here: they are simply said to be bodies sharing

the same matter, differing in form, and interacting to produce coming-to-be and passing-away. Their role as the constituents of the homoiomerous bodies is not explained, and thus their importance to Aristotelian chemistry goes unrecognised. They merely exemplify the general account that has so far been given, and are shown to be engaged in a kind of process which in ch. 14 has to be distinguished from blending. This seems to be because Alexander is interested primarily in the *process* of blending and not the precise character of its end products (see the descriptions at 14. 231. 1-12 and 15. 233. 2-14). His aim is not to describe its place in the Aristotelian physical system as much as to point up those aspects in which it particularly contrasts with the Stoic theory. For this he needs to show only how mixture occurs and in what sense bodies are present in, and separable from, a blend.

Chapter 14

230. 14-24. This general distinction between blending and coming-to-be is an expansion of Aristotle's remarks at *de gen. et corr.* A10 328a 23-31 (cf. 327b 28-31). Its success depends on the validity of the account of blending as a transformation that does not involve the constituents completely changing into another body (see ch. 15 *passim*). Aristotle, for example, speaks of the elements "changing" (μεταβάλλει) to form homoiomerous compounds (334b 25), and of constituents "changing into one another" without losing their own nature (328a 29-30). Again he defines mixture as "the unification of bodies that have undergone qualitative change" (τῶν μικτῶν ἀλλοιωθέντων ἕνωσις—328b 22). Now qualitative change (ἀλλοίωσις) is not blending, for it can only occur in the qualities of a given substance (e.g. *de gen. et corr.* A2 317a 27). Blending must then fall between this and the case of absolute coming-to-be (γένεσις ἁπλῆ) where one substance completely overpowers another (cf. 13. 230. 5-13). But this is awkward, since the distinction between coming-to-be and qualitative change is grounded in the metaphysical and logical distinction between substance and quality. (On the problems that this entails for Aristotle's general physical theory see G. Morrow, "Qualitative Change in Aristotle's Physics," in the *Verhandlungen des 4. Symposium Aristotelicum*, pp. 154-167.) Since there is no intermediate category it is hard to see how there can be an intermediate process. Note the omission of mixture from the lists of

changes in the categories at *de gen. et corr.* A1 314b 26-315a 2, and A4 319b 31-320a 5, although it shares a broad similarity to these changes because its components share in the same matter (cf. *de gen. et corr.* A10 328a 18-23). The difficulty in our text can be seen from the common terminology that Alexander employs to describe coming-to-be and blending. Both, for example, are said to cause a change (μεταβολή)—cf. 230. 6 with 231. 7, 11—and are described as changes (μεταβολή, μεταβάλλειν)—cf. 230. 11-13, 22 (describing coming-to-be) with 231. 20, 22; also see 232. 14, 17, 29, 32 and 233. 13 where this language describes the formation of a blend and the reconstruction of the bodies that have been blended.

In general this difficulty is a result of Aristotle's theoretical demand that bodies be separable from a blend yet not be actually present in it (cf. 327a 35-327b 12 with 327b 26-28). To solve it he was forced to strain his categories for the description of physical change by establishing a theory which claimed that bodies were only potentially present in a blend. In this way he met his demand, but only at the expense of creating these further difficulties as Alexander's account in ch. 15 also shows; see further on 231. 16-19, 232. 20.

230. 14-24 is translated by Moraux (*Alexandre d'Aphrodise*, pp. 40-41). He argues that the distinction upheld here contradicts Alexander's view in the *de anima* (e.g. 8. 22-9. 3) that the soul is produced by the blending of the elements. On the whole question of this apparently unorthodox interpretation see the more extensive discussion by P. Donini, *Atti del Accad. delle Scienze di Torino* 105 (1971) 61-107. Certainly, as we have seen (on 13. 229. 18-30) the discussion of mixture in this treatise has no relevance to the evolution of forms through the blending of the elements.

230. 14-15. "blending is held to be the passing-away of bodies that are preserved" (σωζομένων τινων). Cf. 2. 215. 13-14 and *de gen. et corr.* A10 327b 30-31. Cf. note on the text at 230. 15.

230. 16. "the one increasing and the other passing-away." Cf. *de gen. et corr.* A10 328a 24-26 where such increase is said to occur only when vastly unequal quantities are mixed. The illustration of this at 328a 26-28 as the submersion of a wine drop in innumerable measures of water is apparently directly contradicted by the Stoic theory of total blending (see above, p. 33).

230. 17. "nutriment is not even said to be blended." This probably picks up *de gen. et corr.* A10 327b 14, a text that may have, at least in part, justified the discussion of organic growth in ch. 16.

230. 18. "if both constituents are totally destroyed." Cf. *de gen. et corr.* A10 327b 4-6 where this is one of Aristotle's initial *aporiai* in his discussion of blending.

230. 31. "...lose that superiority in their contrarieties" (ἀποβαλόντα τὰς κατὰ ἐναντιώσεις τὰς ὑπεροχάς). Cf. Alex. *de sensu* 64. 22 and *Quaest.* II. 24. 75. 20-21 for ἀποβάλλειν in this sense; also *Quaest.* II. 15. 60. 7. It is not an Aristotelian term.

231. 1. "retain their own surfaces." "Preservation of surfaces" was, we argued (on 1. 213. 2-13), Alexander's particular way of emphasising the sense in which constituents had to be present in a blend for the Stoics. Cf. on 7. 220. 37-221. 7.

231. 2. "in juxtaposition" (κατὰ τὴν παράθεσιν). Cf. 231. 11, 14; 15. 233. 2-5; and Aristot. *de gen. et corr.* 328a 33, 328b 2. This juxtaposition is simply a consequence of constituents of a blend being easily-divided (see note on 8. 221. 25-28) and is only a preliminary to unification. Aristotle's objections to it as a complete explanation of blending are integrated into Alexander's critique of Democritus and Epicurus in ch. 2, as we have seen (notes on 214. 18-28 and 214. 28-215. 8). For an elaboration of this principle see Aristot. *de sensu* 440a 31ff. where bodies divisible into minima (e.g. the individual in a species) are distinguished from infinitely divisible bodies, and it is claimed that while the former could only be juxtaposed the latter could be blended. Alexander (*de sensu* 64. 2-15) expresses this as a distinction between solid anomoiomerous bodies and moist homoiomerous ones, thus integrating it more closely with the general theory of blending. The relevance of infinite divisibility to blending was however limited for an Aristotelian by the fact that a mixture had to be a unity, and constituents could not be said to be blended in their undivided parts (cf. 8. 222. 13-14).

Simplicius (*de caelo* 660. 18-661. 14) holds a Platonic view that bodies remain juxtaposed while only their qualities are fully blended; cf. Zabarella, *de mistione* 2 for its elaboration by Avicenna. On the attempts of later scholasticism to base Aristotle's theory of blending on a doctrine of corpuscles (*minima naturalia*) see E. J.

Dijksterhuis, *The Mechanization of the World Picture,* pp. 24, 200-206, 277-279.

231. 3. "... a mutual contrariety of equal power" (τὸ ἐναντίωσιν ἔχειν ἰσοκρατῆ πρὸς ἄλληλα). Cf. 231. 6 and Aristot. *de gen. et corr.* A10 328a 29-31. This common-sense notion (cf. p. 33 n. 56 above) may have also held broader implications for Alexander at the level of elemental blending. Simplicius (*de caelo* 83. 30-32) quotes him as saying that "the reason for the preservation and stability (σώζεσθαι καὶ συμμένειν) of the elements, and their not being destroyed, was the equality of their powers (τὴν τῶν δυνάμεων ἰσότητα)." We know that "preservation" and "stability" were terms particularly applied to bodies in prescriptions of providential power over the cosmos (note on 11. 224. 30-32), and the present text may reflect such an account, perhaps apropos *de caelo* A3 where Simplicius quotes him. Note also *de mundo* 396b 34-397a 1 where the preservation (σωτηρία) of the cosmos is said to be due to the "agreement" (ὁμολογία) between the elements not to exceed one another.

231. 10. "Blending, then, can be defined..." This provisional definition of the process of blending omits to describe either the precise nature of the product or to explain in what sense the constituents are present in the blend. The latter point is taken up in ch. 15 but the former is never explored. It is simply assumed that the unified bodies created by blending are uniform (homoiomerous)—see 15. 231. 26-27—without the question of the formation of the homoiomerous bodies (flesh, blood, bone, etc.—cf. *de gen. et corr.* B7) being explored (cf. on 13. 229. 30-230. 14).

Chapter 15

231. 12-29. The separability of constituents from a blend was not discussed by Aristotle, but merely asserted as a consequence of the candidates for blending being substances—see note on 13. 228. 10-25. Alexander is forced to deal with it because of his critique of the Stoics. At 1. 213. 6-13 he claimed that a theory of blending which held that constituents were separable but not completely mixed was superior to the Stoic view that they were totally mixed and separable. Again in ch. 7 he argued that for the Stoics blending must be fusion if it is to be total blending, and will therefore make

constituents inseparable (221. 7-20), but went on to say that if we see that they are separable this theory will be disproved—see 221. 20-25. These hints are now brought to a point as Alexander argues that just as bodies are easily mixed (231. 12-19; cf. 8. 221. 25-28), so they are easily separated (231. 19-29). This establishes a physical criterion for the notion of being in potentiality; it is simply a state both easily achieved and easily lost. Yet it still needs to be characterised more positively and this Alexander attempts with some difficulty at 231. 27-29 (see note on 231. 16-19).

Since this idea of constituents being easily separated is not Aristotelian it needs special attention. Alexander emphasises that "slight assistance" (ὀλίγη βοήθεια, 231. 19, 232. 16; cf. ὀλίγη πέψις [slight fermentation], 232. 6) is required to produce this change, which necessarily occurs easily (cf. ῥᾳδίως, 231. 23, 28). Cf. Galen *de tempr.* I. 9 (p. 32. 16-19 Helmreich) where a broadly similar connection is drawn between the presence of constituents in potentiality in a blend and their easy removal.

Now this agent of assistance is some physical addition to a homoiomerous substance that actualises the bodies that have been blended. The resulting causal process is comparable to that which the Stoics thought occurred when some form of assistance helped bodies to extend themselves in various ways; cf. the argument reported in 4. 217. 17-26. As we saw (pp. 39-40 above) these processes were a species of what the Stoics called "cooperant causes" (συνεργὰ αἴτια), and it would seem certain that Alexander has derived his reasoning in this chapter from that text. For the Stoics certain agents help to extend bodies and so explain, by analogy, the extension of body through body in a blend (see 217. 26-32), while Alexander argues (231. 19-29) that given agents will aid a compound in readily dissolving into its constituents. The same notion of "assistance" is employed, and rather significantly even the verb συνεργεῖν ("to cooperate") is used at 231. 23 (cf. 231. 12) though it does not occur in ch. 4. Also cf. Sext. Emp. *P.H.* III. 15 where a Stoic cooperant cause is said to "contribute a slight force to the easy production of an effect" with 231. 22-24.

This assimilation of a Stoic thought-pattern is a minor example of a common feature of Alexander's relation to the Stoics (see above, pp. 27-28 and on 232. 1-5 below). He grasps the general principle of a Stoic idea and either criticises it or assimilates it, or as in the present case, does both; for at 5. 218. 10-5 he criticises the analogical argument from the common notions reported in ch. 4, while here he has in essence copied it. It is also typical that this assimilation is entirely implicit.

As for the validity of his reasoning, it is neither weaker nor stronger than the Stoic argument in ch. 4. An analogy is drawn between cases where constituents of blends are easily separated and the general theory that bodies are reconstituted from a blend without coming-to-be, because they are present in it in potentiality, and yet when reconstituted are not identical with the bodies that had been present in the blend. Like the Stoic examples of extension in ch. 4 Alexander's various cases of separation at 231. 29-232. 7 are "suasions" (πίστεις); only if we accept the independent validity of the analysis of blending in chs. 13 and 14, and the intelligibility of the notion of being in potentiality, can we be persuaded that the present account elucidates its nature. This is quite apart from the irrelevance of these examples to the blending of *opposites* which at 13. 229. 17-18 is said to be an essential part of the process.

The argument of this chapter is not in itself, then, adequate to prove its thesis, just as Stoic arguments from the common notions were, I believe, similarly deficient and only intended to support theories established independently on more rigorous grounds (see *Symb. Os.* 48 [1973] 58-60, and above at p. 41, n. 99). Alexander's argument is not as rigorous as some of the analogical reasoning that we find in Aristotle (on this see M. Hesse, *Phil. Quart.* 15 [1965] 328-340). Nor, indeed, is it as informative as analogies that we find elsewhere in Alexander's works; cf. *de an.* 6. 16-20, 9. 14-23, which relate parts of a theory rather than attempting to prove a preestablished theory. The closest parallel to the present text is *de an.* 4. 11-18 which compares the relation of matter and form to imprint and wax (cf. Aristot. *de an.* B1 412b 7). Similar analogical arguments are used at 16. 237. 5-218. 10; cf. notes *ad loc.*

231. 16-19. "a body not in actuality any of the bodies that have been mixed..." This describes in metaphysical terms the product of blending characterised in physical terms at lines 13-16. It is the basis for a similar description of the separation of constituents at 231. 24-29, where it is inferred that the bodies separated are not the original constituents but the product of an easily-achieved change into a similar body not identical with them. The

inherent obscurity of this point is unfortunately increased by the unsatisfactory text at 231. 27-29.

While we can grant the evident ease of the physical process by which constituents blend and are separated, it is not clear that this in itself explains the ontological status of constituents throughout the change. Their being in potentiality in a blend fulfills the requirement that constituents be neither destroyed nor juxtaposed (cf. also 232. 20-31), but this state is not given any clear independent characterisation. Philoponus, for example, compared it to the state of a geometer, who in potentiality could do geometry, but had this power in a diminished form through being drunk (*de gen. et corr.* 188. 21-26). Some such account is required because the sense in which constituents are in potentiality in a blend is quite novel. It is neither like the case where a student is potentially a geometer, or where an inactive geometer is potentially one; for the first would mean that the constituents did not exist at all, and the second that they had not fully merged their identity in the blend. See further Joachim on *de gen. et corr.* A10 327b 22-31. None of this is explored by Alexander, although it became a major topic in later scholasticism as Zabarella's *de mistione* shows. He is content to offer a set of analogies with actual physical processes (231. 30-232. 18) without independently explaining the concepts which these analogies clarify. A similar conceptual weakness occurs in his general distinction between blending and coming-to-be (see on 14. 230. 14-24), and in his uncritical claim that the constituents of blends are substances with forms (see on 232. 29 below).

231. 21-22. "for they do require some addition to be actualised, but not absolute coming-to-be and change" (προσθήκης γὰρ δεῖταί τινος εἰς τελειότητα, ἀλλ' οὐ παντελοῦς γενέσεώς τε καὶ μεταβολῆς). When the bodies are reconstituted they reach a τελειότης (lit. a "completeness"; "Vollendung" [Rex]). Now this could be taken in a strictly physical sense as when Aristotle says at *Meteor.* 379b 20 that a body "is completed" (τετελείωται) by fermentation. Interestingly Alexander *ad loc.* (*Meteor.* 186. 34-35) describes this as a body coming to be in actuality, exactly the phenomenon both here and in the other cases where similar language is used (232. 29, 32, and cf. 232. 12 where agents are said to bring bodies in potentiality to completion [τελειοῦντα]). This means that τελειότης carries a metaphysical sense, and indeed we find it used by Alexander to mean

form (e.g. *de an.* 23. 17-18) and actuality—ἐνέργεια (*de an.* 43. 7-8). Hence τέλεος ("complete") at 14. 230. 23, 25 and 231. 10 would mean both physical and metaphysical completeness. But cf. further on 232. 29.

231. 29-33. "a heated stone cast into milk." For this case cf. Aristot. *Meteor.* 381a 7 and 384a 21-25 with Alexander *ad loc.* Also see Galen *nat. fac.* I. 15 (II. 58K) where it is used as an analogy in a discussion of the passage of blood through the kidneys. It is probably copied from Galen by Athenagoras *de ress.* 6; see L. Barnard, *Latomus* 31 (1972) 426.

232. 2-5. "the action of the sponge." This example would seem to be borrowed from an account of the Stoic tripartition of blends: see Stob. *Ecl.* I. 17, p. 154. 8-11 W (= *SVF* II, p. 153. 21-23), and Philo *de conf. ling.* 184 (= *SVF* II, p. 153. 36-39). Sambursky, *Isis* 49 (1958) 332-333 has a brief discussion of it. It is one example of Alexander's dependency on, and transformation of, the former text (see above pp. 55-65). Cf. also Nemesius *de nat. hom.* 3, p. 128 Matth. where this example serves to illustrate the preservation of constituents in a blend below the level of perception, a theory attributed to Democritus by Alexander at 2. 214. 18-28.

232. 6-10. "the onset of a slight fermentation in must." There is a general reference to the heating of must (γλεῦκος) at Aristot. *Meteor.* 380b 31-32.

232. 9a. "just above." I.e. at 10. 223. 20-25. Here (at 232. 8 and 9) I translate πνεῦμα (pneuma) as "air", its meaning in that earlier "realistic" argument where in the context of the general polemic against pneuma I transliterated it (with inverted commas) "pneuma"; see note on 10. 223. 19-27.

232. 20. "the body that was affected by them ... came to be" (ἡ γένεσις τῷ ταῦτα πάσχοντι σώματι). That is, the compound is created by the interaction of the constituents and can therefore be described, albeit loosely, as coming-to-be by being acted on by them. Similar language is used to describe the formation of pneuma as a compound from fire and air at 10. 224. 20; see also 232. 13 where Alexander speaks of the process (γένεσις) of separation, and cf. *de an.* 19. 27, 29, 32. This usage reflects the fact that blending has to be described as a form of change, yet all the relevant vocabulary

(γίγνεσθαι, μεταβολή) is appropriated by the Aristotelian doctrine of absolute change; see further on 14. 230. 14-24.

232. 23. "as we see when Air is separated from Water." Water (Moist and Cold) changes into Air (Moist and Hot) when the Hot overcomes the Cold; see *de gen. et corr.* B4 331a 19-20 and 29-31 for the reverse change. Cf. on 16. 234. 7 below.

232. 24-25. "neither are bodies present in actuality dissociated...' (οὔτε γὰρ ἐνεργείᾳ ἐνυπάρχοντα διακρίνεται). That is, if bodies are juxtaposed then they are present in actuality. Cf. Aristot. *Phys.* Δ4 212b 3-6 which distinguishes parts in a whole being (a) "in actuality, as with a heap" (κατ' ἐνέργειαν ὥσπερ σωρός), from (b) those in potentiality (κατὰ δύναμιν) in a homoiomerous body.

232. 26-27. "in these cases." When a mass of juxtaposed bodies is reduced it remains the same in form although diminished, and Water, part of which is converted into Air, still retains its identity as Water.

232. 29. "changing into the actuality of which it was deprived..." (μεταβάλλον εἰς τὴν τελειότητα, ἧς ἀφῃρέθη). τελειότης, which is translated as "actuality," is, as we saw (note on 231. 21-22) a complex term. If it means form here, as it sometimes can, then Alexander could be implying that constituents lose their forms in a blend, a process usually described as passing away (cf. 9. 223. 5-6). It must, then, mean actuality (ἐνέργεια) or physical completeness, so that this statement would be compatible with the idea that forms are retained in the blend. As we have seen (on 13. 229. 18-30) by neglecting the question of the ontological status of constituents, *qua* substances with forms, in a blend, Alexander does not probe further the character of a form in this diminished state. For such discussion we have to turn to a later scholasticism—concisely summarised in Zabarella's *de mistione.*

It is notable that in ch. 16 Alexander has a great deal to say about the preservation of form throughout the process of growth. This is presumably because Aristotle had explicitly solved the problem of how a substance preserves its identity throughout this particular form of change in terms of the concept of form in *de gen. et corr.* A5. In A10, the chapter on mixture, form gives way to being in potentiality in the account of a substance's retention of identity. Alexander is therefore only as precise as his authority allows him to be. Cf. also on 16. 235. 14-236. 5.

232. 31-233. 2. The theory of blending must accord with (1) the basic principles of Aristotelian physics—presumably the theory of the elements and powers, the concepts of matter and form, and actuality and potentiality; (2) "how bodies change and come to be," referring presumably to the distinction drawn in ch. 14 between absolute coming-to-be and the particular change involved in blending; and (3) the common preconceptions about blending—cf. 215. 31-32 where these were invoked as a general criterion for any theory of blending. (3) ought also to apply to the analogical arguments at 231. 29-232. 18, given the relation that they have to Stoic argument from the common notions in ch. 4 (see note to 231. 12-29). Alexander however is probably only referring to general principles such as the preconceptions that a mixture be unified (215. 10-11), and the natural notion that body does not go through body (218. 17-18). As ch. 4 itself shows he does not seem to have appreciated the role of the common notions in an argument, and is content to take them over in the loose sense of general principles—see on 5. 218. 11. It is also, as we have seen, questionable how far Alexander's account is satisfactorily related to (1) and (2); on the former see notes on 13. 228. 10-25, 229. 18-30, and 15. 231. 16-19, and on the latter on 14. 230. 14-24.

233. 2-14. This text is rather irrelevant here. It simply points to the empirical verifiability of blending in its initial stage when bodies divide one another, and belongs with the general comments on this at the beginning of ch. 15—231. 12-18. It adds nothing new, though it reemphasises Alexander's concern to establish an empirically based theory, evident also from the analogical argument for the nature of constituents' presence in a blend at 231. 30-232. 18. Cf. the reference to surface contact at 233. 11-12 with the general principle that constituents are juxtaposed, i.e. in surface contact, before being fully blended; see 14. 231. 1-2, 11 and note on 231. 2.

Chapter 16

The principal justification for this long chapter being in the *de mixtione* is that it deals with the paradox of body going through body. Alexander says that some thinkers have tried to explain organic growth in this way (233. 14-24) and must be refuted. This he does very briefly at 234. 23-32 in an argument that elaborates Aristotle's point (at *de gen. et corr.* A5 321b 15-16) that growth cannot occur by body since this will mean that two bodies occupy the same place (cf. also *de gen. et corr.* A10 327b 14 where Aristotle makes the very general point that nutriment is not mixed with a body). The remainder of the chapter is a long paraphrase of Aristotle's discussion of growth in A5 with only two further brief references to the notion of body going through body at 236. 12-14, 238. 20-23. There is also a brief reference to the fact that growth is not blending or coming-to-be (234. 5-9) which links it to the discussion in ch. 14 (cf. 230. 17), and the analogical arguments of ch. 15 (231. 30-232. 18) are closely followed in form at 237. 5-238. 10. Apart from this, the present chapter must be considered an appendix to the *de mixtione*. As we have argued above (pp. 83-84) it is certainly not anti-Stoic, and in view of the independent discussion of "body going through body" in the exegetical tradition must owe its place here to Peripatetic concerns.

The commentary, as with chs. 13-15, will deal principally with Alexander's contribution rather than the Aristotelian theory of growth itself on which Joachim's commentary on *de gen. et corr.* A5 goes into great detail. See also the notes in W. J. Verdenius and J. H. Waszink, *Aristotle on Coming-to-be and Passing-away*. Like chs. 13-15 this discussion is undoubtedly closely related to Alexander's lost commentary on the *de generatione et corruptione* (see Additional Note at the end of this chapter). In a long paraphrastic note on A5 of the Aristotelian work Philoponus includes all the salient points of our text in roughly the same order (*de gen. et corr.* 102. 31-108. 17); some parallels are noted in the commentary, though I have not made a detailed comparison. Like much of that work it was undoubtedly derived from Alexander (see Kroll, *RE*, IX-ii, col. 178).

Other relevant Alexandrian texts are the discussion of the nutritive soul in *de anima* 33. 13-38. 11 apropos Aristot. *de an.* B4; *Quaest.* I. 5, a more abstract discussion than that offered here at 235. 14-238. 10 of why growth occurs in the form and not the matter; the so-called *de augmento*, a Latin text translated from the Arabic by Gerard of Cremona (see Cranz, *Catalogus*, I, 88-89), which provides an orthodox defence of Aristotle's theory. It is ultimately derived from Alex. *Quaest.* I. 5. On the relevant Arabic manuscripts see A. Dietrich, *Nachr. Gött. Akad.* (1964), p. 98, and A. Badawī, *La Transmission de la philosophie grecque au monde arabe*, p. 96. The heterodox theory of growth attributed to Alexander by

Averroes (*Middle Commentary on the de generatione et corruptione*, tr. S. Kurland, p. 38) must be based on a misunderstanding.

233. 15-22. The contrived nature of this polemic is further evident if we compare the link established here between nutriment's being assimilated throughout the whole of a body and bodily interpenetration, and that established in ch. 8 between total mixture and complete division. In the present case Alexander himself admits that growth occurs in all parts and uses this to attract, as it were, the theory of bodily interpenetration (cf. also 234. 23-29 and 236. 12), whereas in ch. 8 he compels the theory of total blending to admit a compatibility with the paradox of a complete division of constituents.

233. 25. "our problem." I.e. the claim that growth involves body going through body. Cf. 234. 33-34 where Alexander also seems to be justifying the relevance of his lengthy exposition by referring to this topic.

233. 26-27. "only bodies with their own nutritive faculty are nourished..." (τρέφεται δὲ ὅσα τὴν θρεπτικὴν ἔχει δύναμιν ἐν αὑτοῖς). This nutritive faculty is not mentioned in *de gen. et corr.* A5, though it is implied (see Joachim on 322 31-33). It is that basic faculty of the soul possessed by all animals (see *de an.* B4 415a 23-24, cf. *de gen. an.* 740b 34). In his discussion of *de gen. et corr.* A5 Philoponus deals with it at length—see 107. 32-108. 3, 117. 22-23, 29-32. See also below 234. 11, 14; 236. 20-21, 26; 237. 24. At *de an.* 35. 9-17 Alexander draws a distinction between the nutritive faculty and the faculty of growth within the vegetative soul (τὸ φυτικόν)—e.g. the old are nourished but do not grow (35. 12-13). This is not of central importance for the general theory of growth enunciated in this chapter; the two are linked at 234. 11 (cf. also 233. 22, 24-25) and distinguished in very general terms at 236. 18-26.

233. 30-234. 15. This is based on *de gen. et corr.* A5 321a 17-321b 10, though without the aporetic character of the Aristotelian text.

234. 7. "the change from Water to Air." For this example see *de gen. et corr.* A5 321a 10-17. Alexander has added the case of blending (234. 9) because of its relevance to the main theme of this treatise; cf. also 15. 232. 20-31 where it is distinguished from this process of change.

234. 17-23. "It is not by the incorporeal." This is derived from the brief Aristotelian argument against growth by the incorporeal at *de gen. et corr.* A5 321a 6-7: "If then by the incorporeal there will be separable void (χωριστὸν κενόν)—but it is impossible that there be separable matter for a magnitude." Taking only the first clause Alexander has interpreted "separable void" to mean the non-occupancy of place by the cause of growth. Although this argument is only applied to the body that causes growth (τὸ αὐξῆσαν, 234. 21) it would equally apply to the body that grows, if it were said to be incorporeal.

234. 23-32. At *de gen. et corr.* A5 321b 15-16 Aristotle places three restrictions on any theory of growth: (1) that the body [that grows] is not void; (2) that there are not two bodies in the same place; and (3) that growth does not occur through something incorporeal. (2) is dealt with here by referring to the familiar paradox of body going through body (234. 23-29); cf. Philop. *de gen. et corr.* 103. 10, 15. On (1) see 234. 29-32 where the body that is void is clearly the one that is nurtured (τὸ τρεφόμενον, 234. 30). (1) is also expressed in more general terms at 5. 219. 3-9, and can be derived from Aristot. *Phys.* 214b 8-9 where the view that growth requires the void is refuted. See also Philop. *de gen. et corr.* 104. 14-17 and cf. above, pp. 78-79. (3) is the correlative of (1), and is paraphrased by Alexander at 234. 17-23 above (see note *ad loc.*). Philoponus (*de gen. et corr.* 103. 10-15) groups both (1) and (3) together as entailing the introduction of the void.

Joachim's note on *de gen. et corr.* 321a 5-9 goes into more detail on this topic, and I have drawn on it here.

235. 1-12. This is a more elementary version of *de gen. et corr.* A5 321b 16-19.

235. 1. "whole units" (τὰ ὅλα). Growth is in all parts (see 233. 16, 19-21, 234. 1-2); therefore whatever grows must grow as a whole.

235. 12-14. "preserved and stable" (σώζεσθαι τε καὶ μένειν). Cf. 234. 10 and 236. 23, and see note on 11. 224. 30-32 where we show that this language is particularly applied to a body's form, and probably carries providential overtones. Its use in the present text anticipates the argument beginning at 235. 17 which shows that growth must be in the form.

235. 14-236. 5. Cf. *de gen. et corr.* A5 321b 19-28. At *Met.* Z16 1040a 5-16 the parts of animals are ruled out as substances. Joachim (on *de gen. et corr.* 321b 19-22) accuses Alexander of interpreting Aristotle as talking about form and matter in abstraction from the living organism. Yet he does make references to the nutritive faculty of organisms (cf. on 233. 26-27), which is more than Aristotle does in *de gen. et corr.* A5. Joachim (on 322a 31-33), wrongly I believe, takes the term δύναμις (322a 28-29) to refer to the nutritive soul, but in the context it is only an alternative for εἶδος (cf. *de gen. et corr.* A10 327b 31) and used as it is with reference to the analogy of an expanding duct (cf. on 237. 28-238. 10) has no biological connotation. The abstractness of Alexander's account is no greater than Aristotle's and, like his authority, he treats form throughout this chapter as the form of a given quantity and not as the nutritive soul of an organism, the view for which Joachim (on 321b 16-17) argues speculatively.

235. 33. "prevents its total destruction." Cf. Alex. *Quaest.* I. 5. 13. 11-16 and Philop. *de gen. et corr.* 106. 3-11 for the *aporia* that if growth is in form then it is conceivable that all the matter will be dispersed and a separate form left behind. Both solve it by stating essentially the same theory of controlled dispersal outlined here at lines 29-33.

236. 2. "would abolish the principle" (ἀναιρήσει). Cf. on 6. 219. 20-22.

236. 14-26. This should be compared with *de gen. et corr.* A5 322a 25-28 where Aristotle distinguishes nutriment from growth in its being because it produces growth only in so far as it is in potentiality "so much flesh" (ποσὴ σάρξ), but is nutriment where it is only potentially flesh. Alexander has simplified this point. For the general distinction between the processes of growth and nutrition see also *de an.* B4 416b 11-17, and *de gen. an.* B4 744b 33-36.

236. 14-15. "For matter is not at all stable in a preexistent state..." (οὐδὲ γὰρ τὴν ἀρχὴν προϋπάρχον ὑπομένει). προϋπάρχειν ("preexist") seems to have been a quasitechnical term to describe the state of the substrate to which nutriment is added. See 236. 17, Philop. *de gen. et corr.* 118. 23, and cf. on 237. 8-9. The only precise Aristotelian precedent would seem to be *Rhet.* 1419b 22 (ἡ τῶν σωμάτων

αὔξησις ἐκ προϋπαρχόντων ἐστίν), though the verb is used in a more general discussion of change at *de gen. et corr.* A3 317b 17.

236. 27-28. "...which transforms and assimilates food and absorbs it for the body being nourished" (τῆς ἀλλοιούσης τε τὴν τροφὴν καὶ ἐξομοιούσης καὶ προσκρινούσης τῷ τρεφομένῳ). As a result of this repetitious triad (for other examples see on 3. 216. 15-16) we are forced here and at 237. 17 and 238. 11-13 to distinguish ἐξομοίωσις (literally "making like," and therefore rendered "assimilation") from πρόσκρισις otherwise translated as "assimilation" but which has a root-meaning of "separating towards" or "dispersing in." There is no difference in the processes envisaged: at 238. 11-13 they define one another, and are used together at Alex. *de an.* 35. 1.

236. 32-237. 5. This description of proportional growth and the subsequent illustration is based on *de gen. et corr.* A5 321b 23-322a 4. The details of the forward shift of parts (προωθεῖν) that occurs in order to preserve the same shape (237. 1-5, anticipating the analogy at 237. 5-15) is a scholastic addition. We find this description throughout Philoponus' account—see *de gen. et corr.* 107. 29-30, 108. 6, 30, 33, 109. 4, and in a biological version of the present text at 117. 22-118. 2. (On the general process of assimilating nutriment see also Alex. *de an.* 33. 16-34. 26.) It seems to have developed under the influence of the passage from the *de caelo* B14 on the sphericity of the earth used to illustrate proportional growth in the next passage (237. 5-15) where προωθεῖν ("push-forward") occurs (see 297a 28, cf. 297b 13). See note on 237. 5-15.

236. 29-30. "so it maintains it along with its own shape." Shape is the physical manifestation of a body's form. Although the form of the homoiomerous bodies will not be specifiable as a particular shape, the total effect of growth can be measured by reference to an animal's shape. Form and shape (σχῆμα, μορφή) are often closely related by Aristotle (e.g. *de gen. et corr.* A5 321b 27-28) though in his discussion of growth he does not emphasize the preservation of shape as do his commentators; cf. also Philop. *de gen. et corr.* 105. 15-16.

237. 4. "an equal addition." That is, an equally distributed addition and not the additional nutriment referred to at 236. 18 which is equal to the dispersed flesh and therefore only nourishes *without* causing the very growth under discussion in the present passage.

237. 5-238. 10. On this analogical reasoning see note on 15. 231. 12-29. Though not derived from Stoic sources like those at 15. 231. 30-232. 18 the validity of these analogies is again no greater than that of the Stoic argument from the common notions in ch. 4. They merely sustain a theory already established in abstract terms (at 235. 17-236. 5) by clarifying the particular process of growth prescribed by that theory. They are *a posteriori* suasions rather than directly informative analogies.

237. 5-15. Alexander's illustration here is based on part of Aristotle's discussion of the sphericity of the earth at *de caelo* B14 297a 8-297b 17. There it was argued that if we conceived of the earth being generated the heavy particles would all move towards the centre of the universe and so all be at a constant distance from it (297a 8-21). This would apply even to the case of particles of different size, since they would all have the same impulse to the centre and the larger would drive the smaller ahead of it (τὸ γὰρ πλεῖον ἀεὶ τὸ πρὸ αὑτοῦ ἔλαττον προωθεῖν ἀναγκαῖον μέχρι τοῦ μέσου τὴν ῥοπὴν ἐχόντων ἀμφοῖν—297a 27-29; cf. 237. 10-13). There would be no difficulty even if a heavier weight than the earth were added to it in one hemisphere (cf. 237. 5-9), for by the same principle of impulsion towards the centre the sphericity of the earth would be maintained and the centre of the earth would not be displaced from the centre of the universe (297a 30-297b 14). For other examples of a commentator reworking Aristotelian illustrations, see above, p. 80; also on 13. 229. 3, 15; 232. 1-5.

237. 6-7. "both in its own nature and by its own middle" (τῷ ἰδίῳ μέσῳ). The first expression is explained at 237. 13-14 (cf. Aristot. *de caelo* 297a 16-17 and Alex. *de fato* 185. 28-31). The second may refer to the fact that because of its particular composition a given mass of earth will not move to the exact centre of the universe but to the centre peculiar (ἴδιος) to it. This is the explanation forwarded by Alex. ap. Simplic. *de caelo* 546. 15-23 for the earth not being exactly spherical. Earth is "non-uniform" (ἀνομοιομερής) and therefore different quantities of it will have impulses towards slightly different centres: "heavy bodies aim to occupy the middle by the middle of their individual impulse (τῷ τῆς ῥοπῆς τῆς οἰκείας μέσῳ), not by the centre of the magnitude [of the earth]" (546. 20-21).

237. 8-9. "this would not sink through the preexisting earth" (οὐ διὰ τῆς προϋπαρχούσης γῆς διαδύεται). The use of προϋπάρχειν here and at line 10 is influenced by the earlier account of growth; see above on 236. 14-15. For the contrast between "sinking-through" and proportional growth made in the context of a general description of growth see Philop. *de gen. et corr.* 117. 24-118. 2. διαδύεσθαι is the commentator's term for the process that would lead to an additional weight displacing the earth from the centre of the universe and not combining with it to form a single body with one centre; it has no Aristotelian precedent.

237. 26-27. "by the agency of nature and the nutritive faculty" (ὑπὸ τῆς φύσεώς τε καὶ τῆς θρεπτικῆς δυνάμεως). Cf. Philop. *de gen. et corr.* 108. 8-12 where the assimilative power of the nutritive faculty is explained as the action of nature in maintaining the shapes of an organism's parts by effecting a proportional distribution of nutriment.

237. 28-238. 10. "wine moving along a channel." This example is almost certainly a supplement to *de gen. et corr.* A5 321b 24-25 where the absorption of water in a measure that retains its shape is offered as an analogy for the proportionality of growth. This must also be true of Philoponus' more elaborate example of a duct of skin (σώλην δερμάτινος) expanding and contracting when different amounts of water are present in it but always maintaining the same shape; see *de gen. et corr.* 105. 22-26, 107. 27-31, and 108. 12-14. At Aristotle *de gen. et corr.* A5 322a 27-31 we have the case of a duct (αὐλός) absorbing liquid only in virtue of Joachim's emendation of the ἄϋλος ("immaterial") of the manuscripts. The word *tibia* occurs in a sixteenth-century translation, but ἄϋλος was the reading in the manuscripts known to Alexander and Philoponus; see the latter *ad loc.*

Philoponus (*de gen. et corr.* 106. 12-17) offers a further illustration of growth in form by referring to the way that a shadow thrown across a moving river preserves its shape while the water flows.

237. 29. "because of its elasticity and pliancy" (δι' ὑγρότητα καὶ μαλακότητα). As Philoponus emphasises (*de gen. et corr.* 108. 3-9, 117. 27-29) it is through their moistness (ὑγρότης) that organic bodies assimilate food. Perhaps in Alexander's commentary this was also stressed when the present example was used. However in

the analogy here ὑγρότης describes not only the composition of the channel but its flexibility in absorbing nutriment.

238. 14-15. "the ultimate nutriment—blood in redblooded creatures and its analogue in other things..." (ἡ ἐσχάτη τροφή, ὅπερ ἐστὶν ἐπὶ τῶν ἐναίμων τὸ αἷμα, ἐπὶ δὲ τῶν ἄλλων τὸ ἀνάλογον). This could be based on Aristot. *de somn.* 3. 456a 34-35; also cf. *de gen. an.* B4 740a 21-22. The "other things" are bloodless creatures (τὰ ἄναιμα). On "ultimate nutriment" cf. also *part. an.* Δ4 678a 7-9, and Alex. *de an.* 94. 26-95. 6.

238. 15-16. "by passing through the vessels that belong to them" (διὰ τῶν ἀγγείων φερομένη τῶν καθηκόντων ἐπ' αὐτά). For καθήκειν ἐπί τι in this sense see Alex. *de sensu* 41. 2-5. For ἄγγεια = φλέβες ("blood-vessels") see on 5. 218. 31.

238. 17-20. "and we must conceive of the process..." For these two examples see *de gen. et corr.* A5 322a 9-11. Alexander separates the example of fire from the point that Aristotle derives from it (322a 11-16)—that nutriment and the growing flesh must be together, because if they were apart new flesh would grow, and this would be coming-to-be, just as an already existing fire may either be kindled (growth) or logs may be set on fire (coming-to-be). Alexander only makes this point in general terms at lines 10-13.

238. 24. "remarkable theory" (θαυμαστὸν δόγμα). Cf. 227. 4 and 213. 6-7.

Additional Note: Alexander's lost commentary on the *de generatione et corruptione*. The evidence for this work's survival in the Greek tradition is gathered by Zeller, p. 790 n. The principal source is Philoponus whose commentary is the only one on this Aristotelian treatise extant in Greek; he cites it 35 times. Cf. also the title to Alex. *Quaest.* II. 22.

A. Wartelle (*Inventaire des manuscrits Grecs d'Aristote et ses commentateurs*, 1963) reports the existence of three manuscripts of this work: Athens, National Library, Nos. 1173 and 1301 (Wartelle Nos. 38 and 43), and Athos, Library of Vatopédi, No. 20 (Wartelle No. 173). I have been informed by the Athens library that the description of the first two mss. is incorrect, and I would assume that the third, an eighteenth century ms., is similarly misreported.

The only extant part of this commentary is the Arabic fragment at Chester Beatty ms. 3702 fol. 168b (see Peters, *Aristoteles Arabus*, p. 38 n. e); for the text and an English translation see A. A. Ghorab, *Islamic Philosophy and the Classical Tradition* pp. 81-82. This material is not related to any of the themes of the present study.

TEXTS FROM THE *DE MIXTIONE* IN *STOICORUM VETERUM FRAGMENTA*

de mixtione	*SVF*
1. 213. 2-8	II. 481
3. 216. 1-14	II. 470
3. 216. 8-14	III. p. 258 (Sosigenes)
3/4. 216. 14-218. 10	II. 473
6. 219. 16-20	II. 466
7. 220. 37-221. 3	II. 481
8. 221. 16-19	II. 474
10. 223. 25-224. 4	II. 441
10. 224. 14-26	II. 442
11. 224. 32-225. 10	II. 310
11. 225. 18-27	II. 1044
11. 226. 10-21	II. 1047
11. 226. 24-30	II. 1048
12. 226. 34-227. 17	II. 475
16. 233. 14-24	II. 735
16. 234. 23-32	II. 735[2]

BIBLIOGRAPHY I

1. ANCIENT AUTHORS

(a) *Collections*

Commentaria in Aristotelem Graeca. 23 vols. Berlin, 1883-1907. (*CAG*).
Doxographi Graeci. H. Diels, ed. Berlin, 1879 (reprinted 1958).
Epicurea. H. Usener, ed. Leipzig, 1887 (reprinted 1966).
Die Fragmente der Vorsokratiker. H. Diels and W. Kranz, eds. 3 vols. 6th ed., Berlin, 1951-2 (reprinted 1968).
Stoicorum Veterum Fragmenta. H. von Arnim, ed. 3 vols. Stuttgart, 1903 (reprinted 1964).
Vol. IV, Indices. M. Adler. Stuttgart, 1924 (reprinted 1964).
Supplementum Aristotelicum. 3 vols. Berlin, 1885-1903. (*SA*).

(b) *Texts and Editions of Individual Authors*

Albinos. *Epitome* (ed. P. Louis). Paris, 1945.
Alexander of Aphrodisias. *In Analytica Priora* (ed. M. Wallies). *CAG* II-i (1883).
——. *In Metaphysica* (ed. M. Hayduck). *CAG* I (1891).
——. *In Meteorologica* (ed. M. Hayduck). *CAG* III-ii (1890).
——. *In De Sensu* (ed. P. Wendland). *CAG* III-i (1901).
——. *In De Sophisticis Elenchis* (ed. M. Wallies). *CAG* II-iii (1898).
——. *In Topica* (ed. M. Wallies). *CAG* II-ii (1891).
——. *De anima liber cum mantissa* (ed. I. Bruns). *SA* II-i (1887).
——. *De anima liber secundus una cum commentario de Mistione* (trans. A. Caninius). 3rd ed. Venice, 1555.
——. *Scripta minora reliqua* (*Quaestiones, de fato, de mixtione*) (ed. I. Bruns). *SA* II-ii (1892).
——. *The Refutation by Alexander of Aphrodisias of Galen's Treatise on the Theory of Motion* (trans. with introduction, notes and an edition of the Arabic text by N. Rescher and M. Marmurra). Islamabad, 1965.
Anonymus Londiniensis. Ed. H. Diels. *SA* III-i (1893).
Aristoclis Messenii Reliquiae (ed. H. Heiland). Diss., Giessen, 1925.
Aristotle. *De Anima* (ed. R. D. Hicks). Cambridge, 1907 (reprinted Amsterdam, 1965).
——. *Aristotle on Coming-to-be and Passing-away* (ed. H. H. Joachim). Oxford, 1922.
——. *Aristotle on Coming-to-be and Passing-away: Some Comments* (W. J. Verdenius and J. H. Waszink). 2nd ed. Leiden, 1968.
——. *Generation of Animals* (with an English translation by A. L. Peck). London-Cambridge, Mass., 1953.
——. *Metaphysics* (ed. Sir D. Ross). Oxford, 1924.
——. *Meteorologica* (with an English translation by H. D. P. Lee). London-Cambridge, Mass., 1952.
——. *Meteorologica* (ed. J. L. Ideler). 2 vols. Leipzig, 1836.
——. *Parva Naturalia* (ed. Sir D. Ross). Oxford, 1955.
——. *Physics* (ed. Sir D. Ross). Oxford, 1936.

Atticus. *Fragments* (ed. J. Baudry). Paris, 1931.
Averroes. *Aristotle's de generatione et corruptione, Middle Commentary and Epitome* (trans. S. Kurland). Cambridge, Mass., 1958.
Calcidius. *Commentaria in Platonis Timaeum* (ed. J. H. Waszink), *Plato Latinus* IV. London-Leiden, 1962.
Cicero. *Academica* (ed. J. S. Reid). London, 1885.
Diogenes Laertius. *Lives of Eminent Philosophers* (with an English translation by R. D. Hicks). 2 vols. London-New York, 1925.
——. *Vitae Philosophorum* (ed. H. S. Long). Oxford, 1964.
Eudemus. Ed. F. Wehrli, in Vol. VII of *Die Schule des Aristoteles*. 2nd ed. Basel, 1969.
Galen. *De tempramentis* (ed. G. Helmreich). Stuttgart, 1904 (reprinted 1969).
——. *Scripta Minora* (edd. I. Marquardt, I. Mueller, G. Helmreich). 2 vols. Leipzig, 1884, 1891.
Ps.-Galen. *De qualitatibus incorporeis* (ed. J. Westenberger). Diss., Marburg, 1906.
Hierokles. *Ethische Elementarlehre* (*Pap. 9780*) (ed. H. Von Arnim), *Berliner Klassikertexte* IV. Berlin, 1906.
Marcus Aurelius. *The Meditations* (ed. A. S. L. Farquharson). 2 vols. Oxford, 1948.
Nemesius of Emesa. *De Natura Hominis* (ed. C. F. Matthaei). Halle, 1802 (reprinted Hildesheim, 1967).
"Ocellus Lucanus." Ed. R. Harder, *Neue Philologische Untersuchungen* I. Berlin, 1926.
Philoponus. *In de anima* (ed. M. Hayduck). *CAG* XV (1897).
——. *In de generatione et corruptione* (ed. H. Vitelli). *CAG* XIV-ii (1897).
——. *In Physica* (ed. H. Vitelli). *CAG* XVII-XVIII (1888).
Philostratus. *Lives of the Sophists* (with an English translation by W. C. Wright). London-New York, 1922.
Plotinus. *Schriften* (trans. R. Harder, ed. R. Beutler and W. Theiler). 6 vols. Hamburg, 1956-1970.
Plutarch. *Moralia* (ed. M. Pohlenz, rev. R. Westman). Vol. VI, Fasc. 2. Leipzig, 1959.
Sextus Empiricus. *Opera* (ed. H. Mutschman, rev. J. Mau). 3 vols. Leipzig, 1958-1961.
——. *Sextus Empiricus* (with an English translation by R. G. Bury). 4 vols. London-Cambridge, 1933-1949.
Simplicius. *In de anima* (ed. M. Hayduck). *CAG* XI (1882).
——. *In de caelo* (ed. I. L. Heiberg). *CAG* VII (1894).
——. *In Categorias* (ed. C. Kalbfleisch). *CAG* VIII (1907).
——. *In Physica* (ed. H. Diels). *CAG* IX-X (1882).
Stobaeus. *Eclogae Physicae et Ethicae* (ed. C. Wachsmuth). 4 vols. Berlin, 1908 (reprinted 1958).
Themistius. *In de anima* (ed. R. Heinze). *CAG* V-iii (1890).
——. *In de caelo* (ed. S. Landauer). *CAG* V-v (1902).
——. *In Physica* (ed. H. Schenkl). *CAG* V-ii (1900).

2. MODERN AUTHORS

Abel, K.-H. Review of J. Gould, *The Philosophy of Chrysippus*, *Gnomon* 44 (1972) 645-651.
Apelt, O. "Die schrift des Alexander von Aphrodisias über die mischung," *Philologus* 45 (1886) 82-99.
Aubonnet, M. J. See under Galperine.

Babut, D. *Plutarque et le Stoïcisme*. Lyons, 1969.

Badawī, A. *La Transmission de la Philosophie Grecque au Monde Arabe*. Paris, 1968.

Baeumker, C. *Das Problem der Materie in der Griechischen Philosophie*. Münster, 1890 (reprinted Frankfurt, 1963).

Barth, P. *Die Stoa*. 6th ed. rev. by A. Goedeckemeyer. Stuttgart, 1946.

Bazán, B. C. "Le 'De Intellectu' attribué à Alexandre d'Aphrodise," *Revue Philosophique de Louvain* 71 (1973) 468-487.

Birley, A. *Septimius Severus: The African Emperor*. London, 1971.

Bowersock, G. W. *Greek Sophists in the Roman Empire*. Oxford, 1969.

Bowie, E. L. "Greeks and their past in the second sophistic," *Past and Present* 46 (1970) 3-41.

Bréhier, É. *Chrysippe et l'Ancien Stoïcisme*. Rev. ed. Paris, 1951.

——. *La Théorie des Incorporels dans l'Ancien Stoïcisme*. 3rd ed. Paris, 1962.

Brinkmann, A. "Ein Schreibgebrauch und seine Bedeutung für die Textkritik," *RM* 57 (1902) 481-497.

Brown, H. V. B. "Avicenna and the Christian Philosophers in Baghdad," pp. 35-48 in *Islamic Philosophy and the Classical Tradition*. Ed. S. M. Stern, et al. Oxford, 1972.

Browning, R. "An Unpublished Funeral Oration on Anna Comnena," *Proceedings of the Cambridge Philological Society* n.s. 8 (1962) 1-12.

Cherniss, H. *Aristotle's Criticism of Presocratic Philosophy*. Baltimore, 1935 (reprinted New York, 1964).

Christensen, J. *An Essay on the Unity of Stoic Philosophy*. Copenhagen, 1962.

Cranz, F. E. "Alexander of Aphrodisias" in *Catalogus Translationum et Commentarium*. Ed. P. Kristeller. Vol. I. Washington, D.C., 1961; II, 1971.

——. "The Prefaces to the Greek Editions and Latin Translations of Alexander of Aphrodisias, 1450-1575," *American Philosophical Society Proceedings* 102 (1958) 510-546.

DeLacy, P. H. "Stoic Views of Poetry," *AJP* 69 (1948) 269-280.

Dietrich, A. "Die Arabische Version einer unbekannten Schrift des Alexander von Aphrodisias," *Nachr. der Gött. Akad.* (1962) 90-104.

Dijksterhuis, E. J. *The Mechanisation of the World Picture*. Trans. C. Dikshoom. Oxford, 1961.

Donini, P. L. "L'anima e gli elementi nel de anima di Alessandro di Afrodisia," *Atti dell'Accademia delle Scienze de Torino* 105 (1970-1971) 61-107.

Dörrie, H. *Porphyrios' "Symmikta Zetemata"*. (*Zetemata* 20) Munich, 1959.

——. "Ὑπόστασις, Wort- und Bedeutungsgeschichte," *Nachr. der Gött. Akad.* (1955) 36-83.

Duhem, P. *Système du Monde*. Vol. I. Paris, 1913 (reprinted Paris, 1954).

Düring, I. *Aristotle in the Ancient Biographical Tradition*. Göteborg, 1957.

——. "Von Aristoteles bis Leibniz: einige Hauptlinien in der Geschichte des Aristotelismus," *Antike und Abendland* 4 (1954) 118-154.

Edelstein, L. *Ancient Medicine: Selected Papers of Ludwig Edelstein*. Edd. O. Temkin and C. L. Temkin. Baltimore, 1967.

——. *The Meaning of Stoicism*. Cambridge, Mass., 1966.

——. "The Philosophical System of Posidonius," *AJP* 57 (1936) 286-325.

Évrard, E. "Jean Philopon, Son 'Commentaire sur Nicomaque' et ses Rapports avec Ammonius," *REG* 78 (1965) 592-598.

Festugière, A. J. *L'Idéal religieux des Grecs et l'Évangile*. Paris, 1932.

——. "Modes de composition des commentaries de Proclus," *MH* 20 (1963) 77-100.

Fuhrmann, M. *Das Systematische Lehrbuch*. Göttingen, 1960.

Furley, D. J. "Lucretius and the Stoics," *Bulletin of the Institute of Classical Studies, U. of London* no. 13 (1966) 13-33.
——. *Two Studies in Greek Atomism*. Princeton, 1967.
Galperine, M. "Le Thème du mélange dans le Stoïcisme et le Néo-platonisme," *Actes du VIIe Congrès de l'Association Guillaume Budé* (Paris, 1964), pp. 298-300. Comment by M. J. Aubonnet, *ibid.*
Ghorab, A. A. "The Greek Commentators on Aristotle," pp. 77-88 in *Islamic Philosophy and the Classical Tradition*. Ed. S. M. Stern, *et al.* Oxford, 1972.
Goldschmidt, V. *Le Système Stoïcien et l'Idée de Temps*. Paris, 1953.
——. "ΥΠΑΡΧΕΙΝ et ΥΦΙΣΤΑΝΑΙ dans la philosophie Stoïcienne," *REG* 85 (1972) 331-344.
Gould, J. B. *The Philosophy of Chrysippus*. ("Philosophia Antiqua," Vol. XVII.) Leiden, 1970.
Grant, R. M. "Greek Literature in the Treatise *De Trinitate* and Cyril *Contra Julianum*," *JTS* 15 (1964) 265-279.
Graeser, A. "A propos ΥΠΑΡΧΕΙΝ bei den Stoikern," *Archiv für Begriffsgeschichte* 15 (1971) 299-305.
——. *Plotinus and the Stoics: A Preliminary Study*. ("Philosophia Antiqua," Vol. XXII.) Leiden, 1972.
Grimal, P. "La critique de l'Aristotélisme dans le *de vita beata*," *REL* 45 (1967) 396-419.
Guthrie, W. K. C. *A History of Greek Philosophy*. 3 vols. Cambridge, 1962-1969.
Hadot, P. "Zur Vorgeschichte des Begriffes 'Existenz,' bei den Stoikern," *Archiv für Begriffsgeschichte* 13 (1969) 115-127.
Happ, H. *Hyle: Studien zur Aristotelischen Materie-Begriffe*. Berlin-New York, 1971.
——. "Weltbild und Seinslehre bei Aristoteles," *Antike und Abendland* 14 (1968) 72-91.
Hesse, M. "Aristotle's logic of analogy," *Phil. Quart.* 15 (1965) 328-340.
Horne, R. A. "Aristotelian Chemistry," *Chymia* 11 (1966) 21-27.
Howald, E. "Das philosophiegeschichtliche Compendium des Areios Didymos," *Hermes* 55 (1920) 68-98.
Hunt, H. K. "The importance of Zeno's physics for an understanding of Stoicism during the later Roman Republic," *Apeiron* 1 (1967) 5-14.
Ilberg, J. "Ueber die Schriftstellerei des Klaudios Galenos," *RM* 44 (1889) 207-239, 47 (1892) 489-514.
——. "Aus Galens Praxis. Ein Kulturbild aus der Römischen Kaiserzeit," *Neue Jahrbücher* 15 (1905) 276-312.
Jannone, A. "Aristote et la physique stoïcienne," *Actes du VIIe Congrès de l'Association Guillaume Budé* (Paris, 1964), pp. 284-287.
Joachim, H. H. "Aristotle's Conception of Chemical Combination," *JP* 29 (1904) 72-86.
Jones, W. H. S. *The Medical Writings of Anonymus Londiniensis*. Cambridge, 1947.
Kahn, C. "On the Terminology for *Copula* and *Existence*," pp. 141-158 in *Islamic Philosophy and the Classical Tradition*. Ed. S. M. Stern *et al.* Oxford, 1972.
Krämer, H. J. *Platonismus und Hellenistische Philosophie*. Berlin-New York, 1971.
Lilla, S. R. C. *Clement of Alexandria: A Study in Christian Platonism and Gnosticism*. Oxford, 1971.

Long, A. A. "Aristotle's Legacy to Stoic Ethics," *BICS* 15 (1968) 72-85.
——. Ed. *Problems in Stoicism*. London, 1971.
——. "Stoic Determinism and Alexander of Aphrodisias' *De Fato* (I-XIV)," *AGPh* 52 (1970) 247-268.
Lynch, J. P. *Aristotle's School: A Study of a Greek Educational Institution*. Berkeley-Los Angeles-London, 1972.
MacDiarmid, J. "Theophrastus on the Presocratic Causes," *HSCP* 61 (1953) 85-156.
Martin, Th.-H. "Questions connexes sur deux Sosigènes, l'un astronome et l'autre peripatéticien, et sur deux peripatéticiens Alexandre, l'un d'Égée et l'autre d'Aphrodisias," *Annales de la Faculté des Lettres de Bordeaux* 1 (1879) 174-187.
Marrou, H. I. *A History of Education in Antiquity*. Trans. G. Lamb. New York, 1964.
Martorana, A. L. "Il Maestro di Alessandro di Afrodisia," *Sophia* 36 (1968) 365-367.
Merlan, P. *Monopsychism, Mysticism, Metaconsciousness: Problems of the Soul in the Neoaristotelian and Neoplatonic Tradition*. The Hague, 1963.
Mette, H. J. Review of M. Pohlenz, *Die Stoa*, *Gnomon* 23 (1951) 27-39.
Montanari, E. "Per un'edizione del Περὶ Κράσεως di Alessandro di Afrodisia," *Accademia Toscana di Scienze e Lettere "La Colombaria"* (Florence, 1971), Vol. 36, pp. 17-58.
Moraux, P. *Alexandre d'Aphrodise, Exégète de la Noétique d'Aristote*. Liège-Paris, 1942.
——. "Alexander von Aphrodisias *Quaest.* 2. 3," *Hermes* 95 (1967) 159-169.
——. *D'Aristote à Bessarion: Trois exposés sur l'histoire et la transmission de l'aristotélisme grec*. Laval, 1970.
——. "Aristoteles, der Lehrer Alexanders von Aphrodisias," *AGPh* 49 (1967) 169-182.
——. "Einige Aspekte des Aristotelismus von Andronikos bis Alexander von Aphrodisias," pp. 203-208 in *Antiquitas Graeco-Romana ac tempora nostra* (Acta congressus internationalis habiti Brunae 1966). Prague, 1968.
——. "L'exposé de la philosophie d'Aristote chez Diogène Laërce (V. 28-34)," *Revue Philosophique de Louvain* 47 (1949) 5-43.
——. *Les Listes Anciennes des Ouvrages d'Aristote*. Louvain, 1951.
Morrow, G. "Qualitative Change in Aristotle's Physics," pp. 154-167 in *Naturphilosophie bei Aristoteles und Theophrast*. Ed. I. Düring. Heidelberg, 1969.
Mueller, I. "Stoic and Peripatetic Logic," *AGPh* 51 (1969) 173-187.
Müller, A. *Die Griechischen Philosophen in der Arabischen Überlieferung*. Halle, 1873.
Nutton, V. "Galen and Medical Autobiography," *Proceedings of the Cambridge Philological Society* n.s. 18 (1972) 50-62.
Ogereau, F. *Essai sur le Système Philosophique des Stoïciens*. Paris, 1885.
Orth, E. "Les Oeuvres d'Albinos," *Antiquité Classique* 16 (1947) 113-114.
Owens, J. "The Grounds of Universality in Aristotle," *American Philosophical Quarterly* 3 (1966) 162-170.
Pearson, A. C. "Stoica Frustula," *JP* 30 (1907) 211-222.
Peters, F. E. *Aristoteles Arabus: The Oriental Translations and Commentaries on the Aristotelian Corpus*. ("New York University, Department of Classics: Monographs on Mediterranean Antiquity," Vol. II.) Leiden, 1968.

Pines, S. "Omne quod movetur necesse est ab aliquo moveri: A Refutation of Galen by Alexander of Aphrodisias and the Theory of Motion," *Isis* 52 (1961) 21-54.

——. "Un Texte Inconnu d'Aristote en Version Arabe," *Archives d'histoire doctrinal et littéraire du moyen âge* 34 (1959) 295-299.

Plezia, M. *De Andronici Rhodii studiis aristotelicis*. Krakow, 1946.

Pohlenz, M. *Kleine Schriften*. Ed. H. Dörrie. 2 vols. Hildesheim, 1965.

——. Review of K. Reinhardt, *Kosmos und Sympathie*, *GGA* 184 (1922) 161-187 (= *Kleine Schriften*, I, 199-232).

——. "Plutarchs Schriften gegen die Stoiker," *Hermes* 74 (1939) 1-33 (= *Kleine Schriften*, I, 448-479).

——. *Die Stoa: Geschichte einer geistigen Bewegung*. 2 vols. Göttingen, 1948.

Praechter, K. Review of *CAG*, *Byzantinische Zeitschrift* 18 (1909) 516-538.

Reesor, M. "The Stoic Concept of Quality," *AJP* 75 (1954) 40-58.

Reinhardt, K. *Kosmos und Sympathie: Neue Untersuchungen über Poseidonios*. Munich, 1926.

Rex, F. *Chrysipps Mischungslehre und die an ihr geübte Kritik in Alexander von Aphrodisias de mixtione*. Diss., Frankfurt, 1966.

Reith, O. *Grundbegriffe der Stoischen Ethik*. (*Problemata* 9.) Berlin, 1933.

Rist, J. M. *Stoic Philosophy*. Cambridge, 1969.

Robert, L. "Inscriptions d'Aphrodisias," *Antiquité Classique* 35 (1966) 397-432.

Rodier, G. "Corrections au texte du περὶ μίξεως d'Alexandre d'Aphrodisias," *Revue de Philologie* 17 (1893) 10-13.

——. *Études de Philosophie Grecque*. Paris, 1957.

Sambursky, S. "Conceptual Developments in Greek Atomism," *Archives Internationales d'Histoire des Sciences* 11 (1958) 251-261.

——. "Note on John Philoponus' Rejection of the Infinite," pp. 351-353 in *Islamic Philosophy and the Classical Tradition*. Ed. S. M. Stern, *et al.* Oxford, 1972.

——. "On Some References to Experience in Stoic Physics," *Isis* 49 (1958) 331-335.

——. *The Physical World of the Greeks*. London, 1956.

——. *Physics of the Stoics*. London, 1959.

Sandbach, F. H. "ΕΝΝΟΙΑ and ΠΡΟΛΗΨΙΣ in the Stoic Theory of Knowledge," *CQ* 24 (1930) 44-51 (= pp. 22-37 in *Problems in Stoicism*. Ed. A. A. Long).

——. "*Phantasia Kataleptikė*," pp. 9-21 in *Problems in Stoicism*. Ed. A. A. Long.

——. "Plutarch on the Stoics," *CQ* 34 (1940) 20-25.

Schmidt, H. *Der Hermino Peripatetico*. Diss., Marburg, 1907.

Seeck, G. A. *Über die Elemente in der Kosmologie des Aristoteles: Untersuchungen zu "de generatione et corruptione" und "de caelo."* (*Zetemata* 34.) Munich, 1964.

Simon, H. and M. *Die Alte Stoa und Ihr Naturbegriff*. Berlin, 1956.

Solmsen, F. *Aristotle's System of the Physical World: A Comparison with his Predecessors*. Ithaca, N. Y., 1960.

——. *Cleanthes or Posidonius? The Basis of Stoic Physics*. ("Mededelingen der Koninklijke Nederlandse Akademie van Wetenschappen, Afd. Letterkunde, Nieuwe Reeks," Vol. XXIV, No. 8.) Amsterdam, 1961. (= *Kleine Schriften*, I, 436-460.)

——. *Kleine Schriften*. Vol. I. Hildesheim, 1968.

Sokolowski, R. "Matter, Elements and Substance in Aristotle," *JHP* 7 (1970) 263-288.

Spanneut, M. *Le Stoïcisme des Pères de l'Église*. Paris, 1957.

Théry, G. *Autour du Décret du 1210: II. Alexandre d'Aphrodise*. Le Saulchoir Kain, 1926.

Thillet, P. "Une traité inconnu d'Alexandre d'Aphrodise sur la Providence dans une version inédite," *Actes du Premier Congrès International de Philosophie Médiévale: L'Homme et son destin* (Louvain, 1960), pp. 313-324.

Todd, R. B. "Alexander of Aphrodisias and the Alexandrian *Quaestiones* II. 12," *Philologus* 116 (1972) 293-305.

——. "Alexander of Aphrodisias *de mixtione* 11. 226. 13: An Emendation," *Hermes* 101 (1973) 278-282.

——. "Chrysippus on Infinite Divisibility (Diogenes Laertius VII. 150)," *Apeiron* 7 (1973) 21-29.

——. "*Epitedeiotes* in Philosophical Literature: Towards an Analysis," *Acta Classica* 15 (1972) 25-35.

——. "The Stoic Common Notions: A Reexamination and Reinterpretation," *Symbolae Osloenses* 48 (1973) 47-75.

——. "ΣΥΝΕΝΤΑΣΙΣ and the Stoic Theory of Perception," *Grazer Beiträge* 2 (1974) 251-261.

Trabucco, F. "Il Problema del *De Philosophia* di Aristocle di Messene e la sua dottrina," *Acme* 11 (1958) 97-150.

Troost, K. *Zenonis Citiensis de rebus physicis doctrinae fundamentum ex adiectis fragmentis*. Berlin, 1891.

Ueberweg, F. *Grundriss der Geschichte der Philosophie*. Revised K. Praechter. 13th ed. Tübingen, 1953. (Cited "Ueberweg-Praechter.")

——. *Die Philosophie des Altertums*. Pt. I of *Grundriss der Geschichte der Philosophie*.

Usener, H. Review of *CAG*, *GGA* 26 (1892) 1001-1022 (= *Kleine Schriften*, III [Leipzig, 1914], 193-214).

Verbeke, G. "Aristotélisme et Stoïcisme dans le *De Fato* d'Alexandre d'Aphrodisias," *AGPh* 50 (1968) 73-100.

——. *La Doctrine du Pneuma du Stoïcisme à S. Augustine*. Paris-Louvain, 1945.

Vlastos, G. "Equality and Justice in Early Greek Cosmologies," *CP* 62 (1947) 156-178.

——. "Zeno of Elea," *Encyclopaedia of Philosophy* (New York, 1967), VIII, 369-379.

Walden, J. H. *The Universities of Ancient Greece*. New York, 1909.

Walzer, R. *Galen on Jews and Christians*. Oxford, 1948.

Weil, E. "Remarques sur le 'matérialisme' des Stoïciens," *Mélanges A. Koyré* (Paris, 1964), II, 556-572.

Wolfson, H. A. *The Philosophy of the Church Fathers*. Vol. I: *Faith, Trinity, Incarnation*. 2nd ed. Cambridge, Mass., 1964.

——. "Plato's Pre-existent Matter in Patristic Philosophy," pp. 409-420 in *The Classical Tradition: Essays in Honor of H. Caplan*. Ithaca, 1966.

Zabarella, J. *de rebus naturalibus libri xxx*. Frankfurt, 1607 (reprinted Frankfurt, 1966).

Zeller, E. *Die Philosophie der Griechen in Ihrer Geschichtlichen Entwicklung*. 3 vols in 5. 3rd ed. Leipzig, 1876-1889. (Vol. III-i [1880] cited "Zeller.")

Zumpt, C. G. "Ueber den Bestand der philosophischen Schulen in Athen und die Succession der Scholarchen," *Abhandlungen der Akademie*, Berlin (1843), pp. 27-123.

BIBLIOGRAPHY II

ALEXANDER OF APHRODISIAS: A SELECT BIBLIOGRAPHY

This listing is designed primarily as a supplement to the general introduction to Alexander in Part One and to Pt. I of the Bibliography. It is not exhaustive but I hope complete enough to serve as a useful guide.

1. *General Studies*

Major encyclopaedia articles are: A. Gercke, *RE* I-ii (1894) 1453-1455; A. C. Lloyd in *Encyclopaedia of Philosophy* (New York, 1967) I, 73; and P. Merlan in *Dictionary of Scientific Biography* (New York, 1970), I, 117-120. Among the histories of philosophy see Zeller, III-i³, 789-801; Ueberweg-Praechter, pp. 564-565 and p. 179 of the supplement; and P. Merlan in *The Cambridge History of Later Greek and Early Medieval Philosophy* (reprinted Cambridge, 1970), pp. 117-123. See also F. Cranz's articles in the *Catalogus Translationum* (listed in Bibliography I) on the Latin translations of Alexander. There is some general material also in G. Movia, *Alessandro di Afrodisias tra naturalismo e misticismo* (Padua, 1970), as well as Moraux, *Alexandre d'Aphrodise* and G. Théry, *Décret du 1210* (both in Bibliography I). Movia's work also contains an extensive bibliography.

2. *Treatises and Commentaries*

de anima. P. L. Donini, "Il De Anima di Alessandro di Afrodisia e Michele Efesio," *Rivista di Filologia e di Instruzione Classica* 96 (1968) 316-323, and his article cited in Bibliography I; J. Zahlfleisch, "Die Polemik Alexanders von Aphrodisias gegen die Verschiedenen Theorien des Sehens," *AGPh* 8 (1895) 373-386, 498-509, 9 (1896) 149-162. See also Movia, Sect. 1 above, and Moraux, *Alexandre d'Aphrodise*, in Bibliography I.

Quaestiones. I. Bruns, "Studien zu Alexander von Aphrodisias II, *Quaestiones* II. 3," *RM* 45 (1890) 138-145, and "Studien zu Alexander von Aphrodisias III, Lehre der Vorsehung," *RM* 45 (1890) 223-235; P. Merlan, "Zwei Unterschuchungen zu Alexander von Aphrodisias II, Der Platoniker bei Alexander von Aphrodisias *Quaestiones* II. 21, p. 65. 17-71. 2 Bruns," *Philologus* 113 (1969) 88-91. Also see Bibliography I under Moraux, *Hermes* 95 (1967).

de fato. Translations: J. G. Schulthess, *Abhandlungen von dem Schicksal und von der Freiheit des Willens* (Zûrich, 1872); J. F. Nourrisson, *De la liberté et du hasard. Essai sur Alexandre d'Aphrodisias suivi du Traité du destin et du libre pouvoir aux empereurs* (Paris, 1870); A. Fitzgerald, *Alexander of Aphrodisias On Destiny* (London, 1931).

Textual Studies: O. Apelt, "Die Kleinen Schriften des Alexander von Aphrodisias," *RM* 49 (1894) 59-71; H. Von Arnim, "Textkritik zu Alexander von Aphrodisias," *Wiener Studien* 22 (1900) 1-10; G. Rodier, "Conjectures sur le texte du *De Fato*," *Revue de Philologie* 25 (1901) 66-71; H. Langerbeck, "Zu Alexander von Aphrodisias *de fato* X," *Hermes* 64 (1936) 473-474; R. Hackforth, "Notes on Some Passages of Alexander Aphrodisiensis *de fato*," *CQ* 40 (1946) 37-44; P. L. Donini, "Note al *Peri Heimarmenés* di Alessandro di Afrodisia," *Rivista di Filologia e di Instruzione Classica* 97 (1969) 298-313.

Special Studies: I. Bruns, "Studien zu Alexander von Aphrodisias I, Der Begriff des Möglichen und die Stoa," *RM* 44 (1890) 613-630; R. A. Pack, "A Passage in Alexander of Aphrodisias relating to the theory of tragedy," *AJP* 58 (1937) 418-436; P. Merlan, "Zwei Untersuchungen zu Alexander von Aphrodisias I, Eine eigenartige Erklärung des *eph hemin*," *Philologus* 113 (1969) 85-88. See Bibliography I under Verbeke, *AGPh* 50 (1968) and Long, *AGPh* 52 (1970).

Commentaries. P. Wilpert, "Reste verlorener Aristotelesschriften bei Alexander von Aphrodisias," *Hermes* 68 (1940) 369-396; G. Volait, *Die Stellung des Alexander von Aphrodisias zur Aristotelischen Schlusslehre* (diss., Bonn, 1907); V. C. Coutant, *Alexander of Aphrodisias, Commentary on Bk. IV of Aristotle's Meteorologica* (diss., Columbia, 1936).

Fragments. E. G. Schmidt, "Alexander von Aphrodisias in einem altarmenischen Kategorien-Kommentar," *Philologus* 110 (1966) 277-286. See Bibliography I under Grant and Thillet.

3. *Alexander's Interpretation of Aristotle's Doctrine of the Intellect*

A. Gunsz, *Die Abhandlung Alexanders von Aphrodisias über den Intellekt* (diss., Berlin 1887); A. Elfes, *Aristotelis doctrina de mente humana ex commentariorum Graecorum sententiis eruta: I. Alexandri Aphrodisiensis et Joannis Grammatici Philoponi commentationes* (diss., Bonn, 1887); H. Kurfess, *Zur Geschichte der Erklärung der aristotelischen Lehre vom sogennanten* νοῦς ποιητικός *und* παθητικός (diss., Tübingen, 1911); O. Hamelin, *La Théorie de l'Intellect d'après Aristote et ses commentateurs* (Paris, 1953); P. Merlan, *Monopsychism, Mysticism, Metaconsciousness* (in Bibliography I), ch, 2 *passim*.

On the later history of Alexander's interpretation: P. Wilpert, "Die Ausgestaltung der aristotelischen Lehre vom Intellectus agens bei griechischen Kommentatoren und in der Scholastik des 13. Jahrhunderts," *Beiträge zur Gesch. der Phil. und Theol. des Mittelalt.*, Supplband III-i (1935) 447-462; E. P. Mahoney, "Nicoletto Vernia and Agostino Nifo on Alexander of Aphrodisias: An Unnoticed Dispute," *Rivista critica di storia della filosofia* 23 (1968) 268-296. See also Bibliography I under Moraux, *Alexandre d'Aphrodise*, and Bazán.

4. *Alexander and Plotinus*

P. Merlan, "Plotinus Enneads 2.2.," *TAPA* 74 (1943) 179-191; H. R. Schwyzer, "Plotinos," in *RE* 21 (1951) cols. 573-574; P. Henry, "Une Comparaison chez Aristote, Alexandre, et Plotin," *Les Sources de Plotin* (Fondation Hardt, "Entretiens," V: Vandoeuvres-Genève, 1960), pp. 427-449; A. H. Armstrong, "The Background of the Doctrine 'That the Intelligibles are not outside the Intellect'," *Les Sources de Plotin* (Fondation Hardt, "Entretiens," V), pp. 414-425; J. M. Rist, "On Tracking Alexander of Aphrodisias," *AGPh* 48 (1966) 82-90; H. J. Blumenthal, "Plotinus Ennead IV. 3. 20-21 and its Sources—Alexander, Aristotle and Others," *AGPh* 50 (1968) 254-261; F. P. Hager, "Die Aristotelesinterpretation des Alexanders von Aphrodisias und die Aristoteleskritik Plotins bezüglich der Lehre von Geist," *AGPh* 46 (1964) 174-187.

5. *Alexander Arabus*

Two recent works of reference (fully cited in Bibliography I) are F. E. Peters, *Aristoteles Arabus*, and A. Badawī, *La Transmission de la philosophie grecque au monde arabe* at pp. 94-99 and 121-165.

Special studies: J. Finnegan, "Texte arabe du Περὶ Νοῦ d'Alexandre d'Aphrodise," *Mélanges de l'Université Saint Joseph* (Beyrouth) 33-2 (1956) 157-202; J. Van Ess, "Über einige neue Fragmente des Alexander von Aphrodisias und des Proklos in arabischer Übersetzung," *Der Islam* 42 (1966) 148-168; H. Gaetje, "Zur Arabischen Überlieferung des Alexander von Aphrodisias," *Zeitschrift des Deutschen Morgenländischen Gesellschaft* 106 (1966) 255-278; *idem*, "Die Arabische Übersetzung der Schrift des Alexander von Aphrodisias über die Farbe," *Nachr. der Gött. Akad.* (1967) 341-382; *idem*, *Studien zur Überlieferung der aristotelischen Psychologie im Islam* (Annales Universitatis Saraviensis, Bd. II, Heidelberg, 1971). See also Bibliography I under Dietrich, Pines, and Thillet.

INDICES

1. INDEX OF NAMES

(a) *Ancient Authors*

(b) *Modern Authors* (*Select*)

2. INDEX OF GREEK TERMS

3. INDEX LOCORUM

(References in brackets are to the *SVF* volume and text)

Sem. Griekse Letterkunde
BI